3rd International Conference on Efficient Building Design 2018

Materials and HVAC Equipment

Beirut, Lebanon
4 - 5 October 2018

ISBN: 978-1-7138-0598-4

Printed from e-media with permission by:

Curran Associates, Inc.
57 Morehouse Lane
Red Hook, NY 12571

Some format issues inherent in the e-media version may also appear in this print version.

Printed with permission by Curran Associates, Inc. (2020)

ISBN: 978-1-7138-0598-4

American Society of Heating, Refrigerating and Air-Conditioning Engineers (ASHRAE)
1791 Tullie Circle
N.E. Atlanta, GA 30329
USA

Phone: (404) 636-8400
Fax: (404) 321-5478

ashrae@ashrae.org

Additional copies of this publication are available from:

Curran Associates, Inc.
57 Morehouse Lane
Red Hook, NY 12571 USA
Phone: 845-758-0400
Fax: 845-758-2633
Email: curran@proceedings.com
Web: www.proceedings.com

TABLE OF CONTENTS

Controlling Sunlight Entering the Indoor Space through Windows using Dual Solar Screens

Esam M. Alawadhi, PhD
Kuwait University

ABSTRACT

Sunlight is utilized in a sustainable building to minimize the energy usage and maximize the comfort of the occupants. Sunlight provides the comfort to the occupants by naturally illuminating the indoor space, and reduces energy consumed by artificial lightings. However, sunlight increases the energy for cooling, and discomfort visual condition. Hence, controlling the sunlight entering the indoor space using shading devices for windows is essential. The present work introduces the idea of dual perforated solar screens is introduced to control the sunlight entering the indoor space. Two screens with square holes are installed externally on a west-facing façade window. The perforation ratio of screens is 25%, and all experimental measurements are conducted in June. The holes of the two are perfectly aligned. When no solar screen in installed, measured illuminance in the room reaches as high as 13,000 Lux, and with a single solar screen is installed, the maximum illuminance in the room is decreased 90%. However, there was large contrast in the room. If the second solar screen is installed, the illuminance in the room becomes uniform, and around 200 Lux. Moreover, the illuminance is nearly independent with time, between 2 PM and 5 PM.

INTRODUCTION

Buildings consume large amount of energy in air condition units and lightings. Reducing energy consumption and achieve the comfort for the occupants are two of many requirements for sustainable buildings. To achieve these requirements, sunlight is utilized to reduce energy consumed by artificial lightings, and naturally illuminate the indoor space that provide the comfort the occupants. On the other hand, the amount of sunlight that entering the indoor space should be carefully controlled to prevent high heat gain (J. Lam et al., 2005). Shading devices are commonly used to control sunlight entering the indoor space (A. Freewan, 2014), provide privacy to the occupants, and to reduced energy consumed by artificial lightings (G. Yingming et al., 2017). To control sunlight entering the indoor space, solar screens are considered as a component of a buildings' envelop, and recently, more attention to the solar screens is given (E. Aljofi, 2005). The solar screens are inexpensive, because they can be made from common building materials, and, easy to be integrated in new and historic buildings (A. Ayssa, 1996). Figure 1 shows a solar screen installed on window in Kuwait city. Sherif et al. (2012) studied the effect of the solar screens on reducing energy used for cooling and lighting, and the effect of the screen's holes geometry on energy used reduction was also studied by Sherif et al. (2012). Their results indicate that the solar transmittance is strongly affected by the geometry of the screen.

Figure 1 A solar screen installed on a window in Kuwait city.

In this research, the illuminance distributions is measured and heat gain is calculated for a window with Dual solar screens. Two perforated screens with square holes are installed on a window facing the west direction, and the experiments are conducted in June. Figure 2 shows single and Dual solar screens used in the experiments. The window is located on the façade that got the most of the global solar radiation. The results of the solar screens are compared to window without the solar screen to evaluate the effect of solar screen on the illuminance and heat gain through the window. The perforation ratio of a solar screen is 25%, and if two solar screens are installed, the perforation ratio is decreased to 12.5%. For comparison purposes, window without solar screen are considered in the study.

(a) (b)

Figure 2 The (a) single and (b) double solar screens used in the experiments.

Problem description

The surface of the solar screen is subjected to sunlight, and reflected radiation from surroundings, which are transmitted to the indoor space through the solar screen's holes. Additionally, the window emits heat to the indoor space, and natural convection flow is generated on the indoor surface of the window. The instantaneous heat gain per unit area to the indoor space is the sum of convected heat from the window, radiation emitted by the indoor surface of the window, solar radiation through the solar screens, and radiation emitted by the surroundings (K. Ismail et al., 2009):

$$Q = Q_c + Q_r + Q_s + Q_{sr} \tag{1}$$

Q_C is the heat convection from the window to the indoor space air:

$$Q_c = h_w(T_w - T_i) \tag{2a}$$

where h_w is the heat transfer coefficient along the indoor surface of the window. Tw and Ti are window surface and indoor air temperatures, respectively. Qr is radiation emitted by the indoor surface of the window to the indoor:

$$Q_r = \varepsilon_w \, \sigma \, (T_w^4 - T_i^4) \tag{2b}$$

where ε_w is the emissivity of the window, σ is Stefan-Boltzman constant, and Ti is the indoor surrounding temperature. The temperature of the indoor surroundings is assumed equal to indoor air temperature. Qs is the solar radiation through the holes of the solar screens:

$$Q_s = \tau_w \, A_P \, Q_s \tag{2c}$$

where τ_w is the coefficient of solar transmittance of the window, AP is the perforation ratio of the solar screen, and Qs is the solar radiation on the surface of the solar screen. Qsr is surrounding radiation emitted by the outdoor surroundings through the holes of the screen:

$$Q_{sr} = \varepsilon_o \, \tau_w \, \sigma \, A_P \, (T_o^4 - T_i^4) \tag{2d}$$

where ε_o, and To is the outdoor surroundings emissivity and temperature, respectively. The thickness of the solar screens is 0.5 cm, and the window is double-glazed. The distance between the two screens, and the window and screen, is 10 cm. The two screens have square holes of 10 cm side length. The perforation ratio is defined as the ratio of the holes to the total area of the screen, and the perforation ratio of the solar screens is 25%. All lights are switched off during the experiments, to ensure accurate illuminance measurements. A heat flux sensor is attached to the indoor surface of the window to measure heat flux through it, while the surface temperature of the generated millivolt by thermocouples and heat flux sensor are measured, and logged using a programmable data acquisition device. Solar power meters are used to measure the solar radiation in the site. A light meter is used to measure illumination in the room, and thermography camera is used to capture the indoor surface temperature of the window. The emissivity and coefficient of solar transmittance of the window are ε_w=0.83 and τ_w=0.3, respectively. The emissivity of the outdoor surrounding is ε_o=0.5, and the indoor air temperature is maintained at 23.5°C during the experiments.

RESULTS AND DISCUSSION

Figure 3 shows the outdoor air temperature and global radiation measurements during the day in June. The sunrise and sunset time are 4:59 AM and 6:50 PM, respectively. The minimum and the maximum temperatures are 33.3 oC and 48.65 oC, respectively. From 4 AM to 1:30 PM, the window is completely shaded, and the surroundings are emitting radiation to the window.

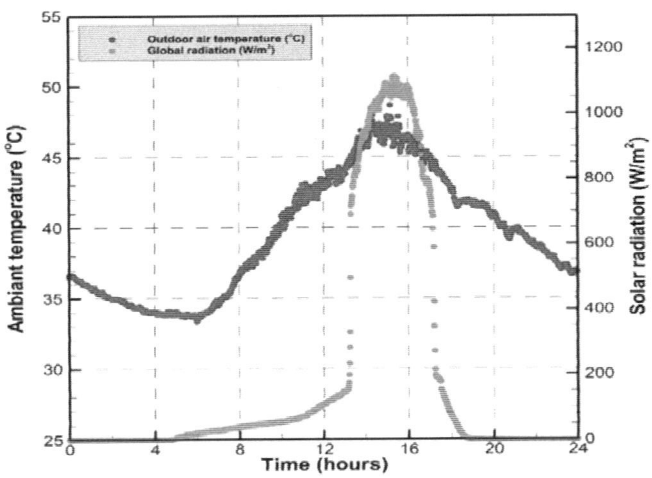

Figure 3 The outdoor air temperature, and global radiation for the month of June.

After 1:30 PM, the sunlight reaches the window. As shown in the figure, the increasing rate of the solar radiation is high, and at 3:20 PM, the solar radiation reaches its maximum value, which is 1114 W/m2. At the sunset time, the solar radiation is decreased at a high rate.

Indoor illuminance

Illuminance is the total luminous flux incident on a surface per unit area, the SI unit for the illuminance is Lux, and one Lux is equal to one lumen per square meter. In a common living room, the illuminance should be is around 350 Lux, and illuminance level higher than 750 Lux is considered to be high (D. Dilaura et. al, 2011). The perforation ratio of the single and Dual solar screens with respect to front view will be changed as the sun position changes, because there is angle between sunlight and a line orthogonal to the solar screen. As a result, the amount of sunlight entering the indoor space will change during the day. Figure 4 shows the sunlight pattern on the room's floor for window with single and Dual solar screens, at 3:00 PM. With a second screen, the amount of sunlight penetrating the room is reduced. Comparing a sunlight pattern of a single solar with a Dual solar screen, the amount of sunlight falling on the floor at 3 PM is reduced by half. A light meter is used to measure of illuminance with time at a single point. The illuminance is measured at 0.75 m from the floor, and 1.5 m from the window. The aim of this measurement is to illustrate the effect of installed single and Dual solar screens on the illuminance in the room.

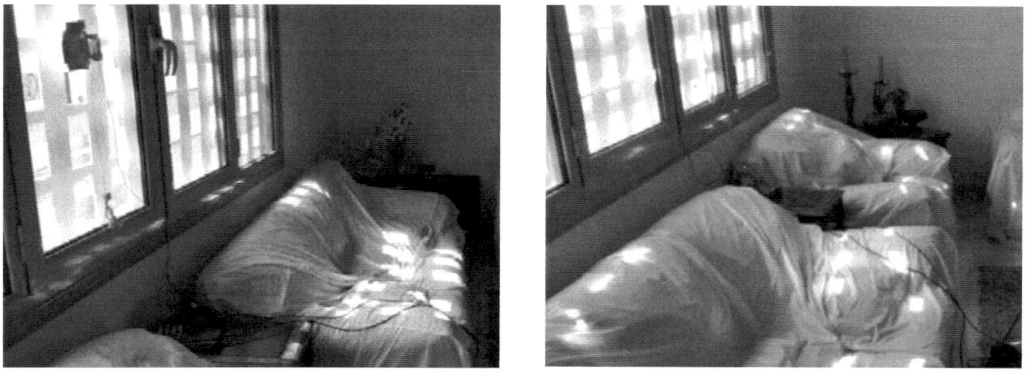

(a) (b)

Figure 4 Sunlight pattern on the floor at 3:00 PM, for window with single solar screen, double solar screen.

Figure 5 shows the indoor illuminance level for window with single solar screen, and Dual solar screens. The illuminance measurements for window without the solar screens is performed, and presented in the figure. As shown, the illuminance in the room is started to increase at 5:00 AM, and the increasing rate is low until 12:00 PM. The source of illuminance is the penetrated sunlight that is reflected from the outdoor surrounding area. Figure 5 indicates that the increasing rate of the illuminance for window without the solar screen is relatively higher than window with the solar screen. At 1:00 PM, the sunlight is started to fall on the window. The illuminance level reach highest value at 4:00 PM, and the illuminance as high as 13,800 Lux, for a window without a solar screen. With a single solar screen, the sunlight entering the room is reduced by more than 75%. Figure 5 indicates that the illuminance in the room is decreased by more than 70%, and the illuminance in the room is above 500 Lux level for just about two hours. With a Dual solar screen, the amount of sunlight entering the room is additionally decreased. For Dual solar screens, the illuminance in the room becomes nearly constant, especially between 3:00 PM and 5:030 PM, and the illuminance is maintained at 200 Lux. At 5:030 PM, the illuminance level is decreasing at a high rate as the sun goes down. The illuminance distribution images enable designers to study the availability, distribution, quality of daylight, and to detect high illuminance spots. Figure 6 shows illuminance distribution extracted from HDR images of at the window region, for window without solar screen, with a single solar screen, Dual solar screens. For a window without solar screen, Figure 6a, the brightness and contrast are very high, and the maximum illuminance in the floor is more than 1,250 Lux.

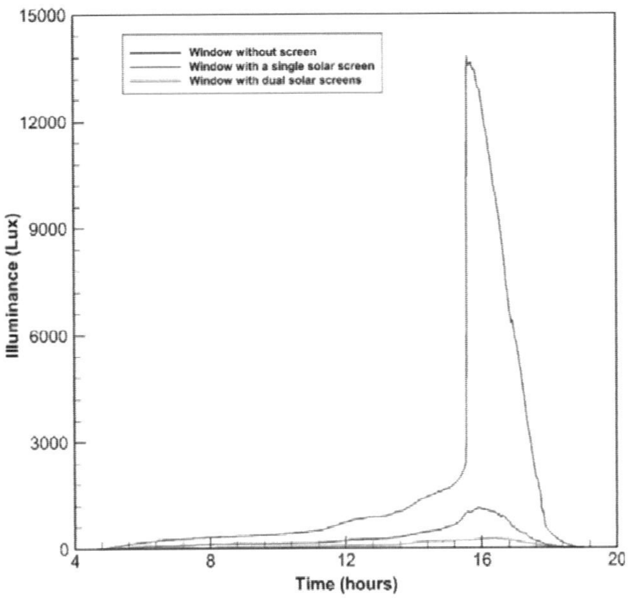

Figure 5 The measured indoor illuminance.

On the other hand, with a single solar screen installed, Figure 6b indicates that the contrast in the room is definitely reduced, but still the room is bright at a region close to the window. The maximum illuminance on the floor is reduced by 50%, from 1250 Lux to 630 Lux. When the second screen is installed, Figure 6c indicates that the illuminance in the room becomes uniform. The maximum illuminance is reduced additionally by 70% to 194 Lux. Hence, the Dual solar screens are capable of making illuminance uniform in the room, and to eliminate the high illuminance spots.

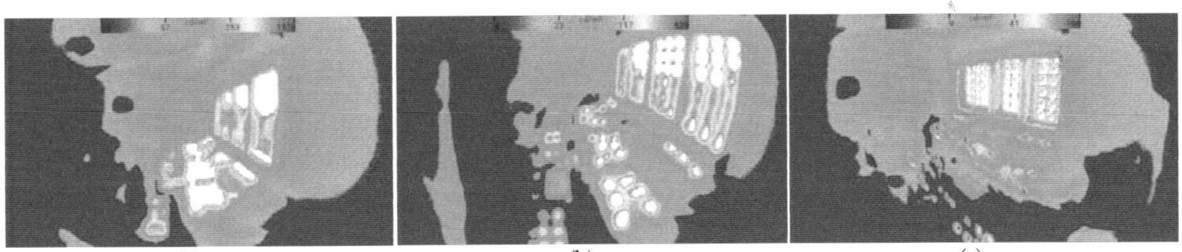

(a) (b) (c)

Figure 6 illuminance distribution extracted from high dynamic range image in the room for the window (a) without solar screen, (b) with a single solar screen, double solar screen.

Heat flux calculations

The heat flux through the window is measured, and presented in Figure 7, for window without solar screen, with a single solar screen, and Dual solar screens. Heat flux is calculated using equation (1). The heat flux at the indoor surface of the window is the heat gain. At 5 AM, heat flux starts to increase to reach its maximum value at 4:15 PM, and then it decreases as the sun goes down. Between 9 PM and 5 AM, the heat gain is due to high surface temperature of the window. Additionally, Figure 7 indicates that installing a single screen causes the heat flux to decrease compared to the window without solar screen. The maximum heat gain for a window without screen is 530 W/m2. If a single screen is installed, the maximum heat gain is decreased 300 W/m2. The solar screen blocks 75% of the sunlight that reduces the heat gain, and reduces the indoor illuminance. If the second solar screen is installed, the heat gain is additionally decreased 200 W/m2. The second screen blocks most of sunlight from entering the indoor space that increases the heat gain, and create better illuminance distribution in the room.

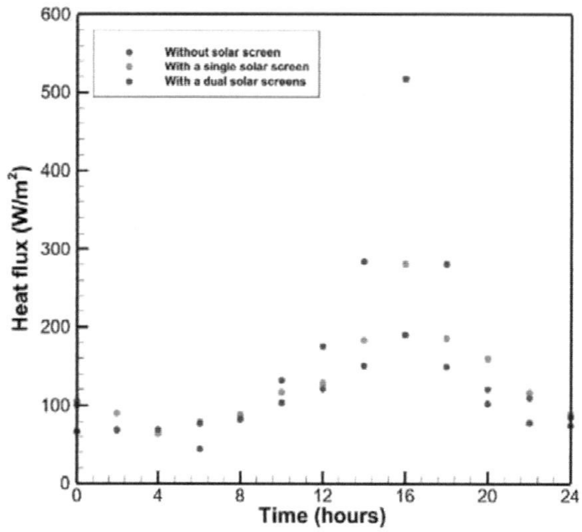

Figure 7 Heat flux gain with time through the window with a solar screen with difference perforation ratios, and window without solar screen.

CONCLUSION

Window allows the sunlight to illuminate the indoor space to create healthy indoor environment. However, excessive amount of sunlight creates a high illuminance spots, and increases the energy consumes by air-condition units. A solar screen made of perforated wooden board is installed on window to reduce the amount of the sunlight entering the indoor space. Experimental measurements are conducted in June to determine the effect of installing solar screen on the illuminance distribution, and heat gain. For a window without solar screen, there is high illuminance in the room, between more 1,000 Lux that makes the room uncomfortable. With dual solar screens, the illuminance level has a maximum value of 200 Lux. Therefore, the visual comfort is attained. If a single screen is installed, the maximum heat gain is decreased by 50%, and installing the second solar screen will additionally decrease the heat gain by 70%.

ACKNOWLEDGMENTS

The work presented was carried out with the financial and technical support of the research administration at Kuwait University under the grant EM01/16.

REFERENCES

Lam, J., Tsang, C., Li, D., Cheung, S., 2005. Residential building envelope heat gain and cooling energy requirements. Energy 30, 933–951.

Yingming, G., Mingdong, X., Huanyue, Z., Xu, G., Ping, L., Nianyu, Z., 2017. Study on distributed individuation lighting model and analysis to energy consumption character. Energy and Buildings 151, 45-52.

Sherif, A., Sabry, H., Tarek., R., 2012. External performance solar screens for daylight in residential desert building: identification of minimum perforation percentages. Solar Energy 86, 1929-1940.

Sherif, A., Sabry, H., Gadelhak, M., 2012. The impact of changing solar screen rotation angle and its opening aspect ratios on Daylight Availability in residential desert buildings. Solar Energy 86, 3353–3363.

Aljofi, E., 2005. The potentiality of reflected sunlight through Rawshan screens. Proceedings of the International Conference Passive and Low Energy Cooling for Built Environment, Santorini, Greece.

Ayssa, A., 1996. The traditional Yemeni window and natural lighting. Renewable Energy 8, 214-218

Ismail, K., Salinas, C., Henriquez, J., 2009. A comparative study of naturally ventilated and gas filled windows for hot climates, Energy Convers. Manage. 51, 1583–1591.

Freewan, A., 2014. Impact of external shading devices on thermal and daylighting performance of offices in hot climate regions. Solar Energy 102, 14-34.

Dilaura, D., Steffy, G., Mistrick, R., Houser, K., 2011. The Lighting Handbook, 10th Edition, Illuminating Engineering Society.

Energy Comparison of Air Conditioning Split System Vs. Solar Absorption Systems with Optimization for a Prototype Educational Building

Hesham Safwat, PhD
ASHRAE member

Fady Anees

ABSTRACT

This paper aims to study solar energy in Egypt and its potential utilization in solar thermal cooling (absorption technique). The paper includes a detailed proposed solar absorption cooling system for one building located at the British University in Egypt with maximum cooling load of 141.3 ton. Of refrigeration. The building cooling loads at different months were calculated using the hourly analysis program (HAP). A detailed design using a design software (TRNSYS) is introduced to model the system and for optimization. A cost analysis is finally included to compare the running costs of both the proposed cooling system and the current system. The paper shows that more than 80% energy savings can be achieved using the solar thermal cooling system compared to ordinary split air-conditioning units.

INTRODUCTION

Egypt has been witnessing a huge power dilemma for the last ten years. This problem began to escalate during the past summer seasons when frequent breakdowns of power during peak hours started. In some areas, breakdowns lasted for continuous 18 hours a day (Raslan, 2013). Oil, natural gas and coal count for 94% of total energy consumption in Egypt. While solar and wind energy count for less than one percent of the total energy consumption (Egypt/ International energy data research and analaysis, 2015). Meanwhile, Energy subsidies have reached 21% of the total state budget which is a huge burden for the government. The installed power generation capacity of 31 GW is not capable of meeting the high peak demand during peak hours in summer (Burger, 2015). This is mainly due to high cooling loads from big scale refrigeration and central air-conditioning and cheap, low efficiency air conditioning units in domestic applications. Electric power demand is expected to increase annually by 5-6% (Burger, 2015). Also, the government has decided to gradually phase out all the energy subsidies within five years (Sarant, 2105). This will cause a noticeable increase in energy costs for all citizens. Although this may seem so disturbing and unpromising, those consequences have actually urged the government, companies and individuals to start looking for other sources of energy and utilizing them. Renewable energy in Egypt seems to be the future and the one option available eventually.

Hesham Safwat is a lecturer in the Department of Mechanical Engineering, British University in Egypt, Cairo, Egypt. **Fady Anees** is a teaching assistant at the British university in Egypt Cairo, Egypt.

SOLAR COOLING

Solar energy is a great source for air-conditioning and refrigeration applications. There are various cycles that have been developed over the past years to utilize solar energy for cooling purposes. Basically, there are two main types of solar cooling systems; photovoltaic cooling systems and solar thermal cooling systems. This paper only deals with absorption solar thermal cooling systems.

Scoop of work:
The paper includes a detailed proposed absorption cooling system for one building located at the British University in Egypt with 840 Kw cooling load for research purposes. A preliminary study is first introduced to discuss the feasibility of such a design for the building and then a detailed optimized design using a design software (TRYNSYS) is introduced. A cost analysis is finally included to compare the running costs of both the proposed cooling system and the current system.

CASE STUDY

Methodology

The cooling loads were first calculated by HAP software (Hourly Analysis program) certified by ASHRAE. After that the cooling system was modeled by TRNSYS software to simulate the cooling process and obtain results such as the solar fraction. Both loads and simulations were conducted from the first of May to mid-October.

Building loads and data

Building H is a building with 2586 m2 area served by the cooling plant and 32 zones. The loads are calculated by HAP by the peak coincident load method; which means that all heat sources cannot reach their peaks at the same time. The actual circumstances and conditions of the building such as the number of people, lighting, electric equipment and ventilation were taken in consideration during the load calculation. Figure (1) shows the cooling loads in tons of refrigeration at different times of the day (14th of each month) from May to September. The maximum cooling load is 141.3 ton of refrigeration as shown in figure 1.

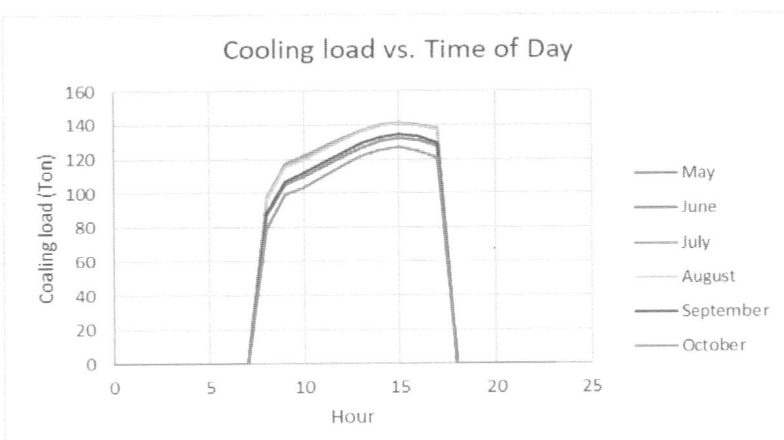

Figure 1 Cooling load vs. Time of Day at during summer

3.3 System modeling on TRNSYS:

The system was modeled on TRNSYS as shown in figure (5). It consists of two subsystems: Solar thermal cycle and the Absorption cooling cycle. The system consists of the following components with their equations:
Evacuated tubes solar thermal collectors (type 72): They are similar in the TRNSYS code to the flat plate design but it reads an external file with transverse and logitidunal IAMs. the thermal efficiency equation is as following:

$$\eta = a_0 - a_1\left(\frac{\Delta T}{I_T}\right) - a_2\frac{(\Delta T)^2}{I_T}$$

$$\Delta T = T_I - T_a$$

Thermal Storage tank (type 4): the tank is working at mode 1 (load enters at the bottom and hot water enters at the top of the tank) the equations used in this model are showin in figure (2) taken from TRNSYS software mathematical reference:

$$M_i C_{pf}\frac{dT_i}{dt} = \begin{cases} (\dot{m}_1 - \dot{m}_3)\, C_{pf}(T_{i-1} - T_i) & \dot{m}_1 \geq \dot{m}_3 \\ (\dot{m}_3 - \dot{m}_1)\, C_{pf}(T_{i+1} - T_i) & \dot{m}_1 < \dot{m}_3 \end{cases}$$

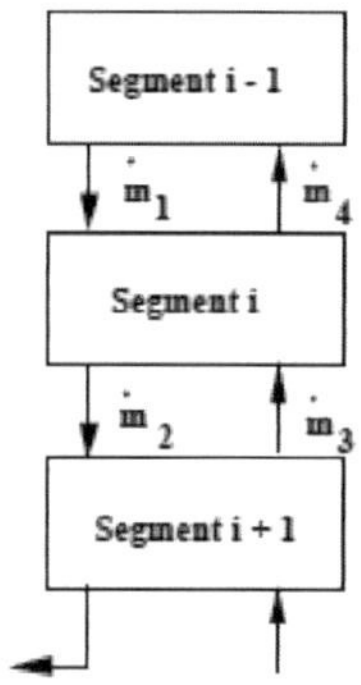

Figure 2 Thermal storage tank flow equations with diffrent segments inside

Auxiliary heater (type 6): the auxuilary heater works such that if the output flow temperature is less than the setpoint temperature (95°C), it receives a control signal from the controller to actuate the heater. the auxuilary heater equations used are shown in figure (3) (taken from TRNSYS software mathematical reference):

If $T_i \geq T_{set}$, $\dot{m}_i \leq 0$, or $\gamma = 0$ then

$T_o = T_i$, $\dot{m}_o = \dot{m}_i$, $\dot{Q}_{loss} = 0$, $\dot{Q}_{fluid} = 0$, and $\dot{Q}_{aux} = 0$

Otherwise, an energy balance on the steady-state heater reveals:

$$T_o = \frac{\dot{Q}_{max}\,\eta_{htr} + \dot{m}\,C_{pf}\,T_i + UA\,T_{env} - \frac{UA\,T_i}{2}}{\dot{m}\,C_{pf} + \frac{UA}{2}}$$

$$\dot{m}_o = \dot{m}_i$$

$$\dot{Q}_{aux} = \dot{Q}_{max}$$

$$\dot{Q}_{fluid} = \dot{m}_o\,C_{pf}\,(T_o - T_i)$$

Figure 3 Auxuilary heater equations

absorption chiller (type 107): the absorption chiller equations used are shown in figure (3) taken from TRNSYS software mathematical reference:

$$\dot{Q}_{remove} = \dot{m}_{chw} Cp_{chw} \left(T_{chw,in} - T_{chw,set}\right)$$

$$f_{DesignLoad} = \frac{\dot{Q}_{remove}}{Capacity_{Rated}}$$

$$Capacity = f_{FullLoadCapacity} * f_{NominalCapacity} Capacity_{Rated}$$

$$\dot{Q}_{hw} = \frac{Capacity_{Rated}}{COP_{Rated}} f_{DesignEnergyInput}$$

$$T_{hw,out} = T_{hw,in} - \frac{\dot{Q}_{hw}}{\dot{m}_{hw} Cp_{hw}}$$

$$T_{chw,out} = T_{chw,in} - \frac{MIN(\dot{Q}_{remove}, Capacity)}{\dot{m}_{chw} Cp_{chw}}$$

$$\dot{Q}_{cw} = \dot{Q}_{chw} + \dot{Q}_{hw} + \dot{Q}_{aux}$$

$$T_{cw,out} = T_{cw,in} + \frac{\dot{Q}_{cw}}{\dot{m}_{cw} Cp_{cw}}$$

$$COP = \frac{\dot{Q}_{chw}}{\dot{Q}_{aux} + \dot{Q}_{hw}}$$

Figure 4 absorption chiller equations

The Absorption chiller used in this simulation is a Single-Effect, Low Temperature Hot Water Hermetic Absorption Liquid Chiller with a nominal cooling capacity of 155 ton of refrigeration. The Following table (table 1) represents some performance data obtained from the chiller's catalogue:

Table 1 chiller performance data

	Chilled water	Cooling Water	Hot Water
Flow rate (Kg/s)	23.5	35.2	21.4
Temperature (oC)	6.67	29.4	95

other model components: weather data, solar collector pump, controller for the solar thermal pump, , control unit for the auxiliary heater, , hot water pump, cooling tower, cooling water pump, cooling load data, integrators, Calculator and unit conversion tools.

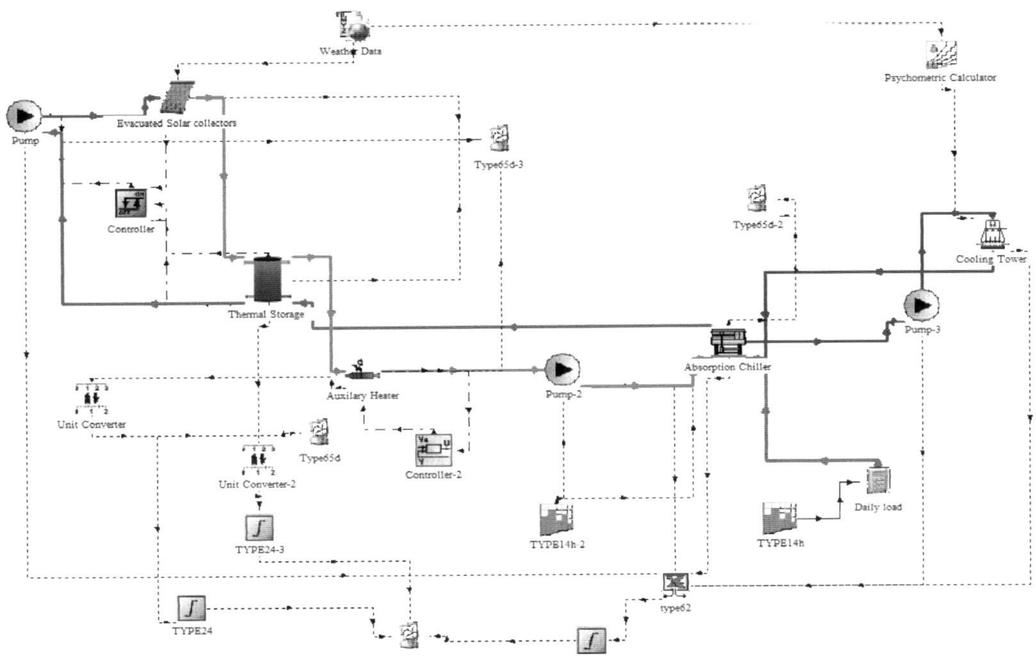

Figure 5 TRNSYS import diagram

RESULTS AND DISCUSSION

System optimization

Several factors can affect the solar fraction (sun energy gained fraction of the total heat energy required) some of those factors will be studied and the TRNSYS simulation will be used for optimization to determine the best value fort those factors. So, the aim is to increase the solar fraction which means increasing the system efficiency or decreasing auxiliary heat needed.

Tilt angle (the inclination angle of the solar thermal collectors): Tilt angle has a great effect on harnessing solar energy by the collectors and being best exposed to direct irradiance. From the simulation it was decided that in order to gain the highest sun energy possible, the tilt angle should be adjusted monthly. Table (2) shows the tilt angles that the system will be adjusted at each month to obtain the highest sun fraction:

Table 2 tilt angles

Month	May	June	July	August	September	October
Tilt angle (°)	6	7	16	28	40	45

Solar collector's area: It is known that increasing the solar collector area, will increase the energy gained from the sun. However, in our case the collectors are installed on top of the roof of the building. So, the collector area is limited. Accordingly, the collector for the solar thermal collectors is 500 m^2

Storage tank volume: The effect of the storage tank volume was tested by the simulation. The data of solar fraction were recorded as shown in figure (6) for May with the optimized tilt angle for May. It can be seen that first increasing the storage tank volume increases the solar fraction. However, after a specific value (0.5 m^3) the surface area increase of the tank will result in an increase in the dissipiated energy to the environemt (heat losses). Accordingly, an optimum value for the tank (0.5 m^3) is used.

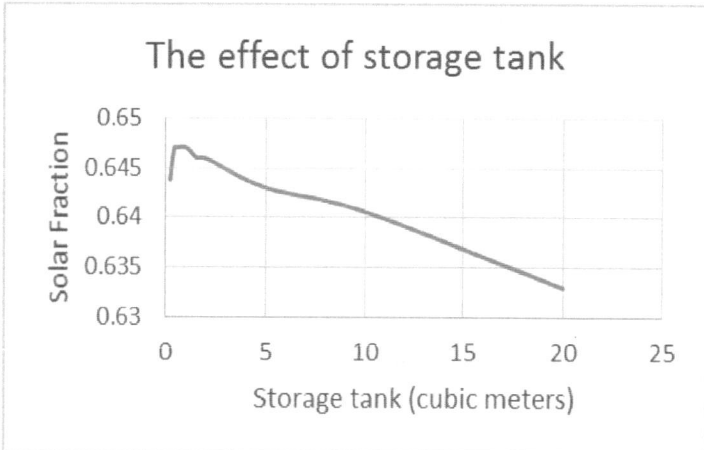

Figure 6 the effect of storage tank volume on solar fraction.

Collector mass flow rate: The collector mass flow rate also has a limited effect on the solar fraction as shown in figure (7). The figure shows that the solar fraction increases (i.e. the auxiliary heating decreases) by increasing the mass flow rate to some level (50000 Kg/hr). Beyond that level, the solar fraction starts decreasing again. However, the change is not signifigant. It should be noted that the electricty input to the pump is not taken in consideration here. However, its effect should not be much but it is expected to increase by increasing the flow rate. Hence, a flow rate of 45000 Kg/hr is chosen.

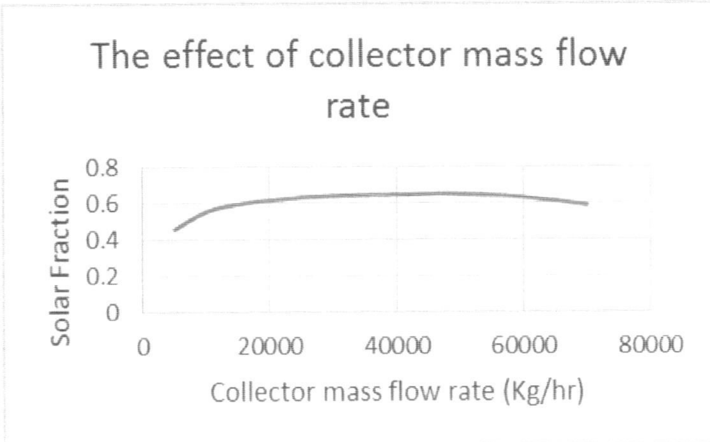

Figure 7 The effect of collector mass flow rate

Cooling tower: the simulations has showed that the chiller would need a cooling tower with a capacity of 1,111 Kw heat rejection rate which is equal to 253 tons of refrigeration given the fact that 1 ton of refrigeration for the cooling tower equals 4.39 Kw.

Other factors that might affect the coefficient of performance of the chiller (COP) are determined by the manufacturer and presented in table 1. It should be noted that the interactions effect between those different factors have not been studied. A further study is needed to determine how the interaction between different levels can affect the solar fraction.

System performance

The previous optimum values for each factor was used in the simulation to determine the amount of auxiliary energy needed and the solar fraction for each month. Figure (8) shows solar heating and auxiliary heating for each month using the previously optimized values. The solar fractions and electricity for months from May to mid-October are as following:

Table 3 solar fraction for each month

Month	May	June	July	August	September	October
Solar Fraction	0.65	0.56	0.51	0.47	0.47	0.41

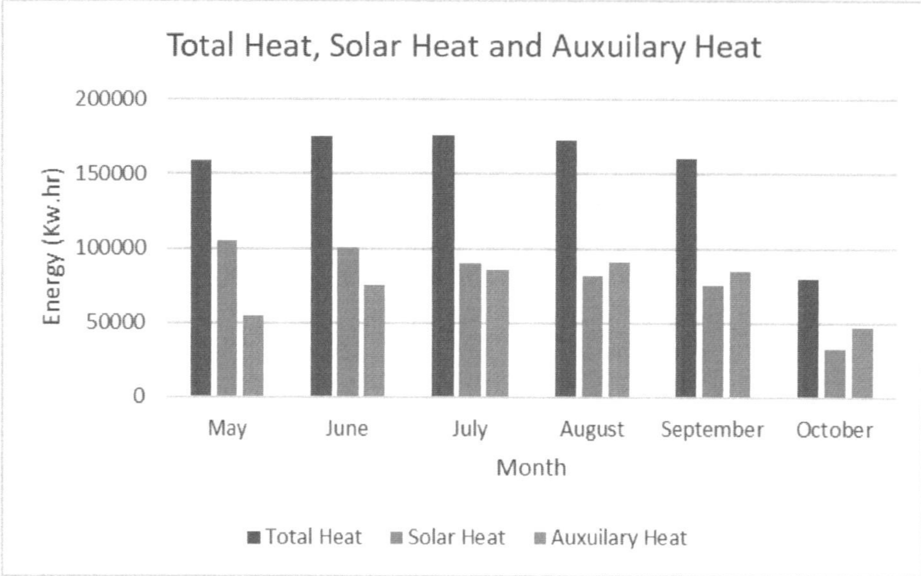

Figure 8 total heat and auxiliary heat for solar thermal cooling system

Energy savings

The proposed system runs on solar energy and auxiliary energy which use natural gas as energy source. The used natural gas heating value is 39.115 MJ/ m³ with a price of 2.49 USD/ MMBTU (L.E. 0.78) per m³ (About EOG, 2016) The following table represents the amount of natural gas required for each month and the costs of natural gas.

Month	May	June	July	August	September	October
Auxiliary energy (MJ)	196268.4	270547.2	307638	326289.6	306475.2	169804.8
Natural gas required (m3)	5017.727	6916.712	7864.962	8341.802	7835.235	4341.168
Natural gas cost for proposed system (L.E.)	3913.827	5395.036	6134.671	6506.606	6111.483	3386.111

The current installed system running costs are based on the assumption that the vapor compression ac unit consumes 1500 wat for each 1 RT. the price for electricity is L.E. 0.84 per Kw.hr in 2016. The following table shows the power consumption for each month and the electricity costs for the currently installed system:

Month	May	June	July	August	September	October
Cooling load per month (ton of refrigeration.hr)	35043	38613	38730	38421	36204	33774
Electricity demand per month (Kw.hr)	52564.5	57919.5	58095	57631.5	54306	50661
Electricity cost (L.E.)	44154.2	48652.4	48799.8	48410.5	45617	42555.2

The following figure shows the data in the previous tables:

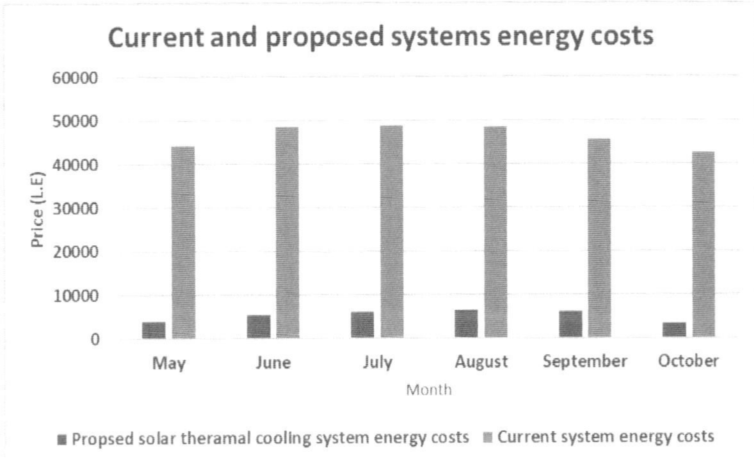

Figure 9 Current and proposed systems energy costs

The figure shows that the solar thermal cooling system will always cost less than 15% compared to the currently installed system. Which means that Energy savings of more than 85% can be obtained during the year. The following table shows the energy costs savings:

Month	May	June	July	August	September	October
Energy savings percentage	91.136	88.91106	87.4289	86.5595	86.60263	92.04302
Energy savings (L.E)	40240.3	43257.34	42665.1	41903.9	39505.6	39169.1

Total Energy costs savings per year= L.E. 246741.4

CONCLUSION

Solar thermal absorption chiller seems a perfect option for building H in the British University in Egypt. Solar Energy covers about 50% of the total energy used for driving the chiller. The remaining energy is mainly gained through natural gas which resembles most of the operating costs but still cheaper than electricity. It should be kept in mind that providing a natural gas line to the British University in Egypt may have additional costs that may eventually result in rejection of the proposed system. However, if natural gas is provided, Energy savings of about L.E. 245,000 will be saved annually. The design was optimized mainly in terms of solar collectors' tilt angle, thermal storage tank through calculating the solar fraction at different values. However each factor was studied independently while fixing other variables. Future studies should include the interaction between these factors and should also concentrate on the relationship between the thermal storage tank and the solar collectors' area.

NOMENCLATURE

Evacuated tube solar thermal collectors

a_0 [-]	Intercept (maximum) of the collector efficiency
a_1 [kJ/h-m^2-K]	Negative of the first-order coefficient in collector efficiency equation

a_2 [kJ/h-m²-K²]	Negative of the second-order coefficient in collector efficiency equation
I_T [kJ/h-m²]	Global radiation incident on the solar collector (Tilted surface)
T_a [°C]	Ambient (air) temperature
T_i [°C]	Inlet temperature of fluid to collector

auxuilary heater

Cpf [kJ/kg.K]	fluid specific heat
mI [kg/hr]	inlet fluid mass flow rate
mo [kg/hr]	outlet fluid mass flow rate
Qaux [kJ/hr]	required heating rate including efficiency effects
Qfluid [kJ/hr]	rate of heat addition to fluid stream
Qloss [kJ/hr]	rate of thermal losses from heater to environment
Qmax [kJ/hr]	maximum heating rate of heater
Tenv [C]	temperature of heater surroundings for loss calculations
Ti [C]	fluid inlet temperature
To [C]	fluid outlet temperature
Tset [C]	set temperature of heater internal thermostat
UA [kJ/hr]	overall loss coefficient between the heater and its surroundings during operation
γ [-]	external control function which has values of 0 or 1
η htr [0..1]	efficiency of auxiliary heater

Thermal storage tank

Mi	mass of fluid in the ith section
Cpf	specific heat of the tank fluid
Ti	temperature of the ith tank segment

Absorption chiller

Capacity [kJ/hr]	Mass flow rate of air entering the "fresh air" side of the device.
$f_{FullLoadCapacity}$ [0..1]	Fraction of the device's full load capacity during operation under current conditions.
$f_{NominalCapacity}$ [0..1]	Fraction of the device's nominal capacity during operation under current conditions.
$Capacity_{rated}$ [kJ/hr]	Rated cooling capacity of the device
Q_{remove} & [kJ/hr]	Amount of energy that must be removed from the chilled water stream in order to reach the set point temperature
$T_{chw,set}$ [°C]	"Chilled water" stream set point.
f_{design} [0..1]	Fraction of design capacity at which the machine is currently

operating.

COP_{rated} [-]	Machine's rated Coefficient of Performance.
$f_{DesignEnergyInput}$ [0..1]	Fraction of design energy Input currently required by the machine.
$T_{hw,out}$ [°C]	Temperature of fluid exiting the "hot water" stream
$T_{hw,in}$ [°C]	Temperature of fluid entering the "hot water" stream
Q_{hw} & [kJ/hr]	Energy removed from the "hot water" stream
m& $_{hw}$ [kg/hr]	Mass flow rate of the "hot water" stream fluid
Cp_{hw} [kJ/kg.K]	Specific heat of the "hot water" stream fluid.
$T_{chw,out}$ [°C]	Temperature of fluid exiting the "chilled water" stream
$T_{chw,in}$ [°C]	Temperature of fluid entering the "chilled water" stream
Q_{chw} & [kJ/hr]	Energy removed from the "chilled water" stream
m& $_{chw}$ [kg/hr]	Mass flow rate of the "chilled water" stream fluid
Cp_{chw} [kJ/kg.K]	Specific heat of the "chilled water" stream fluid.
$T_{cw,out}$ [°C]	Temperature of fluid exiting the "cooling water" stream
$T_{cw,out}$ [°C]	Temperature of fluid entering the "cooling water" stream
Q_{cw} & [kJ/hr]	Energy added to the "cooling water" stream
m& $_{cw}$ [kg/hr]	Mass flow rate of the "cooling water" stream fluid
Cp_{cw} [kJ/kg.K]	Specific heat of the "cooling water" stream fluid.
Q_{aux} & [kJ/hr]	Energy draw of parasitics (solutions pumps, controls, etc.)
COP [-]	Coefficient of Performance for the device.
Mi	mass of fluid in the ith section
Cpf	specific heat of the tank fluid
Ti	temperature of the ith tank segment

References

A.Allouhi, T.Kousksou, A.Jamil, P.Bruel, Y.Mourad, & Y.Zeraouli. (2015). Solar driven cooling systems: An updated review . *Elsevier, RenewableandSustainableEnergyReviews.*

About EOG. (2016, JUNE). Retrieved from EGYPT OIL & GAS WEB PORTAL: http://www.egyptoil-gas.com/about-us/

Absorbtion Refrigeration. (n.d.). Retrieved from Oregon State Univeirsity.

ac-system. (n.d.). Retrieved from aireserv: http://aireserv.com/expert-tips/ac-system

Ahmed AbdelKarim, A. F. (2015). *Design of a distrcit conditioning system.*

Bermejo, P., Pino, F., & Rosa, F. (2010). *Solar absorption cooling plant in Seville, Sol. Energy.* Elsevier.

Best, R., & Rivera, W. (2014). A review of thermal cooling systems. *ScienceDirect.*

Burger, A. (2015, June 9). Egypt's Renewable Energy Drive Gains Steam. *Renewable Energy World Magazine*, 2.

(2015). *Egypt/ International energy data research and analaysis.* U.S. Department of Energy. Washnigton : eia Beta.

Erickson, D. (2007). 100 ton absorption chiller/heat pump demonstrates the real cost of saving energy. *ASHRAE Annual Meeting.* Long Beach, CA: SCOPUS.

F, A., KalogirouSA, AliY, & SopianK. (n.d.). Simulation and optimization of a LiBr solar absorption cooling system with evacuated tube collectors. *Elsevier, Renewable Energy.*

Figueredo, G., Bourouis, M., & Coronas, A. (2008). Thermodynamic modeling of a twostage. *Elseveier,Applied Thermal Engineering.*

Ghafoor, A., & Munir, A. (2014). Worldwide overview of solar thermal cooling technologies. *elsevier.*

Ghafoor, A., & Munir, A. (2015). Worldwide overview of solar thermal cooling technologies. *Elsevier, RenewableandSustainableEnergyReviews.*

Ghoneim, E. (2009). *A Remote Sensing Study of Some Impacts of Global Warming on the Arab Region.*

Grossman, G. (2002). Solar-powered systems for cooling: dehumidification and air conditioning . *Elsevier, Solar Energy.*

Guide to solar thermal energy. (n.d.). Retrieved from Solar Tribune: http://solartribune.com/solar-thermal-power/

How Air Conditionig Works. (n.d.). Retrieved from How stuff works: http://home.howstuffworks.com/ac3.htm

Khalil, A., Mubarak, A., & Kaseb, S. (2010). Road map for renewable energy research and development in Egypt. *Journal of advance research.*

Kohlenbach, P., & Jakob, U. (2014). *Solar Cooling- The Earth Scan Expert Guide To Solar Cooling Systems.* Abingdon, Oxon: Routledge.

(n.d.). *MODEL FOR DOUBLE-EFFECTcx.*

Nakahara, N., Miyakawa, Y., & Yamamoto, M. (1977). Experimental study on house cooling and heating with solar energy using flat plate collector. *Elsevier, Solar Energy.*

Raslan, R. (2013, May 14). *Empowering Egypt: Challenges.* Retrieved November 2, 2015, from Carboun: http://www.carboun.com/energy/empowering-egypt/

Sarant, L. (2105, Feburary 18). *The Rise of Solar Energy in Egypt.* Retrieved from Middle East Institute: http://www.mei.edu/content/article/rise-solar-energy-egypt

Sarbu, I., & Sebarchievici, C. (2013). Review of solar refrigeration and cooling systems. *elsevier.*

Solar hot water collectors. (n.d.). Retrieved from Greenspec: http://www.greenspec.co.uk/building-design/solar-collectors/

Srikhirin, P., Aphornratana, S., & Chungpaibulpatana, S. (2001). A review of absorption refrigeration technologies. *Elsevier, Renewable and Sustainable Energy Reviews.*

Tierney, M. (2007). Options for solar-assisted refrigeration trough collectors and double-effect chillers. *Elsevier, Renewable Energy.*

VASILESCU, C., HERA, D., & FERREIRA, C. I. (n.d.). *MODEL FOR DOUBLE-EFFECT ABSORPTION REFRIGERATION CYCLE.* TECHNICAL UNIVERISTY OF CIVIL ENGINEERING OF BUCHAREST, Romania.

Contributing Algorithms of Energy Efficiency and Renewable Energy in the Residential, Commercial and Industrial Sectors

Georges El-Jamal **Dr. Chantal Maatouk** **Dr. Mazen GHANDOUR**
Dr. Adel Mourtada **Dr. Fouad Kaddah**

ABSTRACT

Renewable energies (RNE) and energy efficiency (EE) are considered as twin pillars in the energy strategic battle to decrease fossil energy consumption. The combined potential of energy efficiency and renewable energy to reshape energy future is substantial. The energy consumption profile of an industry process or facility is considered to be an energy footprint and a main axis in its life

The aim of this paper is to highlight to the existing energy profile of a facility or industry process and to elaborate different steps to reduce it by exposing an innovative algorithm (methodology), contributing between EE and RNE, to achieve a low energy profile in residential, commercial and industrial sectors

INTRODUCTION

Energy is a major significant dilemma that harasses the world. The day of depletion of oil, which runs life on earth, is very close, so replacing this type of energy with another cheaper, clean and lasting become a strategic question. Is renewable energy (RNE) the solution? Nowadays this energy is unable to completely replace the fossil types for different reasons, where, immaturity of the technologies, difficulties of control, mismatching between production and consumption, lack of awareness of consumers, are the main reasons, but the contribution of energy efficiency (EE) and (RNE) can be the main axis in solving the dilemma, where the first keeps energy demand under its minimum level and the second reduce fossil energy demand and emissions, in this context, many tools have been developed to prepare the necessary studies to illustrate the application.

RENEWABLE ENERGY

Wind power:

The record of the year 2016 for wind power has a global capacity of 487 GW and a price of 0.10-0.12 $/kwh

Hydropower

An estimated 1246 GW power plants were installed worldwide and 4100 TWh of electricity was generated during 2016

[1]G. EL-JAMAL, PhD student at Saint-Joseph and Lebanese University, Beirut, Lebanon, (e-mailgfjamal@yahoo.com)
[2]Chantal Maatouk Professor at Saint Joseph University, Beirut, Lebanon, (e-mail chantal.maatoukriachi@usj.edu.lb)
[3]Mazen GHANDOUR Professor at Lebanese University, Beirut, Lebanon, (e-mail drmazengh@yahoo.com
[4]Adel Mourtada Professor at Lebanese University, Beirut, Lebanon, (e-mailecotech@inco.com.lb)
[5]Fouad Kaddah Professor at Saint Joseph University, Beirut, Lebanon, (e-mail fouad.kaddah@usj.edu.lb

Solar energy

By the end of 2016, global solar thermal capacity reached an estimated 420 GW_{th}. and 180 GW for PV

Biomass:

Biomass can be generated by combustion to create heat, or to transform it to biofuel. By the end of 2016, global bio power capacity reached an estimated 112 GW

Geothermal energy:

Geothermal coming from crust rock heated up to 700 °F (370 °C). In 2016 delivered approximately 220 TWh,

ENERGY EFFICIENCY

There are many motivations to ameliorate EE and to minimize energy consumption where is looked as a solution to the increasing of carbon dioxide emissions, so many strategies and applications were developed for this issue.

Technologies and concepts applied in residence and commerce sectors

Many actions can be fulfilled in buildings (residential and commercial) to applied EE as follows

1- Reacting in building's orientation, location, , to benefit from passive solar heating, cooling and lighting to reduce energy used by 20%.
2- Reducing U-value by applying thermal insulation for the envelope of building and proper placement of windows and skylights will save energy by 40%.
3- Using smart heating and cooling systems with high COP and EER will reduce the profile of energy consumption by 25%.
4- Lighting with led light uses 90% less energy than incandescent bulb and life span longer about 30 times and also reduces cooling load by 5 to 7%.
5- Controlling and monitoring the buildings by using BMS for mechanical and electrical equipments will save up to 15% of energy used.
6- Adopting home and office energy efficiency rated equipment, which use 30 to 70% less energy than non rated apparatus

Technologies & concepts applied in the industrial sector

Best available techniques (BAT) are actually well matured methods of application of EE in industry sectors and are summarized below:

1- Maintaining equipment is the first key in energy saving and high performing machineries.
2- Adopting the cogeneration plants deliver more than 50% energy efficiency
3- Recovering 5 to 10% of energy from steam boiler exhausts.
4- Insulating steam and condensing return pipes, avoiding leakage of steam reduce 25 % of the energy used for steam production.
5- Using Voltage Optimization VO system, minimize energy consumption by 15% and ameliorate power quality by balancing phase voltages and screening harmonics.
6- Employing Variable Frequency Drive, VFD, technologies in motors (pumps, fans…etc) permit to users to save 18% of energy.
7- Applying process integration enriches the unification of the processes and acts in the interactions between the different operating phases from the start to the end rather than optimizing them separately. The process integration should envelop five main scopes, the optimization use of raw materials, process operations, EE, emissions minimization, mass and energy integration. The integrating of process concepts can be applied in various fields. one of main method of process integration is Pinch method, The intervention of different technologies of process integration can be achieved a 40% of energy savings

Minimum Energy Required and Energy Losses

The total energy consumed in any industry process is shared between two main energy parts as illustrated in Figure.1:

The Minimum Energy Required MER which is defined as the minimum energy theoretically necessary and sufficient to manufacture the product. Two types of MER exist:

1. MER Product: is the minimum energy absorbed by the product to be changed from raw material to final product.. It depends on product quality
2. MER Process: the minimum energy consumption of process.

Energy Losses energy produced but not used as MER, they considered inactive. It's divided into two parts:

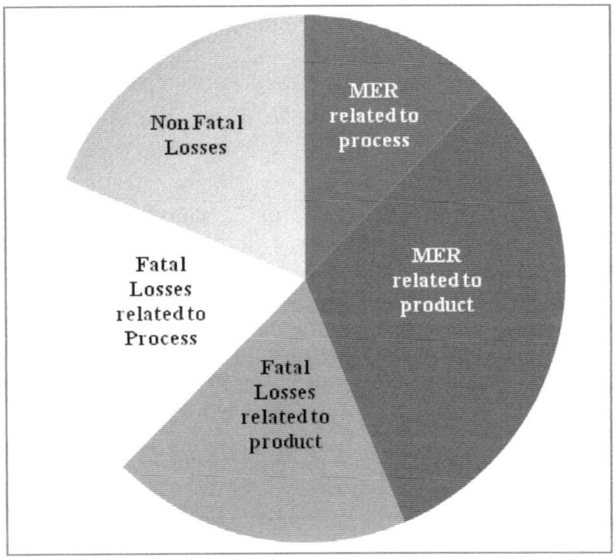

Figure 1 Distribution of energy consumption in industry

1. **Fatal losses** are related to the transformation of the product, these losses require change in process to minimize it.
2. **Nonfatal energy losses, does not require a change in the manufacturing process.**

METHODOLOGY AND BASED DESIGN

Review of existing tools

Many tools have been developed to prepare the necessary studies to illustrate the application of EE and adoption of RNE. The following information shed light on some software that deal with studies

HOMER, Computer program, helps designing RNE hybrid projects.

RETScreen, A technical and financial of application of EE reports are defined by using the RETScreen software, with its five phases; energy analysis, cost analysis, carbon foot print, fiscal study, and risk assessment;

Hybrid2, Software used to design a hybrid system, contains many types of electrical features (battery storage wind turbines, photovoltaic, diesel generators,), and prepare economic analysis.

Development of new tools

Two Algorithms were developed, one for residence and commerce sector, the second for industrial sector. The two algorithms have the target to show the actual energy profile and to calculate the different energy efficiency and renewable profiles by introducing two types of multi criteria decision methods to select the best renewable energy type suitable for the region

Residence and Commerce Algorithm (RCA) (Fig 2) is composed of five main parts:

1. The first part is the calculation of the actual energy consumption profile after the collection of energy data of the different equipment may be exist in building (HVAC, Lighting, hot water, domestic and office equipments, etc).
2. The second part is the energy efficiency application where the previous energy profile will be reduced and a new energy efficiency one will display by acting and adopting a chain of actions above described
3. The third part is the application of multicriteria decision methods to select the best type of renewable energy. The use of Multicriteria Decision Method (MCDM) techniques provides a reliable methodology to rank alternative renewable energy resources. Two methods are used in the algorithm; Analytic Hierarchy Process (AHP) and Preference Ranking Organization METHod for Enrichment Evaluation (PROMETHEE)
   - The above two methods are used in the algorithm in order to determine the priority of one type of renewable energy (wind, hydraulic, solar, geothermal, etc.) over the others and rank the different possible types by using many criteria like cost, environment, operational, site selection, efficiency,...etc

- The purpose of using of two methods is only to confirm the results achieved
- The analytical hierarchy process (AHP) (Saaty, 1982, 1988, 1995) is probably the best known and most widely used MCDM approach. It is based on three principles, namely:
  - Construction of a hierarchy,
  - Priority setting
  - Logical consistency
- PROMETHEE tool is used to analysis the evaluation problem. The tool is divided into tree sub tools:
  - The PROMETHEE I partial ranking of alternatives;
  - The PROMETHEE II complete ranking of the alternatives
  - The GAIA plan visualize graphically

4. The fourth part has a target of calculation of energy efficiency and renewable energy profile after the implementation of the suitable type of RNE, was selected in the previous part of the algorithm.
5. The fifth part is preview for financial reports which contains useful economic indicators. Net present value , Constant annuity, Internal rate of return IRR or TRI, Revenue per unit of expenses (R.U.E), Payback period PBP, Different charts , Saving values of different kind of energy consumption

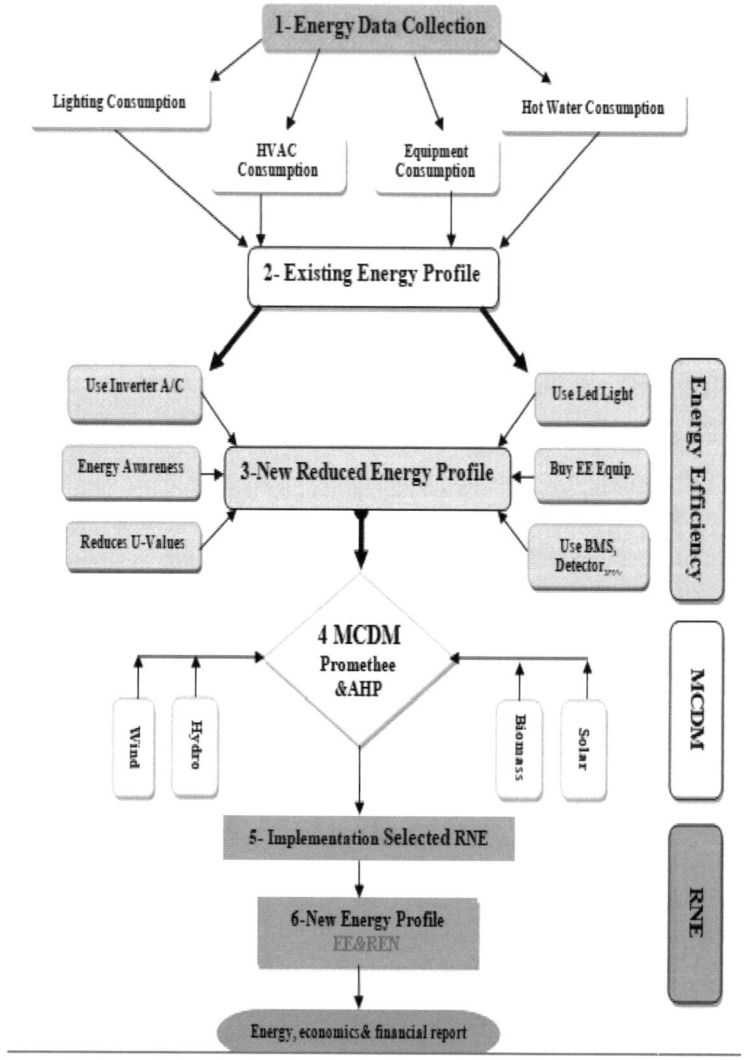

Figure 2 Residence and Commerce Algorithm

The table1 below shows comparison between four algorithms according to four criteria

- Application of Energy efficiency
- Multi criteria decision methods MCDM.
- Implantation of Renewable energy.
- Feasibility study& economic indicators.

Table 1. Comparison table between different algorithms

Factors	RETScreen	HYBRID2	HOMER	RCA
EE	+	--	--	+
MCDM	--	--	--	+
RNE	+	+	+	+
Feasibility Study	+	--	--	+
	+ available	-- not available		

Industry Algorithm Fig.3 is composed also of five main parts

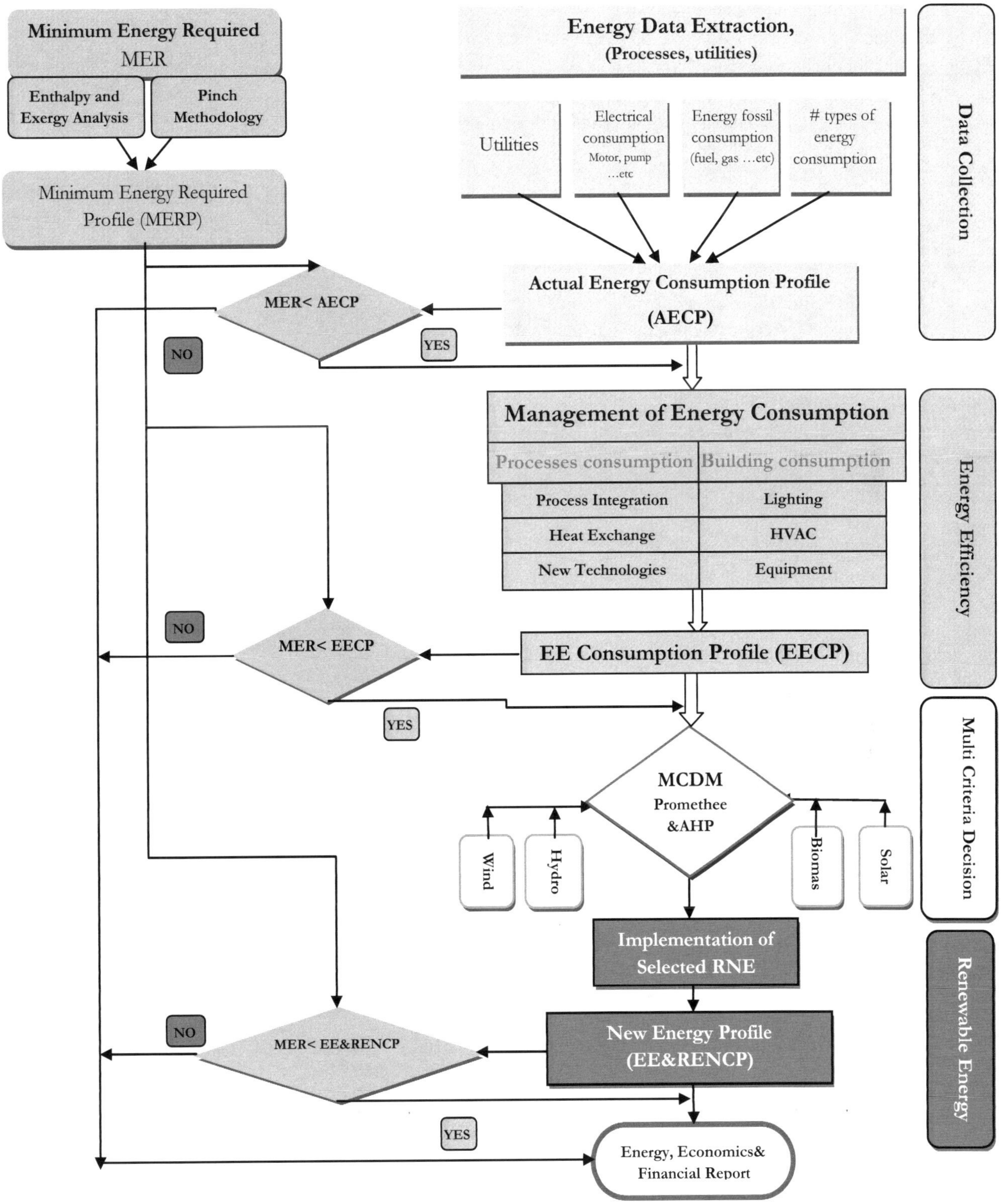

Figure 3 Industry Algorithm

1. First part is the display of actual energy consumption profile which is calculated after energy data extraction of industry process. The data collected mainly are divided into three types:
   - Electrical which run different equipment like motor, lighting, cooling system, pumps etc,
   - Thermal come from fossil oils and produce steam used as heat transfer fluid for equipment
2. The second part is the energy efficiency section, by using the best available technology discussed above, the previous energy profile will be reduced and a new energy efficiency profile will display
3. Third, fourth and fifth parts are similar to the algorithm of different energy profiles of residence and commerce sector

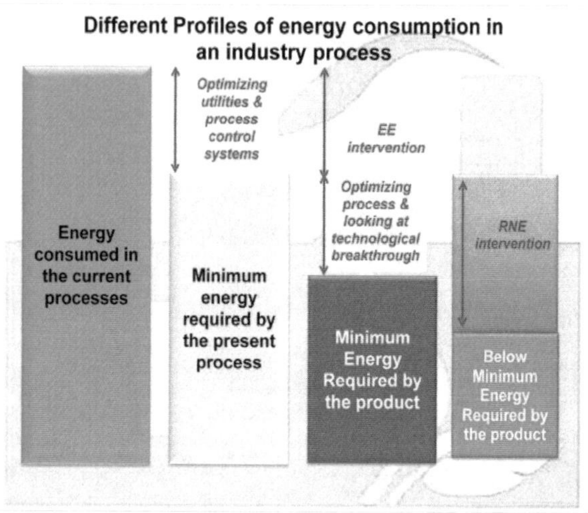

Figure 4. different profiles of energy consumption in industry process

The three energy profiles are tested with the fourth profile, the minimum energy required MER at the end of each part in order to show the percentage of reduction of quantities of energy

The four energy profiles are presented in Fig.4 where the scale relations are displayed between them. The interventions of EE and RNE will reduce the size of profiles

APPLICATIONS, RESULTS AND ANALYSIS

The applications of the two above algorithms lead to the results displayed in table 2 and table 3.

1. Table 2 represents the summary of different energy consumption profiles, actual and after the intervention of energy efficiency and the application of RNE for a residential complex, composed of 3 buildings, each composed of 6 stories, the first has 18 apartments with story area of /450/m², the second and the third have 12 apartments each with story's area of / 350/and/ 325/m².the complex is located in Beirut-Lebanon. where a rationing of 50% is applied
2. -Table 3 shows the summary results of energy consumption profiles of distillation process of rose water in a factory located in Beirut, steam boiler is used heated by fuel #2 with cooling tower in order to condensate the evaporated water with essentials oils

Table 2 Summary table results of applications of Residence and Commerce Algorithm for a residential complex

	Actual situation	EE			RNE			EE+RNE	
Electricity Rationing 50%	**Cons. kWh**	**Cons. kWh**	**Cost $**	**PBP years**	**Cons. kWh**	**Cost $**	**PBP years**	**Total Cost $**	**PBP years**
HVAC Elec. Consumption	186560	79151	210100	9.03	79151				
Equipment Elec. Consumption	119249	119249			104013	39000 PV	11.81		
Lighting Elec. Consumption	150012	8824	40250	1.32	0	21000 LED	10.98	**415350**	**5.3**
Hot water Elec. Consumption	106697	106697			21924	105000 Solar panel	5.44		
Total	**562518**	**313921**	**250350**	**4.65**	**205088**		**6.73**		

Table 3 Summary table of Different Loads for the actual Process and proposed processes for Rose water distillation

	Thermal					Electricity				Energy Used	Energy Cons/Kg
	Total Heating	Total Cooling	Total Energy	Losses Pipes	Make Up Water	Sterilization	Filling	Cooling	RO		
	kwh	kwh	kwh	kwh	kwh	kwh	kwh	kwh	kwh	kwh	kwh
Actual Process	464	357	822	166	7570	14	30.3	424.3	369	9396	20.67
Process Under Pressure	116	5.3	121.3	41.6	111	11	8.7	6.2	92.3	392	0.86
Savings										96%	

The results displayed in Table 2 shows the different energy consumption profiles, actual, after the intervention of energy efficiency and after the application of renewable energy in residential complex where PV and hot water solar panel were applied. The selection of renewable energy type was made by using the multi criteria decision methods. A reduction of 44% of energy consumption has been found after EE intervention and 64% after RNE application and a total payback period of 5.3 years.

Also the results presented in Table3 shows the energy consumption profile of production of rose water by distillation in a factory, by the intervention of process integration methods by using pinch method and adding one heat exchanger between hot and cold streams a 25.5% energy saving was registered. By applying the under pressure process a 96% of energy saving was registered. The application of renewable energy was not studied yet and so the payback period

CONCLUSION

The contribution of energy efficiency and renewable energy in the two discussed algorithms, residence and industry sectors, is well illustrated and display three energy profiles; actual situation, application of EE, implementation of RNE; for buildings or industry processes and have the followings five common steps:

1. Display its actual energy consumption profile.
2. Elaborate the new energy profile by application of energy efficiency.
3. Select the best renewable energy types can be applied by using two types of multicriteria decision methods.
4. Calculate the final energy profile using integration of EE and RNE
5. Preview financial & the useful economic indicators

The participation of energy efficiency and renewable energy to find the solution of the problem of degreasing the consumption of fossils energy has many paths, the above two algorithms explained show a strategic map in issue, the importance of introducing the multicriteria decision methods MCDM in the two algorithms to select the appropriate type of RNE without conducting a feasibility study for all types mentioned energies

NOMENCLATURE

PV = Photovoltaic panel

WIND= Wind power

HE=Hydroelectric

EE = Energy efficiency

RNE. = Renewable energy resources

MCDM= Multi-criteria decision methods

AHP =Analytic Hierarchy Process

PROMETHEE= Preference Ranking Organization Method for Enrichment Evaluation

COP=Coefficient of performance

EER=Energy efficiency rating.

BMS= Building management system.

HVAC= Heating, Ventilation and Air conditioning.

Con. =Consumption

RO= Reverse osmosis

PBP= Pay back periods

REFERENCES

Global Wind Statistics 2016, 10-2-2017, Global wind energy council

Hydropower status report 2017, International hydraulic associations

G.z.Brown and M.Dekay, 2013, the importance of building orientation. Available online www.ecowho.com/articles/6/

N. Gromicko and B. Gromicko, 2012, Structural Design Loads for the Home Inspector. Available on site https://www.nachi.org/building-orientation- optimum- energy.

Reliable plant, U.S. Department of Energy, 2014, Maximize system energy efficiency with proper insulation.

A. Michel, E. Bush, J. Nipkow, C. U. Brunner, 2010, Energy efficient room air conditioners best available technology (BAT), European commission,

LED Light-Emitting Diode (LED) Design, 2016, available on line: www.littelfuse.com.

A. Smith, 2010, Building management system services, Available on site: http:// www.greenconsultancy.com

F. Oosterhuis, 2006, Energy efficient office appliances, Institute for Environmental Studies, Vrije Universiteit De Boelelaan, Amsterdam the Netherlands,

Reference Document on Best Available technology for Energy Efficiency, European commission, February 2009,

B. Darabnia, M. Demichela, 2013, Maintenance an Opportunity for Energy Saving,

ClimateTechbook, 2011, Cogeneration Combined Heat and Power,

Use Feed water Economizers for Waste Heat Recovery, 2012, U.S.Department of Energy,

Steam Distribution and Condensate Return Lines, 2012,U.S. Department of Energy Insulate,

A. Mardapittas, 2016, Voltage-Optimisation-White-Paper, Power star, UK,

Alexander Kamenka, Six tough topics about harmonic distortion and Power Quality indices in electric power systems, 2014

E. Mike, 2012, Adjustable Speed Drives: Drive Up Energy Efficiency,

Bose, B. K, Feb 2009, power electronics and motor drives recent progress and perspective. IEEE Transactions on Industrial Electronics,

N.HALLALE, 2006, brining bright trends in Process integration,

Natural resources Canada, 2003, Pinch Analysis: For the Efficient Use, of energy, water and hydrogen,

B. Linnhoff, D.W. Townsend, D. Boland, G.F., 2007, Second edition, User Guide on Process Integration for the Efficient Use of Energy.

Homer software, 2015, available on site: http://www.homerenergy.com/

RETScreen, software, 2017, available on site: http://retscreen.software.informer.com/4.0/

Hybrid2/software, 2014, available on site: https://www.umass.edu/windenergy/ research/ topics/ tools/ software/hybrid2

G. El-Jamal, sep. 2016, Application of multi-criteria decision- Methods for the selection of renewable energy sources, Lebanese case study, presented at 2nd International Conference Efficient Building Design in AUB

T. L. Saaty, the Analytic Hierarchy Process. New York, McGraw-Hill, 1980

R. Abu Taha and T. Daim, 2009, Multi-Criteria Applications in Renewable Energy Analysis, a Literature Review,

Nijkamp et al, 2004, PROMETHEE and AHP: The design of operational synergies in multicriteria analysis. Strengthening PROMETHEE with ideas of AHP.

Establishment of a Sustainable Energy Action Plan: Case Study of Union of Municipalities

Sabine SAAD,
Student Member ASHRAE
Marwan EL BROUCHE, PhD
Adel MOURTADA, DrEng
Mazen GHANDOUR, DrEng

ABSTRACT

While climate change mitigation and adaptation strategies are the world's most prominent issue today, Lebanon, the Southern Mediterranean country, is trying to develop an integrated plan for urban development, energy supply, and environmental management. For years, the central authority in Lebanon has been unable to provide a steady production and distribution of electrical energy or even invest in the existing renewable resources. This has lead local authorities, or municipalities, to design their own plans despite the lack of accurate data and the absence of supporting measures initiated by the State. In the framework of the Covenant of Mayors (CoM) applications in the South Mediterranean region through the CES-MED program, it has been noticed that several Lebanese applicants wish to commit to the Covenant but are obstructed by many barriers which we shall highlight and evaluate in this paper. The paper also presents a complete case study of the Union of Municipalities of Shouf–Soueijany encompassing eight different municipalities. The case study establishes that the calculation of the total amount of greenhouse gas (GHG) emissions resulting from the overall energy consumption of the 8 municipalities (the Baseline Emission Inventory or BEI) will enable the union to comply with the CoM through the implementation of a Sustainable Energy Action Plan (SEAP). In order to identify the nature of the entities emitting CO2 in the scope of the municipal union, we have resorted to onsite data collection. The unavailability of official reliable statistics and references has made data collection and the determination of a baseline year a significant obstacle to the research.

INTRODUCTION

The Sustainable Energy Action Plan (SEAP) for the Union of Municipalities of Shouf-Soueijany might be one of the few achievements in the South-Mediterranean, and Lebanon in particular. The CES-MED (Cleaner Energy Saving Mediterranean Cities) provided training, technical assistance, capacity building seminars, help desk services and SEAP preparation manuals (in English, Arabic and French) to a considerable number of municipalities in the South-Mediterranean. In Lebanon, 3 cities were selected first (Beirut, Baakline and Qab Elias) and their SEAPs were prepared and supported by CES-MED. Two additional municipalities (Menjez and Ashkout) joined later and prepared their own plans under the project's guidance known as the "Information Kit to develop SEAPs" [1] considered as an effective tool to improve policies and systemize the preparation and financing of SEAPs. The Menjez SEAP was submitted first and approved. It was later followed by Baakline and Qab Elias. The SEAPs of remaining cities have yet to be finalized and submitted. The CES-MED project has designed a self-help guide for municipalities wishing to implement a sustainable energy policy. It is a step-by-step method that helps them achieve their sustainable energy

Sabine SAAD is a PhD student in EDST, Lebanese University, Hadath, Lebanon and in EDSIT, Saint Joseph University, Mar Roukoz, Lebanon.
Adel MOURTADA is a Professor in the Mechanical Department of the Faculty of Engineering III, Lebanese University. ASHRAE Student Branch Advisor.
Marwan El BROUCHE is a Professor in the Department of Preparatory classes, Ecole Superieure d'Ingenierie de Beirut, Saint Joseph University, Lebanon.
Mazen GHANDOUR is a Professor in the Electrical Department of the Faculty of Engineering III, Lebanese University.

goals without full support from CES-MED. Nevertheless, these tools have not always been sufficient as some cities that were selected in the beginning of the program and whose plans were expected by the end of 2015 haven't submitted their plans, yet the CES-MED authorization ended in June 2017. As for the methodology, it has been obstructed by several limitations such as the absence of values regarding the emission factors in the baseline carbon footprint specified to each country. To establish the Baseline Emission Inventory (BEI), local authorities had to rely on inaccurate data that generated inaccurate results. Furthermore, the methodology would not allow any authority wishing to join without any technical assistance from CES-MED to identify and determine the performance of each measure clearly listed by its EE priority and its related investment cost.

Additional cities and villages were supported by the regional program "Sustainable Urban Demonstration Projects" (SUDEP South) to implement sustainable energy actions to address local sustainable development challenges such as energy efficiency, security of energy supply and sustainable economic growth.

However, these projects have failed to contribute to mitigation ambitions and GHG emissions reductions at the national level during the last couple of years. The overall impact of such projects does not have a direct impact on the national achievements in energy strategies and policies, especially in Lebanon. This is particularly obvious in the absence of appropriate laws for mitigation and clean energy production as well as the poor awareness of such issues at the individual, collective and municipal levels.

By dealing with energy consumption and energy bill reduction as strategic priorities for the municipal union, the SEAP established in this study will eventually be considered as a development tool for the union. It suggests eight small and simple actions-solutions that would yield immediate results. The comprehensive target of this SEAP is to accelerate the decarbonization of the scope of the union, strengthen its capacity to adapt to unavoidable climate change impact, and allow its citizens to access secure, sustainable and affordable energy. The quantitative target for this SEAP is to mitigate carbon emission with a total reduction of 54,403 tCO2-eq.

METHODOLOGY, CONSTRAINTS AND BARRIERS

The methodological and technical tools adopted in the framework of COM, mainly targeting their goals of CO2 emissions reduction, energy savings and use of renewable sources, collaborating with 8 local municipalities in the Federation are divided into two categories: (i) Establishment of the BEI and (ii) development of the SEAP.

Establishing the BEI of the Federation will assess its basic situation with respect to the calculation of GHG emissions. It quantifies the amount of CO2 emitted due to energy consumption in the territory of these local authorities in the baseline year chosen 2015 in our case study since it doesn't have data to compile an inventory for 1990 (baseline year recommended by the COM), then it should choose the closest subsequent year for which the most comprehensive and reliable data can be collected. The collecting method applied in this study case is bottom-up data as it is considered at local level. Therefore, some values are evaluated per capita or per household. Then, once the samples were selected in baseline emission inventories database, a reliable dataset will be resulted due to a statistical procedure.

Tackling mitigation and adaptation to climate change approaches are new trends in Lebanon, there is no serious steps toward reducing the energy consumption, decreasing the conventional energy combustions and adopting the renewable clean energy resources. All the energy policies set by the national authority might be unsuccessful when there are no energy conservation laws imposed by the parliament that would help promoting energy conservation initiatives and the electricity generation / distribution is exclusive for one institution with legislative prohibition of any other decentralized decision in this issue. Implementing codes and regulation by local authorities in the building sector with more stringent requirements than national ones, will improve energy efficiency in the building envelope and foster the use of renewable sources for space heating and cooling, contributing then to the total estimated GHG emission reduction. Those are the major constraints and difficulties that might face the implementation of any action toward sustainability. To develop and implement the SEAP of our case study, we use some of the measures developed through different projects and initiatives under the Covenant of Mayors in the European cities as well as in the South

Mediterranean cities. Studying the environmental performance of each as part of a global approach to achieve by the determination of the energy performance, environmental benefit and the related investment cost for each measure. Parallel to these measures, awareness raising, and information campaigns will mobilize public interest in sustainable energy and create broad-based political and social support for the implementation of the SEAP, since it contributes with 26 % to the total estimated GHG emission reduction by 2020 referring to the Joint Research Center (JRC) report [2]. Quantitative indicators such as costs (in $/year), energy savings (in MWh/year), clean energy production (in MWh/year) and estimated CO2-eq emissions reduction (in tons/year) are the instruments to measure the impact of each action clarifying the economical/environmental benefits.

GIVEN FOR THE CASE STUDY

The Union of Municipalities of Shouf-Soueijany includes 9 municipalities Atrine, Ain W Zein, Ainbal, Baaqline, Gharifeh, Jdaideh (Shouf), Kahlouniyeh, Mazraat El Shouf, Semqaniyeh. Baakline will be excluded in this study because this municipality already applied for the COM. The study will hold the 8 other municipalities. This authority, founded in 1979, is in Mount of Lebanon at around 45km away from Beirut. Widespread over 5280 ha, excluding Baakline, it is situated in the heart of the mountain of 820-950 m altitude with 64200 inhabitants distributed in 17000 households occupied all around the year and 1140 households for summer only. It is characterized with a moderated climate with an annual mean temperature around 15°C and an annual rainfall of 1213mm.

The Chouf-Soueijany is rich in renewable energy resources. Many water resources and particularly Bisri river (Awali) ensure the water needs in that area for inhabitants as well as for agriculture. As for the solar potential, the Global Horizontal Irradiance GHI is estimated around 1907 kwh per m2 and the Direct Normal Irradiance DNI is estimated around 2144 kwh per m2 for a Diurnal Air Temperature of 17.7°C [3]. For the waste biomass potential, and referring to the local studies, the waste production is estimated 0.8 kilograms per person per day (kg/p/d) in rural areas to around 1.1 kg/p/d in urban areas, with a national weighted average estimated at about 1.05 kg/p/d [4]. The weak management of waste leads to the burning in the landfill sites to reduce the waste volume and therefore emitting toxic smokes, polluting the soil and the groundwater, disease transmission and GHG emissions. For the biomass upcoming from forests and agriculture, the Chouf-Soueijany has a large potential due to the large green spaces but unfortunately uncontrolled and exposed to fires in hot seasons.

The electricity provided by the local state to this region is limited with periods of interruption of 50% but can reach sometimes 20 hours per day and which is covered by private low efficient, high polluted electric generators and relatively expensive (more than 100USD per household per month). Noting that there is a remarkable absence of any type of renewable energy production to cover the weakness of the local electricity.

BASELINE EMMISSION INVENTORY

The BEI is the instrument allowing the local authority to measure the impact of its actions related to climate change. It will show where the local authority was at the beginning, and the successive monitoring emission inventories will show the progress towards the objective. The activity data that quantify the human activity occurring in the territory of the local authority is recommended to calculate the amount of CO2 emissions by multiplying the emission factor with corresponding activity data. The human activity for the baseline year 2015 is quantified in the Table 1 in all the following sectors: municipal sector (buildings, equipment, facilities), non-residential sector (tertiary, commercial, service...), residential sector, municipal public lighting, transportation (public and private), local electricity production (electric generators and renewable energies locally produced) and is characterized by either electricity consumption or oil/fuel consumption. In Chouf-Soueijany, the activity data are distributed between electricity consumption, gas for cooking and diesel or wood for space heating for municipal, residential and non-residential buildings, diesel or gasoline consumed for the transportation sector, electricity consumption for the public street lighting and diesel consumption for private electrical generations. Noting that the data were too hard to collect for the baseline year leading to many estimations in filling the activity data sheet reported in the Table 1 and giving the total consumption in non-renewable (emitting) energy per sector for the different villages of the federation. The most consuming sector in the territory is the building sector.

The total emissions of GHG is required to complete the BEI by using the conversion factors issued by the ministry of environment in Lebanon for every type of energy and given in Table 2 noting that standard emission factors are used.

By conclusion, the GHG emissions reported in the BEI 2015 dataset have been aggregated per federation sector and sub-sector. The total emissions are 99,300 tCO2-eq per year in Chouf-Soueijany with a preponderant contribution from the buildings sector (48.96%) followed by the private generators (34.56%) as shown on Chart 1. To quantify the expected emissions by 2020 to be considered, the national GHG coefficient of the Guidelines for National Greenhouse Gas Inventories developed by the JRC for the South-MED countries [5] is 1.19 for the baseline year 2015 for Lebanon. This value is multiplied by the baseline GHG emissions of 99,300 to reach a total GHG emissions in the year 2020 of 118,167 tCO2-eq. As a result, residential buildings and private generators represent the main sectors where action should be focused and in which projects that reduce emissions should be developed.

Table1. Human activity in all sectors in Chouf-Soueijany: Energy consumption [MWh/baseline year]

				Energy Consumption [MWh/baseline year] per sector						
	Surface (Hectare)	Population	# of households in summer (max)	Municipal buildings, equipment/facilities	Tertiary (non-municipal) buildings, equipment/facilities	Residential buildings	Private Generators	Municipal public lighting	Private and commercial transport	Total
Ain w zein	500	1700	350	12	6083	2893	840	234	1715	11777
Ainbal	370	600	350	8	106	4209	960	50	3520	8853
Atrine	360	1500	200	4	89	1616	960	70	1626	4365
Gharife	1665	9500	700	8	126	7216	1200	209	5734	14493
Jdeidit chouf	380	30000	5000	18	458	48550	7200	1890	30592	88708
Kahlounyeh	375	2400	270	7	65	2773	840	45	2651	6381
Mazraat chouf	1100	10000	700	36	142	13296	2400	495	8682	25051
Smkanyyeh	530	8500	1500	11	599	23436	12000	216	10183	46445
Percentage of consumptions per sector				0.05%	3.72%	50.46%	12.81%	1.56%	0.05%	31.40%

Table 2. Conversion factors

Type of energy	Conversion factor
EDL	0.735
Private Diesel Generator	1.1
Burning Diesel fuel	0.2637
Burning butane	0.2171
Burning wood	0.39
Burning gasoline	0.25272

SUSTAINABLE ENERGY ACTION PLAN

The main objective of the following action plan is to reduce the amount of CO2 emitted into the atmosphere by 54,403 tCO_2-eq, previously evaluated in the BEI integrated between the baseline year and the target of reduction year 2020, therefore reduced from 118,167 to 63,764 tCO_2-eq.

Many actions should be taken in different sectors to reduce the energy consumption as well as to use renewable clean energy and therefore the CO_2 emissions will be reduced directly. Knowing that the major amount of consumption comes from the Building sector, the national policies promoting energy efficiency, as well as specific local authorities building policies will bring about energy efficiency improvement.

A. Action #1: Replacing the sodium streetlight bulbs by high efficiency LED lamps

This measure will take place in the public lighting sub-sector. Most of the streetlights installed are 150 and 250 watts sodium light bulbs that consumes a huge amount of energy with a lifetime of 20000 to 28000 hours. The aim of this action is to replace these low efficiency lights by the high efficiency 60 watts LED lamps with an 60000 hours lifetime which will reduce the energy consumed by the public lighting sub-sector. This action will reduce the GHG emissions by 1334.5 tCO_2-eq after saving 1454.5 MWh/year of electricity assumed supplied half by EDL and half by the private generators for 12 hours/day all around the year.

B. Action #2: Replacing the incandescent residential light bulbs by high efficiency LED lamps

This measure will take place in the residential building sub-sector. Mostly, all houses use 100 watts light bulbs which they're going to be replaced by 25 watts high efficiency LED lights. This will conserve the energy consumed by residences 6 hours per day for around 5678 MWh/year, considered half supplied by EDL and the other by private generators. The total emissions will be reduced by 5209 tCO_2-eq.

C. Action #3: Replacing the traditional electric water heaters with solar heaters in the households

This measure will take place in the residential building sub-sector. The solar heaters usage is very shy in that region. Thus, installing these systems will eliminate the use of electricity and/or private generators in water heating and will use the renewable clean solar energy. A 200-liter capacity solar heater for a normal household costs 600 $ with a five-year warranty as given by a local retailer and an expected life of twenty years. The total renewable energy produced is estimated around 28672 MWh/year leading to a reduction of 24214 tCO_2-eq/year.

D. Action #4: Integrating photovoltaic panels as a source of electricity in the households stored for the periodic electricity blackout

This measure will take place in the private generators sector. The photovoltaic installation is almost non-existent in that region. Thus, installing these systems will reduce the dependence on private generators during the EDL cutoff hours and will replace it with the renewable clean solar energy. Photovoltaic system in our case is equipped with lithium ions batteries and inverters, it operates during electric blackout period and batteries recharges during normal periods. A 1000 wats photovoltaic system is enough to replace the electric generator in one residence. This action will generate approximately 10268MWh/year and save the atmosphere from 11295 tCO_2-eq/year emitted by the Diesel generators.

E. Action #5: Using the existing biomass in forests for heating

The biomass heating is considered as an efficient renewable energy since a large local area is covered with forests, private and public. They used before to burn every year a large space of dried wood. A biomass trade and logistic center has to be established in the Federation to support use of local wood and its distribution for heating issue since the major inhabitant still use the traditional method (stove) to heat the living rooms. This method will reduce the emissions of fires in the forests directly into the atmosphere as well as it will use the heat of that biomass existed as recovery for residential heating. This reduction is estimated around 4657 tCO_2-eq/year. The residential sub-sector is the main beneficiary of this measure.

F. Action #6: Solar-Powered Public Street Lighting

Public street lighting consumes large amounts of electricity per year to provide sufficient lighting within Chouf-Soueijany internal roads and local highways. Switching to solar power is a solution that would eliminate the need for grid supply especially when designed with sufficient autonomy and power backup. Sector concerned: Public lighting. Total renewable energy produced: 1338 MWh/year. Total emissions reduced: 4081 tCO_2-eq/year.

G. Action #7: House inspection program

The house inspection program is an energy audit that aim to visit the houses and perform efficiency assessment to advice on ways to reduce energy consumption. House inspection will be equipped with necessary tools such as power measurement tools, infrared camera, AC optimization tools and others. In addition, the inspector will provide houses with solutions to avoid thermal losses, infiltration and others. Such inspections lead to better insulation of the buildings (window replacement, better facade insulation, roof insulation) or efficient heating devices, resulting therefore in energy savings in the building sector. This measure will take place in the residential buildings, non-residential and municipal building sectors. Moreover, many small interventions in public buildings will lead to the same results as using LED lamps, installing motion lighting sensor, replacing the air conditions units that consume high energy, fixing of the thermostats of air-conditions, utilize as possible daylight through windows and reduce using artificial lights as much as possible, set PC monitor on sleep mode for maximum two minutes, switch off PC UPS and printer when leaving the work and minimize the usage of printing as possible. This action is assumed to reduce by 10 % the final energy consumption in the residential, the non-residential and municipal building sub-sectors, estimated around 3457 MWh/year and thus will avoid 3172 tCO_2-eq to be emitted per year.

G. Action #8: Forestation

Increasing the number of forest by planting trees improves the land capabilities and potential for carbon sequestration and stocking. It improves the soil conditions, fights deforestation, reduces the impact of CO_2 emissions on the federation and provides sustainable resource for biomass. This measure will take place in the public sector. Assuming that the forestation will cover 0.5% of the area with 1800 trees every hectare, the sequestration will be estimated around 441 tCO_2/year.

By implementing this full set of actions, it will be possible to reach the 2020 target to reduce GHG emissions by 46.1% compared with current trends. With these 8 actions, and referring to the Figure 1, the comprehensive GHG emissions reduction resulting from all proposed sustainability measures sums up to 46.1%, which is 26.1% above the target set by the COM by 2020 with an overall cost of 77 million U.S. Dollars for the whole federation. In Figure 1, the quantitative indicators such as costs (in $/year), energy savings (in MWh/year), clean energy production (in MWh/year) and estimated CO_2-eq emissions reduction (in tons/year) are shown to indicate the impact of each action and to quantify the economical/environmental benefits.

CONCLUSION AND RECOMMENDATIONS

This project will make Chouf-Soueijany a model of sustainable municipalities in Lebanon. It will improve the quality of life towards a healthier, safer and more livable environment through reducing expenditure on energy and GHG emissions. It will promote a low energy consumption housing programs then reducing energy bills and working to raise citizen awareness, reinforce, integrate and promote EE and managing the installation of renewable and sustainable energy systems. This SEAP will be considered as a development tool for the South- Mediterranean region.

As for quantitative overall results table 3:

Table 3. Overall SEAP Results

Total Energy Consumption as per BEI 2015	207140	[MWh/year]
Total Energy Saving per SEAP implementation	22530	[MWh/year]
% of Energy saved per SEAP implementation	10.9%	
Total Renewable Energy produced per SEAP	40279	[MWh/year]
% of Renewable Energy produced per SEAP	19.4%	
Total GHG Emissions as per BEI 2015	99300	[tCO2-eq/year]
the national GHG coefficient of the Guidelines for National Greenhouse Gas Inventories developed by the JRC for the South-MED countries	1.19	
Total CO2 emission in 2020	118167	tCO2-eq
Total GHG Reduction Per SEAP implementation	54404	[tCO2-eq/year]
% CO2 reduction per year after SEAP implementation	46.1	%

A list of recommendations is needed to improve any intervention towards sustainability. For national level, the endorsement of the draft energy conservation law and the draft transport policy by the parliament would help promote energy conservation initiatives in Lebanon particularly as well as in the region. As for local level, municipalities should have the legal right to produce and distribute any type of energy especially the renewables. To avoid errors in data, a methodology shall be developed to build a robust and reliable sample of GHG emission inventories. Behavioral changes are as important as building physics in reducing energy consumption. It is the major policy instrument deployed by local authorities to mobilize public interest in sustainable energy policies and climate change.

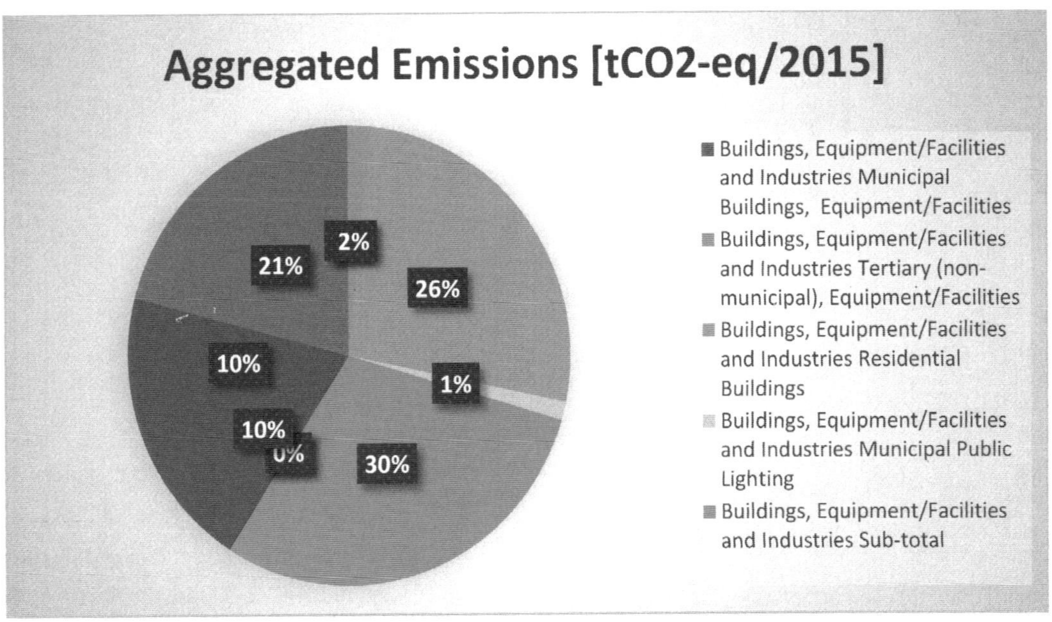

Chart 1. GHG emissions in sub-sectors

Sectors and sub-sectors	Actions (per sector/sub-sector)	Action#	Timeline	Estimated Budget [$]	Estimated Energy Saving [MWh/year]	Energy Saving %	Renewable energy produced [MWh/year]	Renewable Energy Produced %	GHG Reduction [tCO2-eq/year]	GHG Reduction %	Source of reduction
Buildings, Equipment/Facilities and Industries											
Municipal Buildings, Equipment/Facilities											
Tertiary (non-municipal), Equipment/Facilities											
Residential Buildings	Replacing the traditional residential light bulbs by high efficiency LED lamps	2	2018	317450	5677.866	2.74%	0	0.00%	5,209.44	4.41%	EDL-Diesel
	heaters with solar heaters in the households	3	2018-2019	10884000	0	0.00%	28672.44	13.84%	24,213.88	20.49%	EDL-Diesel
	Using the existing biomass in forests for heating	5	2018-2020	192000	11940	5.76%	0	0.00%	4,656.60	3.94%	Carbon Stock
	House inspection program	7	2018-2020	4897800	3457	1.67%	0	0.00%	3,171.80	2.68%	EDL-Diesel
Municipal Public Lighting	Replacing the sodium streetlight lamps by high efficiency LED lamps	1	2018	147610	1454.598	0.70%	0	0.00%	1334.594	1.13%	EDL-Diesel
	Solar-Powered Public Street Lighting	6	2018-2019	7635000	0	0.00%	1337.652	0.65%	4080.94	3.45%	EDL
Industries											
Sub-Total			**2017-2020**	**24073860**	**22529.464**	**10.88%**	**30010.092**	**14.49%**	**42667.251**	**36.11%**	
TRANSPORT											
Municipal Fleet											
Public Transport											
Private and Commercial Transport											
Sub-Total				**0**	**0**	**0**	**0**	**0**	**0**	**0**	
Others											
Private Generators	Integrating photovoltaic panels as a source of electricity in the households stored for the periodic electricity blackout	4	2018-2019	52615000	0	0.00%	10268.493	4.96%	11,295.34	9.56%	Diesel
Public areas	Forestation	8	2018-2020	284310	0	0.00%	0	0.00%	440.681	0.37%	Carbon Stock
Sub-Total				**52899310**	**0**	**0.00%**	**10268.493**	**4.96%**	**11736.024**	**9.93%**	
TOTAL				**76973170**	**22529.464**	**10.88%**	**40278.585**	**19.45%**	**54403.275**	**46.10%**	

References:
[1] CES-MED, 2017. Information Kit, how to prepare a Sustainable Energy Action Plan (SEAP), prepared by CES-MED.
[2] JRC Science and Policy Reports, 2017. Covenant of Mayors in figures: 8-year assessment, JRC Science for policy report, Kona A., Melica G., Bertoldi P., Rivas Calvete S., Koffi B., Iancu A., Zancanella P., Janssens-Maenhout G. Dallemand J.F., Luxembourg: Publications Office of the European Union.
[3] Solar MED Atlas. Online program that gives data for the whole target region covering the South -Mediterranean countries. Database are provided by SOLEMI and Helioclim-3 (SoDa) sources.

[4] Published by the American University of Beirut - Nature Conservation Center (AUB-NCC), 2016. M. Massoud and F. Merhebi. Guide to Municipal Solid Waste Management.

[5] JRC Science and Policy Reports, 2014. How to develop a sustainable energy action plan (SEAP) in South Mediterranean cities guidebook. Saheb Y., Kona A., Maschio I., Szabo S. Luxembourg: Publications office.

Design Optimization for Maintaining Occupants' Outdoor Thermal Comfort

Haneen I. Hamdan, ME **Balsam A. Nehme, ME**

ABSTRACT

Modeling thermal comfort and people's satisfaction with their surrounding environment is grasping a lot of attention. Predicting occupants' tolerability to the ambient conditions helps in optimizing the design of room temperature leading to reduction in the installed cooling/ heating power and thus resultant energy consumption.

For this sake, researchers have developed different indices that predict thermal comfort of occupants present in indoor naturally and mechanically ventilated spaces. The most widely used thermal sensation index is Fanger's PMV (Predicted Mean Vote). While Fanger's PMV allows quantifying the level of comfort of the occupant present in enclosed spaces, it remains inaccurate to be used for assessing the level of thermal comfort of people present in external open spaces such as outdoor pedestrian areas or other spaces specifically designed for outdoor activities. As such, studying outdoor thermal comfort is important in order to optimize the design of the building external construction details as well as the implementation of design measures to the occupied external environment to enhance the occupants' thermal comfort. An outdoor thermal comfort model, referred to as the Standard Predicted Mean Vote (SPMV) was developed.

The purpose of this study is to benefit from the pre-developed SPMV model and develop an in-house tool that predicts thermal comfort of a person present outdoors and directly subjected to the outside environmental conditions such as wind and direct sun radiation. By conducting an hourly assessment of outdoor thermal comfort, the tool allows for identifying critical hours of the day when discomfort due to external conditions might occur. Moreover, the work presents case studies considered and further simulated to assess the impact and optimize the implementation of various potential adequate design measures onto the individuals' thermal comfort in an external environment.

For hourly SPMV calculations as per the model built using Visual basic software, the required weather data characteristics as well as available sky view factors based on the model under consideration are acquired from the Integrated Environmental Solution IES <VE> modeling tool.

INTRODUCTION

Open spaces such as gardens, outdoor restaurants and pedestrian pathways provide physical and visual access to the environment. Research has shown that direct contact with nature aids in reducing stress, promoting relaxation in addition to increasing social interaction and physical activities (Alizera amd Klaus 2013). However, hot summer days impose thermal stress on people present in outdoor spaces and thus necessitate integrating of adequate architectural features and mechanical cooling of the microclimate occupied zones in order to maintain outdoor thermal comfort.

In the recent years, thermal comfort has been considered as a significant parameter in designing occupied zones in order to improve occupants' contentment. Different thermal comfort indices exist such as Fanger's PMV index which is widley used to predict thermal comfort, however, this index is used for occupants present in enclosed and mechanically ventilated spaces (ASHRAE 55). In external spaces, the individual will be directly exposed to his

Hanin Hamdan is a design Mechanical Engineer in the Department of Building Sustainability Unit, Dar al Handasah, Beirut, Lebanon. **Balsam Nehme** is the Head of the Department of Buildings Sustainability Unit, Dar al Handasah, Beirut, Lebanon.

immediate environment and will be subject to variations of sun and shade as well as fluctuations in wind speed, and any other characteristic of the outdoor climate. Standardized Predicted Mean Vore (SPMV) is an index used to assess outdoor thermal comfort (Gagge and Fobelets 1986). This index allows for adjusting the PMV model by replacing the operative temperature by ASHRAE standardized effective temperature (SET) in the heat balance equation.

Using the predeveloped SPMV index, an in-house tool is developed to predict hourly SPMV index and thus optimize the design of microclimate outdoor spaces by implementing adequate architectural features such as shading devices and/or designing for energy efficient cooling systems when necessary.

Throughout this paper, the SPMV index is further explained. Then, the in-house SPMV tool is introduced and its methodology is clarified. Finally, a case study is presented where the use of the in-house outdoor thermal comfort tool aided in optimizing the design of microclimate outdoor zones.

SPMV AND PMV

Thermal comfort is a psychological interpretation of the physiological state of the body and should not be confused with temperature sensation. Our perception and sensation of thermal comfort is mainly determined by the energy exchange between the body and its surroundings. Thus controlling environmental elements which affect this energy exchange is a method to achieve the thermal comfort in any outdoor space. Nevertheless, people's expectations when staying outdoors are different from indoors as greater variability is expected, such as variant in sun and shade, modifications in wind speed and direction, and changes in humidity rates, which assist in tolerating a wider range of outdoor conditions for a longer period of time.

The Predicted Mean Vote (PMV) stands among the most recognized thermal comfort models, and is based on the heat balance between the human body and its surrounding environment. Various environmental factors such as temperature, relative humidity, wind speed, and personal factors such as metabolic rate and clothing levels are taken into account.

However, studies have shown that PMV index is most accurate when used for indoor controlled environments and it is proven that it is insensitive for great fluctuation in the environmental conditions that may occur outdoors. (Gagge and Fobelets 1986). Therefore, the Standard Predicted Mean Vote (SPMV) is introduced to predict thermal comfort of individuals exposed to outdoor conditions.

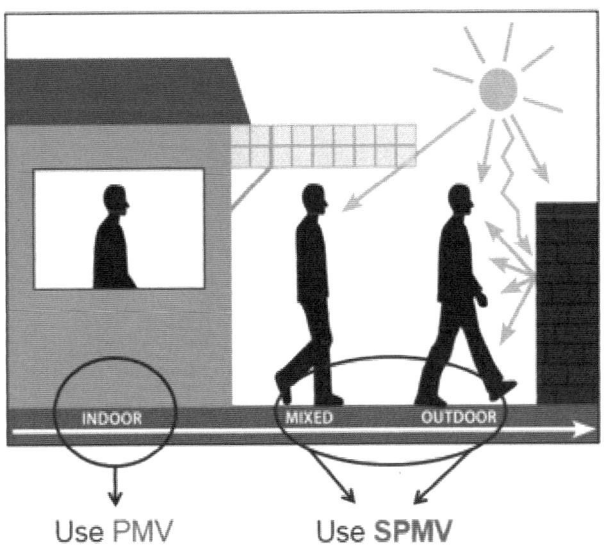

Figure 1 Thermal Comfort Index – SPMV versus PMV punctuation).

SPMV index takes into account the effect of extreme humidity levels, temperatures, solar exposure and individual parameters such as clothing and activity level on the individuals. Its calculation is similar to that of the PMV, however; ASHRAE standard effective temperature (SET) is used instead of the air or operative temperatures. SET is defined as the equivalent air temperature of an isothermal environment at 50% relative humidity, in which the subject wearing the standardized clothing for a given activity will have the same heat losses form the skin and the same skin wittedness.

SPMV is measured on a scale from -4 to +4 where the theoretical peak comfort occurs at a neutral rating (0) with warmer sensations moving up on the scale to +4 (hot), and colder sensation moving down the scale to -4 (cold) as illustrated below.

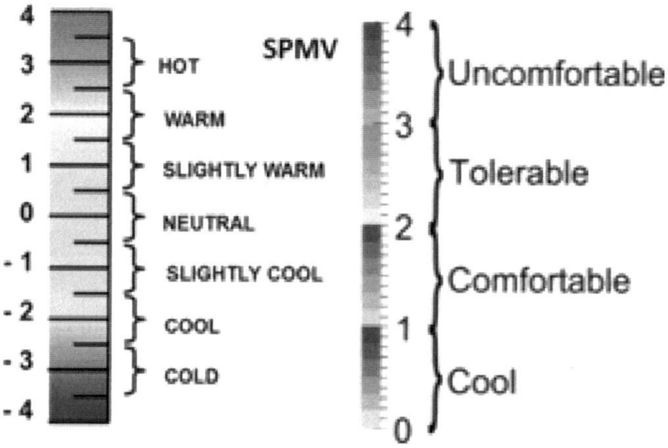

Figure 2 Outdoor thermal Comfort Ranges

SPMV- IN HOUSE TOOL

Thermal and comfort simulation software present in the market today and that are commonly used in engineering applications predict PMV for mechanically ventilated enclosed spaces. These tools lack the ability to predict thermal comfort of individuals present in outdoor spaces. On the other hand, relying only on SPMV equations provided in literature will only be beneficial in case where no mechanical cooling to microclimatic outdoor areas is implemented. For this reason and in an effort to optimize the design for outdoor cooling using SPMV, an in-house tool is generated. This tool combines SPMV equations already found in literature with an energy modeling software in order optimize outdoor cooling and architectural features to provide thermal comfort for individuals.

A visual basic code is developed to calculate the hourly SPMV index as per predeveloped model (Gagge and Fobelets 1986). SPMV equations require various environmental and personal input parameters. Environmental parameters include air temperature, relative humidity, mean radiant temperature, air speed, and direct and diffuse solar radiation of the outdoor zone under study. Usually these environmental inputs are outdoor ambient conditions, however, if mechanical cooling is implemented or shading devices are used, then these parameters will have to be modified. In order to asses the effect of cooling or shading on the microclimate environmental parameters, IES <VE> modeling software is used. In the IES software, the microclimatic outdoor zone where the individuals are present is modeled taking into consideration the effect of any mechanical cooling and any architectural shading feature.

SPMV code will read the outputs from IES <VE> tool that are considered to be the external weather conditions that the individual is exposed to in the outdoor microclimatic space. In addition to environmental factors, the designed has to input the personal factors including the activity level of the person and the clothing ensemble to generate hourly SPMV index.

Figure 3 Sample 3D Model from IES to generate input fro SPMV code

Finally, the inhouse tools will generate hourly contours for the whole year and will aid the designer in identifying the hours of the year where SPMV is beyond tolerable levels and where improvements are to be made either by providing more architectural shades to reduce direct radiation or to further optimize the cooling system.

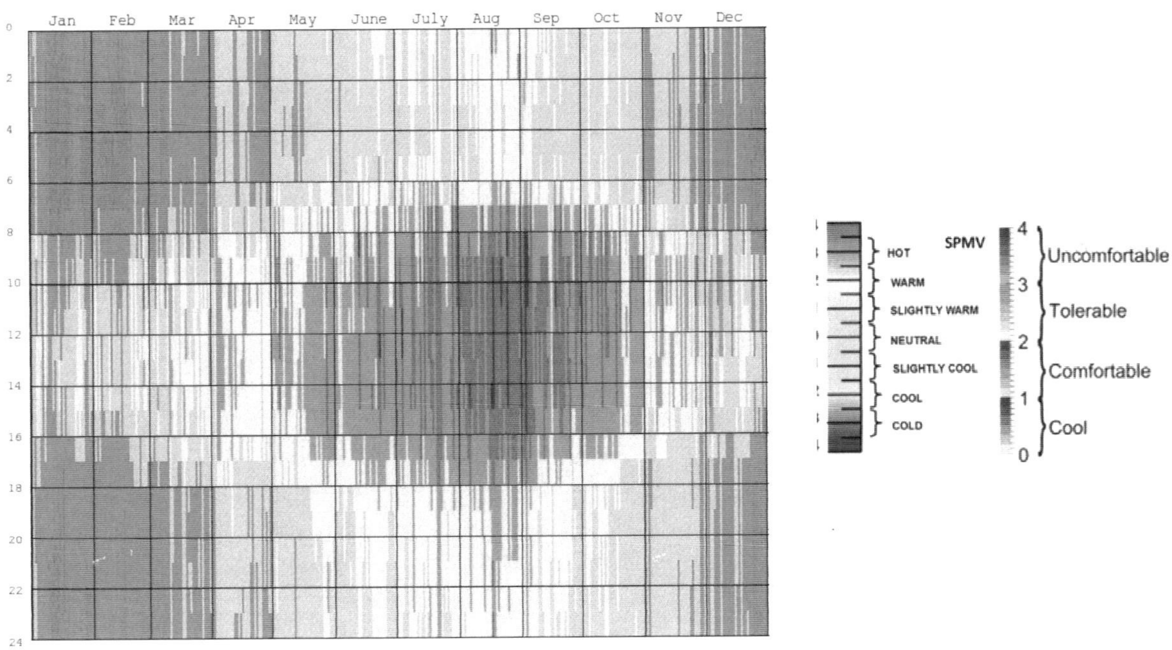

Figure 4 Sample SPMV yearly contour

CASE STUDY

A hotel in Doha includes a number of outdoor Food and Beverage outlets across the promenade area where visitors can enjoy views overlooking the Arabian Gulf. However, due to the hot and humid weather of Doha, visitors will definitely experience discomfort during hot summer days and nights. As such, outdoor cooling is indispensable in order to provide a healthy, vibrant and comfortable outdoor environment that meets visitors' expectations. The aim is to design an optimized cooling system that will provide an adequate microclimate within the F&B outlets.

SPMV in-house tool is used to aid the designers in optimizing the cooling capacity, design supply temperature and flow rate of the mechanical cooling system. In addition, and based on SPMV outcomes, designers may advise the architects to implement shading devices along the occupied outdoor zone.

Figure 5 below provides the SPMV contour for one of the F&B outdoor outlets without the intervention of any mechanical cooling nor shading devices. It is known that during summer days, individuals present in these spaces will be subjected to thermal stress due to elevated air temperature and solar radiation.

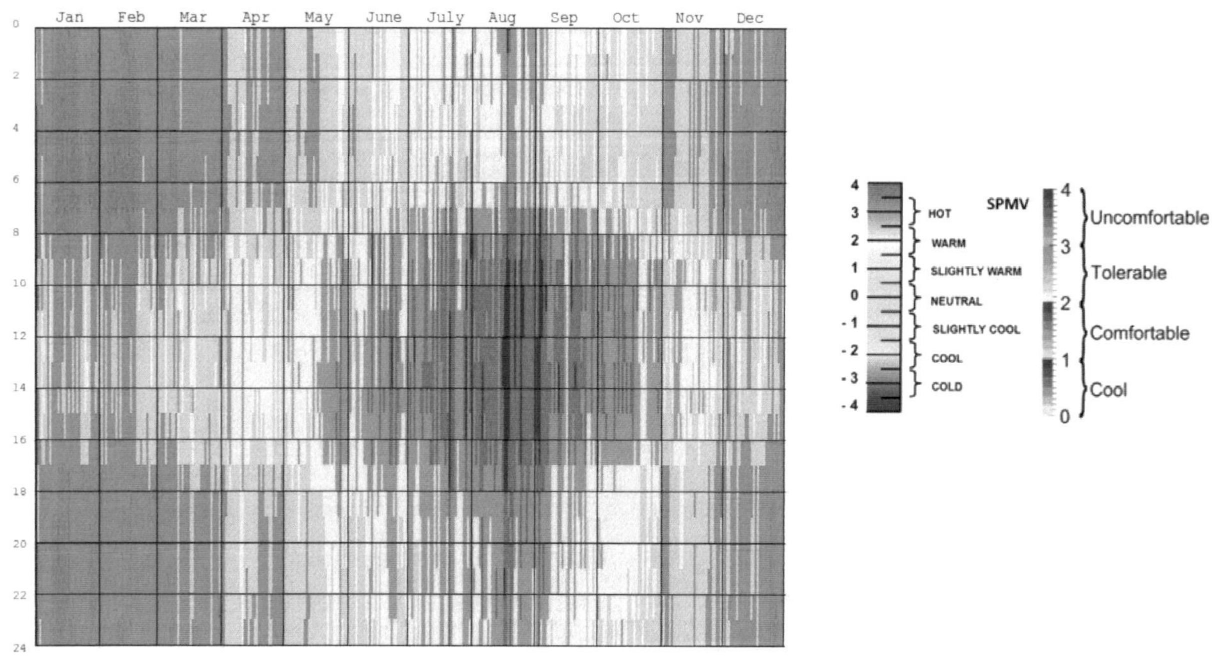

Figure 5 SPMV contour – Basecase: No Shading and No mechanical Cooling

In an effort to overcome this thermal stress, shading is advised to be provided and as such SPMV was reduced as shown in figure 6 below. Providing full shading to the microclimatic outdoor space reduced thermal stress in months of May, June and October. However, it is evident that mechanical cooling is inevitable in months of July, August and September.

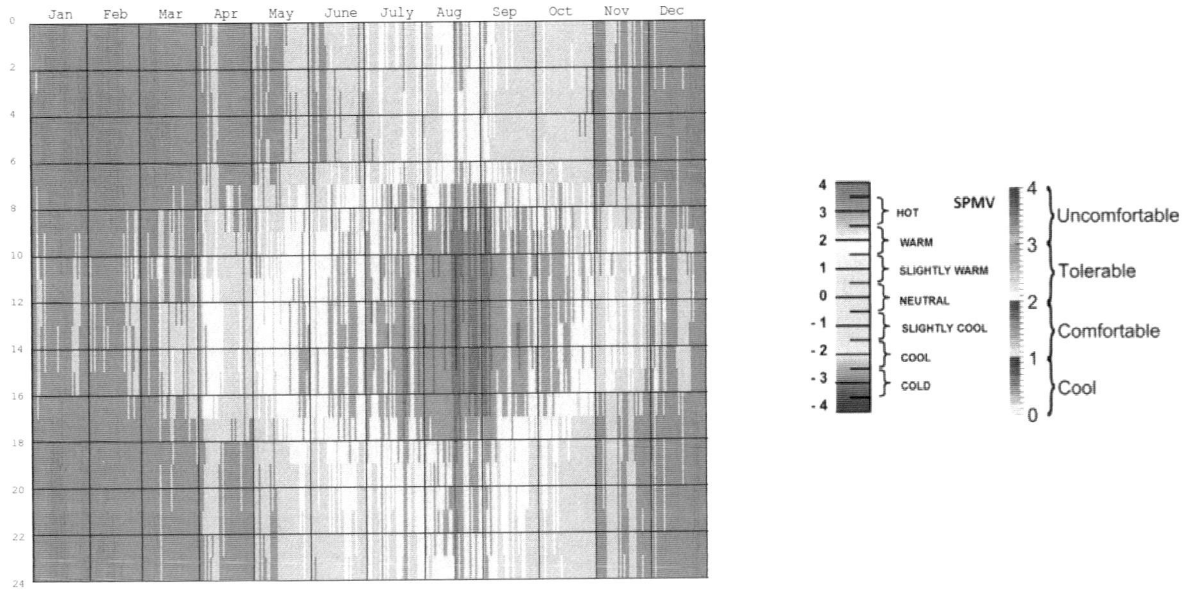

Figure 6 SPMV contour – Case 1: Basecase plus full Shading

As such, a mechanical cooling system was installed, and using IES software, different cooling systems with variant capacities, supply temperature and flow rate where tested to finally reach an optimized design to cool the outdoor space and to provide a tolerable environment and acceptable SPMV levels all year long as shown in figure 7 below.

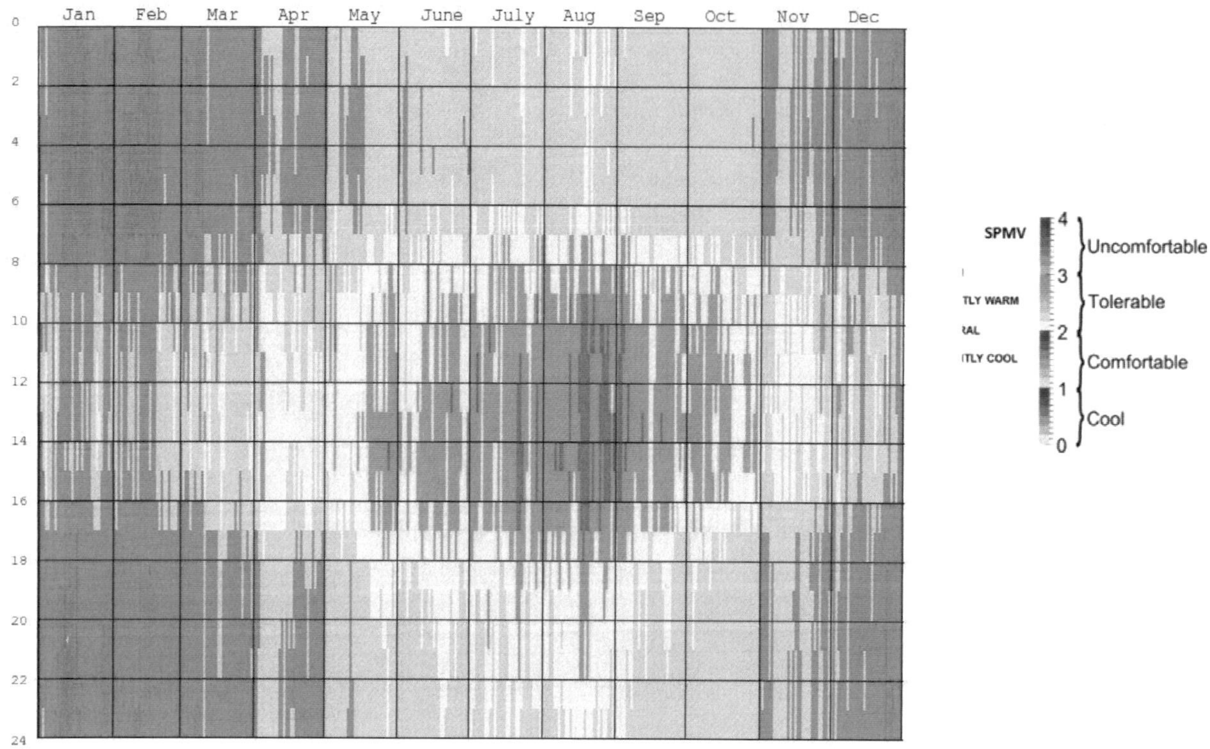

Figure 7 SPMV contour – Case 2: Basecase plus full Shading plus mechanical cooling

CONCLUSION

Outdoor thermal comfort has been lately grasping the attention of designers to optimize energy efficient mechanical cooling design and architectural features of outdoor spaces. SPMV index is used instead of Fanger's PMV to assess thermal comfort in outdoor spaces for the latter is insensitive to outdoor conditions and weather variations. An in-house tool combining pre-developed SPMV model and IES <VE> modeling software is developed. This tool provided an approach that enables the designes to optimize the architectural and mechanical cooling dsign for outdoor spaces present in climates where mechanical cooling is necessary to satisfy indivisuals.

REFERENCES

Alizera Monam, Klaus Rucket, 2013.*The Dependence of Outdoor Thermal Comfort on Urban Layouts.*
ASHRAE Standard 55-2013. T*hermal Comfort Conditions for Human Occupancy*
Gagge, Fobelets, 1986. *A Standard predictive Index of Human Response to the Thermal Environment.*

Transition Engineering the Water-Electricity Nexus Operating in Building Services and Urban Heat Islands – Concept Design – is Airconditioning Really Necessary?

Eric Laurentius Peterson, PhD, PE
Member ASHRAE

ABSTRACT

After firstly determining if passive design cannot be expected to serve at a given development site, it is appropriate to determine what mode of building services would be expected to provide comfortable conditions, as ventilation or evaporative cooling may satisfy without resorting to air-conditioning. Consequently, there are broadly three modes of building services which may be recommended to provide coolth in summer if passive means will not suffice, and similarly another three modes of building services that may be required in to maintain comfort in winter if passive means will not suffice. I suggest that leading winter building service modes are humidification, moderate air-source heat pumping, and icing-heating. The icing-heating condition could be partially managed by defrosting of air-source heat pumps, but may recommend a ground-source or combustion heat-source. For the purposes of the present study there are seven operating modes ranging from direct expansion (DX) air-conditioning in summer to icy-heating in winter.

Balancing comfort and economy, the criteria for the more extreme conditions were taken to be satisfied if the defined criteria are not to exceeded more than 10 days per year – inspired by the Australian approach to setting comfort design conditions with only daily min/max observations – where the average of the tenth highest daily max was used to determine the design dry bulb. So I propose that if mean daily conditions show that air-conditioning is desirable only nine days in a year, then it is not deemed necessary in that location. But if in that location there were at least ten days in a year where evaporative cooling or air-conditioning were desirable, then evaporative cooling would be the suggested system. And finally, passive design may be acceptable for comfort if there were less than ten days in-total in which any of the active modes of cooling were desired. Results include a tabulation of the medians from 157 countries. Larger federations such as USA, Canada, Russia, China, and Australia were not summarized as provincial climatic variability is vast.

The highest drybulb threshold T_{Vhot} arises from ASHRAE 55 standard for comfort and the next two are the most common base temperatures of cooling and heating degree days, and T_{icing} is the upper whole degree associated with freezing up of air-source heatpumps together with ΔT_{ice}=5F of degrees that dewpoint is below drybulb. The highest wetbulb T_{Vhumid} is the author's suggested limit for the suitability of ventilation to manage humidity without air-conditioning, and the lowest webbulb threshold T_{dry} is where the air may be too dry for comfort. The wetbulb T_{humid} is threshold of evaporative cooling.

Author was Inaugural Lecturer of Architectural Engineering, Victoria University, Melbourne, Australia – currently Adjunct Professor visiting University of Leeds.

INTRODUCTION

Bioclimatic analysis (Givoni 1992) predicting achievable indoor conditions has shown that passive or evaporative cooling is ineffective in the summer conditions of Qatar (Sayigh and Marafia 1998), so air-conditioning is an essential building service in that country – although hybridization offers substantial savings in shoulder seasons. In other locations, passive strategy assessment of typical meteorological year (TMY) datasets has been enabled by Climate Consultant Software (Milne, Liggett et al. 2009). This paper suggests that without the benefit of TMY datasets, the minimum/maximum temperature, mean daily dewpoint and pressure summary of the day are sufficient to identify if active heating or cooling are desirable on any particular day – and if passive comfort could be expected in a well insulated building having shaded openable windows that can be operated to maintain comfort.

Global Summary of the Day (GSOD) (Menne, Durre et al. 2012) is maintained by NOAA's National Climatic Data Center. After firstly determining if passive design cannot be expected to serve at a given development site, it is appropriate to determine what mode of building services would be recommended to provide comfortable conditions – as ventilation or evaporative cooling may well satisfy without resorting to air-conditioning. Consequently, I suggest there are broadly three modes of building services which may be recommended to provide coolth in summer if passive means will not suffice, and similarly another three modes of building services that may be required in to maintain comfort in winter. I suggest that leading winter building service modes are humidification, air-source heat pumping, and icing-heating. The icing-heating condition could be partially managed by defrosting of air-source heat pumps, but may recommend ground-source heat pump or combustion heat. For the purposes of the present discussion there are seven operating modes ranging from direct expansion (DX) air-conditioning in summer to icy-heating in winter.

Balancing comfort and economy, the Australian criteria for the more extreme conditions should be taken to be satisfied if the defined criteria are not exceeded more than 10 days in a year (Peterson, Williams et al. 2006). For example, if air-conditioning (level +3 cooling) is recommended only nine days in a year, then it should not be deemed necessary – but evaporative cooling (level +2 cooling) should be recommended if there were ten days in-total recommending air-conditioning or evaporative cooling. If there are less that 10 days recommending air-conditioning and evaporative cooling, but ten days in-total recommending any combination of the three modes of cooling then a ventilation system (level +1 cooling) would be suggested. In the present study I have tallied the selections for each day of the year, choose from seven modes of heating, passive, and cooling as is appropriate based on mean daily dry bulb ($T_{mean\ db}$), coincident wet bulb ($T_{coincident\ wb}$) and coincident dew point ($T_{coincident\ dewpoint}$) derived from GSOD – in decending order from full air-conditioning mode [+3, +2, +1, 0, -1, -2, -3] down to the most extreme heating duty:

+3) Direct expansion vapour compression (**DX**) is suggested if $T_{mean\ db} > T_{hot}$ and $T_{coincident\ wb} > T_{humid}$
– except DX is not required if $T_{mean\ db} \leq T_{Vhot}$ and $T_{coincident\ wb} \leq T_{Vhumid}$

+2) Direct evaporative cooling (**Evap'n**) is suggested if $T_{dry} \leq T_{coincident\ wb} \leq T_{humid}$ and $T_{mean\ db} > T_{Vhot}$

+1) Ventilation cooling and dehumidification (**Vent'n**) is suggested if either of the following are satisfied:
a. $T_{cold} \leq T_{mean\ db} \leq T_{Vhot}$ AND $T_{humid} < T_{coincident\ wb} \leq T_{Vhumid}$
b. $T_{hot} < T_{mean\ db} \leq T_{Vhot}$ AND $T_{dry} \leq T_{coincident\ wb} \leq T_{humid}$

0) **Passive** comfort is readily achieved if $T_{cold} \leq T_{mean\ db} \leq T_{hot}$ and $T_{humid} \leq T_{coincident\ wb} \leq T_{dry}$

-1) Humidification (**Humid'**) is suggested if $T_{cold} \leq T_{mean\ db}$ and $T_{coincident\ wb} < T_{dry}$

-2) Air-source heatpump heating (**Heat**) is suggested if $T_{mean\ db} < T_{cold}$
– except for icy-heating conditions defined in the seventh classification –

-3) Beware that air-source heat pumps are subject to icing (**Icy**), and so consider other sources,
if $T_{mean\ db} \leq T_{icing}$ and $[T_{mean\ db} - T_{coincident\ dewpoint}] \leq \Delta T_{ice}$

I have adopted the thresholds of Table 1 in the forgoing definitions of the seven suggested building services modes, but each of these are the author's nomination and not all are necessarily aligned with ASHRAE standards.

Table 1. Nominal thresholds to classify heating-cooling modes

	Dry Bulb	Wet Bulb	Frost/Dew Point
Very hot and Very humid	T_{Vhot} = 27°C db	T_{Vhumid} = 24°C wb	
hot and humid	T_{hot} = 24°C db	T_{humid} = 19.4 °C wb	
Cold and Dry (heating / humidification desirable)	T_{cold} = 18.3°C db	T_{dry} = 9.4°C wb	
Icy air-source heat pump	T_{icing} = 5°C db		ΔT_{ice} = 5K depression (db – frost or dewpoint)

The highest drybulb threshold (T_{Vhot}) complies with Standard 55 (ASHRAE 2010) at 0.3 m/s ventilation and the next two are the common base temperatures of cooling (T_{hot}) and heating degree days (T_{cold}), while T_{icing} is the upper temperature associated with freezing up of air-source heatpumps together with ΔT_{ice} = 5K of degrees that dewpoint is below drybulb (Peterson 2014). The highest wetbulb threshold T_{Vhumid} is the author's suggested limit for the suitability of ventilation to manage humidity without air-conditioning, and the lowest webbulb threshold T_{dry} is where the air is too dry for comfort. T_{humid} is the 67°F wet bulb limit of effective direct evaporative cooling (Watt 2012) and is concurrent what appears to be 19.4 °C wet bulb threshold in Figure 4-1 "Annual temperature stress by month" p 48, Advanced Energy Design Guide for K-12 School Buildings Achieving Zero Energy (ASHRAE and IESNA 2018).

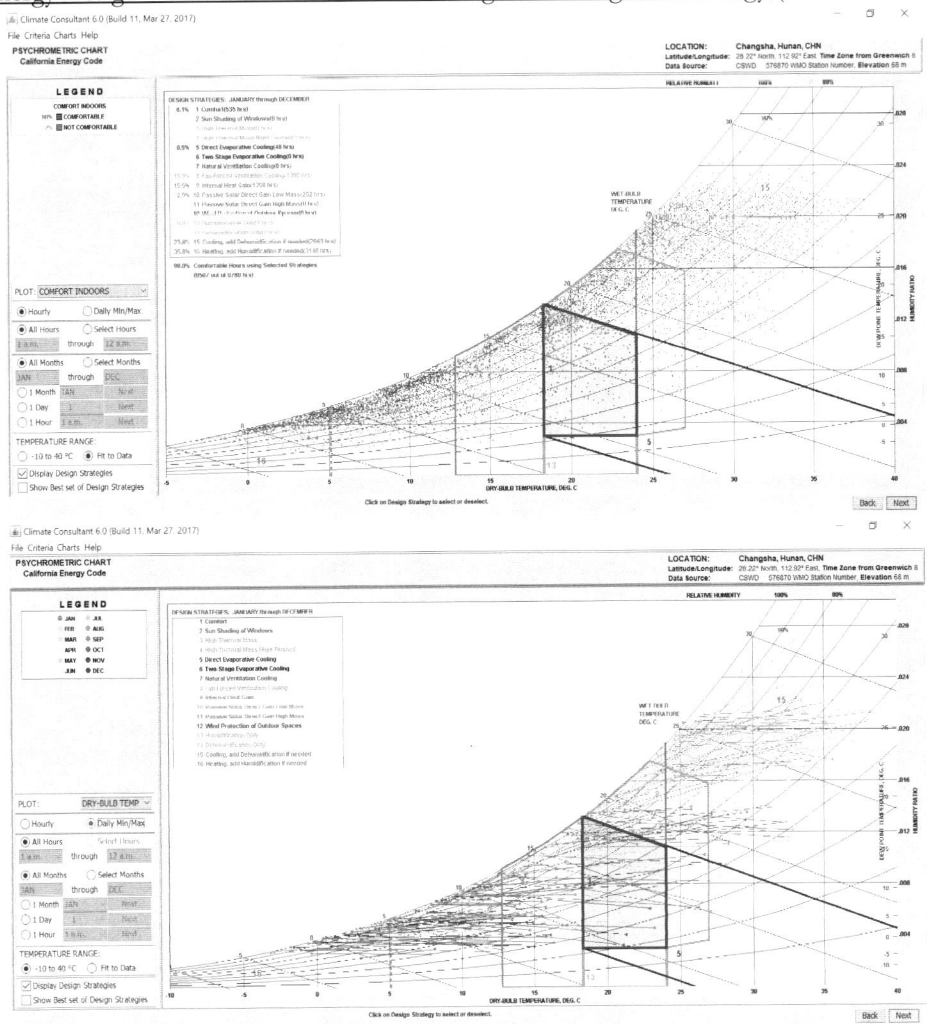

Figure 1 UCLA Energy Tools Group "Climate Consultant" (Milne 2016) showing two views of WMO 576870 CSWD (Yi 2005) TMY file for Changsha, Hunan Province of China. (a) Upper pane of Figure 1 presents 8760 hourly data points and (b) lower pane of Figure 1 presents daily range of 365 days.

Notice in Figure 1b that humidity ratio and dewpoint are relatively constant from the dawn minimum through to the afternoon maximum on any particular day, and so it is reasonable to use GSOD daily observations to estimate coincident wetbulb at the times of minimum, mean, and maximum drybulb observation. Climate Consultant's default is California Energy Code Comfort Model 2013, whereby comfort zone of evaporative cooling is bound by 17.7 °C and 9.3 °C wet bulb. Note that implementation of ASHRAE 55 and/or "Latest Handbook of Fundamentals" evokes a broader evaporative cooling comfort zone – bounded by 20°C and 6.6°C wet bulb. The drybulb comfort zone was adjusted by the author of this paper to be bounded by 18.3°C and 24°C to be comparable with Figure 2 below.

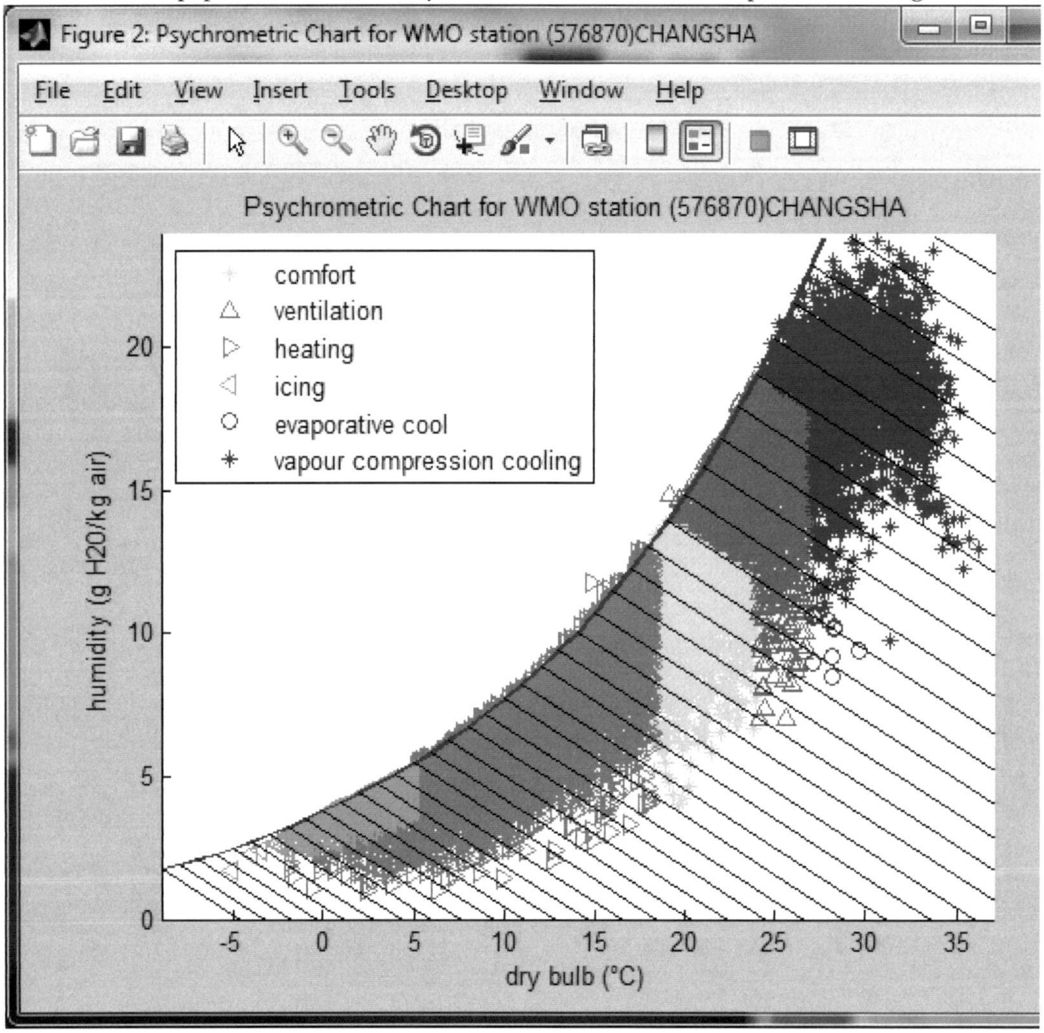

Figure 2 GSOD daily mean data presented on psychrometric plot at the same WMO location as that presented in Figure 1. Example from over 12,000 stations analysed by the author classifying daily modes [+3, ..., -3].

RESULTS

Federations such as USA, Canada, Russia, China, and Australia were not summarized as provincial climatic variability is vast, and so I have excluded any country with more than 99 stations from the present paper. Table 2 summarizes the resulting median findings of 157 smaller countries. The first column is sorted by Federal Information Processing Standards two-letter country codes were used by the US government for geographical data processing, such as World Factbook (CIA 2010). The second column is the number of WMO sites which were analysed in each country to present the following seven columns of results. Results have been adjusted on the basis that there are

365¼ days per annum on average, and because most stations registered observations a number of days less than 365 day in any year. Stations were ignored that registered less than 330 days in a year.

Table 2. Median days per year heating-cooling mode clustered by country

FIPS Code	Sites	-3 Icy	-2 Heat	-1 Humid'	0 Passive	+1 Vent'n	+2 Evap'v	+3 DX	County abrev
AC	2	0	0	0	0	83	0	197	ANTIGUA
AE	9	0	26	0	91	44	21	183	U ARAB E
AF	6	9	136	17	27	33	51	0	AFGHAN
AG	67	0	135	3	54	32	78	10	ALGERIA
AJ	9	39	164	0	54	42	9	21	AZERBA
AL	1	2	189	0	69	47	1	44	ALBANIA
AM	14	105	171	0	29	0	0	0	ARMENIA
AQ	2	0	0	0	0	18	0	323	A SAMOA
AR	61	2	183	0	89	54	1	9	ARGENTI
CN	1	0	0	0	1	130	0	198	COMOS
CO	28	0	0	0	0	30	0	257	COLOMBI
CQ	3	0	0	0	0	16	0	291	N MARIA
CS	6	0	0	0	59	148	0	116	COSTA R
CT	8	0	0	0	24	199	5	50	C AFRICA
CU	43	0	3	0	39	159	0	116	CUBA
CV	5	0	0	0	91	170	0	65	CAPE VER
CW	14	0	0	0	13	145	0	158	COOK I
CY	17	0	154	0	80	48	7	63	CYPRUS
DA	69	95	233	0	21	1	0	0	DENMAR
DO	1	0	0	0	0	111	0	240	DOMINIC
DR	8	0	0	0	1	135	0	199	DOMIN R
EC	15	0	0	0	0	176	0	44	ECUAD
EG	26	0	112	0	82	55	29	52	EGYPT
EI	26	34	321	0	3	0	0	0	IRELAND
EK	2	0	0	0	0	84	0	256	EQUAT G
EN	26	147	179	0	25	3	0	0	ESTONIA
ER	1	0	205	21	100	0	0	0	ERITREA
ES	5	0	0	0	1	72	7	259	EL SALVA
ET	8	0	50	0	204	14	0	0	ETHIOPIA
EU	1	0	0	0	68	141	0	104	EUROPA I
EZ	46	119	197	0	40	4	0	0	CZECH
FG	2	0	2	0	0	106	0	225	FR GUINE
FI	55	159	163	0	17	2	0	0	FINLAND
FJ	16	0	0	0	15	155	0	132	FIJI
FK	5	115	143	0	0	0	0	0	FALKLD I
FM	10	0	0	0	0	0	0	355	MICRON
FO	3	51	275	0	0	0	0	0	FAROE I
FP	12	0	0	0	5	182	0	139	FR POLY
FS	4	70	195	0	0	2	0	0	FR ANTA
GA	1	0	0	0	48	132	11	144	GAMBIA

GB	4	0	0	0	0	103	0	216	GABON
GG	5	67	163	0	65	38	7	7	GEORGIA
GH	2	0	0	0	0	106	1	179	GHANA
GI	1	0	187	0	103	71	1	3	GIBRALT
GJ	1	0	0	0	0	75	0	187	GRENAD
GK	2	13	335	0	12	2	0	0	GUERNSY
GL	37	217	133	0	0	0	0	0	GREENLD
GP	1	0	0	0	1	123	0	176	GUADEL
GQ	1	0	0	0	0	62	0	303	GUAM
GR	34	1	185	0	75	64	4	21	GREECE
GT	8	0	2	0	26	78	0	149	GUATEM
GV	1	0	0	0	0	45	1	268	GUINEA
GY	2	0	0	0	0	7	0	271	GUYANA
HA	1	0	0	0	0	88	0	215	HAITI
HK	1	0	62	0	76	74	0	153	HONG K
HO	16	0	0	0	11	119	0	131	HONDUR
HR	51	49	196	0	71	31	1	10	CROATIA
HU	41	85	187	0	67	17	1	2	HUNGAR
IV	19	0	0	0	6	67	12	121	COTE D'I
IZ	8	2	137	2	41	25	135	3	IRAQ
JE	1	22	324	0	18	0	0	0	JERSEY
JM	2	0	0	0	1	108	0	248	JAMAICA
JO	10	3	156	1	78	49	24	7	JORDAN
JU	1	0	0	0	3	134	0	184	JUAN DE
KE	17	0	87	0	81	3	0	0	KENYA
KG	8	74	160	0	68	33	10	0	KYRGYZ
KN	28	58	192	0	47	48	0	6	N KOREA
KR	3	0	0	0	0	3	0	325	KIRIBATI
KS	56	26	201	0	56	56	0	21	S KOREA
KU	1	0	97	2	53	25	138	50	KUWAIT
KV	2	74	197	0	73	18	2	0	KOSOVO
KZ	90	130	131	0	52	16	6	0	KAZAKIS
LA	8	0	2	0	28	61	1	216	LAOS
LH	89	89	182	0	55	13	1	1	LITHUANI
LI	72	84	173	0	61	15	2	2	LIBERIA
LO	70	84	175	0	61	15	2	2	SLOVAK
LS	35	0	116	1	65	42	21	32	LIECHTEN
LT	33	0	115	1	65	43	28	34	LESOTHO
LU	31	0	114	1	65	44	35	37	LUXEMB
LY	28	0	112	1	65	45	40	41	LIBYA
MA	16	0	0	0	50	149	0	92	MADAGA
MB	1	0	0	0	0	122	0	204	MARTINI
MC	1	0	84	0	58	80	0	142	MACAU
MD	3	93	167	0	73	22	4	3	MOLDO
MF	1	0	0	0	0	141	0	208	MAYOTT
MG	33	104	225	1	26	2	0	0	MONGOL
MH	1	0	0	0	0	63	0	253	MONTSE

MJ	8	14	191	0	61	35	3	14	MONTEN
MK	9	59	175	0	69	18	4	0	MACEDO
ML	9	0	0	0	18	51	78	146	MALI
MO	24	0	167	0	87	49	6	7	MOROCC
MP	6	0	0	0	54	183	0	113	MAURITI
MR	9	0	3	4	30	40	98	101	MAURITA
MT	1	0	180	0	78	81	2	24	MALTA
MU	31	0	8	1	75	46	36	151	OMAN
MY	27	0	0	0	0	37	0	321	MALAYS
MZ	7	0	2	0	90	127	0	59	MOZAM
NC	8	0	2	0	102	170	0	56	NEW CAL
NE	1	0	1	0	27	176	0	102	NIUE
NF	1	0	156	0	151	54	0	0	NORFOLK
NG	14	0	2	6	27	50	100	135	NIGER
NH	9	0	0	0	7	149	0	173	VANUAT
NI	1	0	0	0	4	157	54	107	NIGERIA
NL	57	58	265	0	28	3	0	0	NETHERL
NP	1	0	127	0	48	114	1	1	NEPAL
NR	1	0	0	0	0	8	0	319	NAURU
NS	4	0	0	0	0	89	0	230	SURINA
NT	2	0	0	0	0	74	0	265	NETH A
NZ	38	4	317	0	23	0	0	0	NEW ZL
OO	54	7	0	0	0	0	0	6	ocean
PA	17	0	63	0	74	106	3	72	PARAGU
PC	2	0	74	0	95	139	0	0	PITCAIR
PE	27	0	9	0	17	77	0	7	PERU
PK	48	0	100	0	56	32	13	140	PAKIST
PL	94	112	192	0	42	5	0	0	POLAND
PM	9	0	0	0	0	72	0	265	PANAM
PO	44	0	222	0	97	24	1	0	PORTUG
PP	10	0	0	0	0	14	0	293	PAPUA
PS	1	0	0	0	0	3	0	362	PALAU
PU	1	0	0	0	7	63	31	227	GUINEA
QA	6	0	39	0	64	31	22	130	QATAR
RE	2	0	0	0	67	179	0	96	REUNIO
RI	26	80	177	0	72	23	2	4	SERBIA
RM	3	0	0	0	0	2	0	361	MARSH
RP	60	0	0	0	0	51	0	286	PHILIPPI
RQ	15	0	0	0	0	26	0	189	PUERTO
RW	1	0	8	0	297	34	0	0	RWAND
SA	43	0	73	4	56	36	140	17	SAUDI A
SB	1	150	199	0	8	0	0	0	ST PIERR
SC	2	0	0	0	0	97	0	245	ST KITTS
SE	1	0	0	0	0	72	0	267	SEYCHE
SG	12	0	0	0	17	53	57	193	SENEGA
SH	3	0	199	0	22	20	0	0	ST HELE
SI	18	83	198	0	63	13	0	1	SLOVEN

SN	2	0	0	0	0	20	0	345	SINGAP
SP	65	3	204	0	77	36	3	3	SPAIN
SU	3	0	5	5	31	33	135	74	SUDAN
SV	4	278	195	0	0	0	0	0	SVALBA
SX	3	223	110	0	0	0	0	0	S GEOR
SY	15	6	151	1	53	38	43	33	SYRIA
TD	1	0	0	0	0	101	0	229	TRINID
TH	45	0	4	0	49	86	1	215	THAILA
UP	52	124	163	0	62	14	1	1	UKRAIN
UV	9	0	0	1	13	82	79	148	BURKIN
UY	12	1	175	0	97	63	2	14	URUGU
UZ	22	62	142	0	55	36	45	15	UZBEKI
VC	1	0	0	0	0	40	0	288	ST VINC
VE	52	0	0	0	1	90	0	224	VENEZU
VM	30	0	0	0	1	85	0	223	VIETNA
VQ	8	0	0	0	0	3	0	218	VIRGIN
WA	10	0	93	25	94	60	23	0	NAMIBI
WF	2	0	0	0	0	61	0	289	WALLIS
WQ	1	0	0	0	3	87	0	258	WAKE I
WS	8	0	0	0	0	27	0	275	W SAM
WZ	7	0	71	0	105	71	3	9	SWAZIL
YM	8	0	6	0	40	74	5	151	YEMEN
ZA	2	0	71	0	141	99	16	3	ZAMBIA
ZI	21	0	63	2	117	82	19	2	ZIMBAB

CONCLUSION

Refering to Table 2, it appears that DX air-conditioning would be desirable at least 10 days per year in the median representative of 103 countries. Only Afghanistan, Iraq, Jordan, Kyrgyzistan, Namibia, Zambia, and Zimbabwe have median representatives where direct evaporative cooling is the suggested mode for cooling in summer. The median representative of other countries could be satisfied in summer with fans or passive design, without dissatisfaction more than 10 days in a year.

With regard to winter heating, 37 countries' median representative were exposed icy conditions 10 days pa, and so alternatives to air-source heat pumps should be considered. Meanwhile 40 countries' median representatives might be well served with air-source heat pumps if passive methods cannot be implemented to maintain comfort. The median representatives of Afghanistan, Ethiopa, and Namibia appeared to need winter humidification systems to avoid desiccation 10 day per year. Twenty countries of the countries summarized in Table 2 have median representatives with at least one day pa when humidification is desirable on the basis of main daily conditions available from GSOD.

Twenty countries' median representatives appear in Table 2 that might be well served by air-conditioning in summer, with reverse cycle air-source heat pump function in winter. These countries are Albania, Algeria, Cyprus, Egypt, Greece, Hong Kong, Kuwait, Lesotho, Libya, Liechtenstein, Macau, Malta, Pakistan, Paraguay, Qatar, Saudi Arabia, Syria, United Arab Emirites, and Uraguay.

Only one representative of each country has been presented in Table 1, and conditions will vary at different locations within each country. I recommend that designers seek 8760 hour TMY files close to their project site if available, and then utilize a tool such as Climate Consultant to review the synergistic impact of coincident solar, wind, temperature, humidity, and earth temperature from concept stage – and at later stages use full simulation software.

REFERENCES

ASHRAE (2010). Thermal environmental conditions for human occupancy. Atlanta, Georgia, USA, American Society of Heating, Refrigerating and Air-Conditioning Engineers. **55**.
ASHRAE, A. and U. IESNA (2018). Advanced Energy Design Guide for K-12 School Buildings Achieving Zero Energy. Atlanta, Georgia, USA.
CIA (2010). The World Factbook. See also: https://www. cia. gov/library/publications/the-world-factbook. C. I. A. Office of Public Affairs. Washington, D.C. 20505.
Givoni, B. (1992). "Comfort, climate analysis and building design guidelines." Energy and Buildings **18**(1): 11-23.
Menne, M. J., I. Durre, R. S. Vose, B. E. Gleason and T. G. Houston (2012). "An overview of the global historical climatology network-daily database." Journal of Atmospheric and Oceanic Technology **29**(7): 897-910.
Milne, M. (2016). Climate Consultant v6. 0. Los Angeles, California 90095-1467, Department of Architecture and Urban Design University of California, Los Angeles.
Milne, M., R. Liggett, A. Benson and Y. Bhattacharya (2009). Climate Consultant 4.0 develops design guidelines for each unique climate. American Solar Energy Society Meeting.
Peterson, E. (2014). Climate zone mapping for air conditioners and heat pump devices. S. a. T. a. N. Z. G. The Energy Rating Website - a joint initiative of Australian. Canberra, Australia, Department of the Environment and Energy: 36.
Peterson, E., N. Williams, D. Gilbert and K. Bremhorst (2006). "New air conditioning design temperatures for Queensland, Australia." Ecolibrium **5**(1): 26-36.
Sayigh, A. and A. H. Marafia (1998). "Chapter 1—Thermal comfort and the development of bioclimatic concept in building design." Renewable and Sustainable Energy Reviews **2**(1): 3-24.
Watt, J. (2012). Evaporative air conditioning handbook, Springer Science & Business Media.
Yi, J. (2005). China Standard Weather Data for Analyzing Building Thermal Conditions. ISBN 7-112-07273-3 (13228). T. U. a. Department of Building Science and Technology and C. M. B. Climate Information Center. Beijing, China, China Building Industry Publishing House.

Holistic Approach to Energy Performance of Green Built Environment

Essam E Khalil, PhD, PE
ASHRAE Fellow

ABSTRACT

The global energy crisis coupled with the threats of climate change bring into sharp focus both opportunities and challenges for developing countries. Developed and developing countries have to better address the increasing energy demands of growing economies, as well as address energy poverty issues often highlighted by extreme disparities in income. They also need to deal with the real and potential impacts of climate change and energy efficiency improvement. In addition to these challenges is the global imperative to reduce carbon emissions in order to prevent climate change. While developing nations have thus far been sheltered from obligations to reduce carbon emissions, one cannot anticipate that this situation will continue for long. Holistic approach to energy performance in built environment had been a key tool to improve the overall energy performance. Within this context nations need to follow a very different development path from that established by first world countries. This development path is a low energy, low carbon and generally a resource efficient one.

1. SUSTAINABILITY AND QUALITY OF LIFE

Quality of life has been the domain of development discourse for the past decade. It has been widely recognized that measuring progress in terms of GDP is not sufficient. However, an adequate definition of Quality of life is still missing. Diverse "objective" and "subjective" indicators across a range of disciplines and scales, and recent work on subjective well-being (SWB) surveys and the psychology of happiness have spurred renewed interest. Currently, there are a number of indices proposed and used by different organizations to score and rate cities and countries according to their quality of life as mentioned before. These organizations vary in nature and interests and thus in focus and methodology. The most important indices are Quality of Living by Mercer consultants, Quality of Life index by The Economist Intelligence Unit, Nations' ranking according to quality of life done independently . It is apparent that these different indices focus on a number of common aspects as the main core of quality of life, namely: Housing, Income, Jobs, Community, Education, Environment, Governance, Health, Life Satisfaction, Safety and Work-life balance. Despite these various efforts, an important aspect is still missing in most indices. It is sustainability, whether this level of quality of life can be sustained or not, whether it is affecting the ability of future generations to attain such levels or not. This notion is still underscored, as it is complex to assess. (Khalil et al 2015)

Essam E Khalil is Professor of Energy at Cairo University, Cairo, Egypt

There are other indices that measure and rate the sustainability of a city. One of the most recent sustainability indices is Green City Index. This Index measures the current environmental performance of major cities in different continents, as well as their commitment to reducing their future environmental impact by way of ongoing initiatives and objectives. Quality of life and energy efficiency and renewable energies in the built environment .Despite the involvement of Green City Index in measuring energy efficiency as part of achieving environmental sustainability; it does not explicitly correlate to measuring how much energy is used to attain a certain quality of life for city dwellers. First, the indicators concerning energy efficiency in the Green City Index are extracted from different categories and highlighted. The energy efficiency indicators can be classified into two groups: the first group relates directly to the energy used to achieve a good quality of life, while the other group relates indirectly to the matter and is more concerned with sustaining this quality of life from the point of using energy efficiently. Error! Reference source not found. shows this classification of indicators.

Table Error! No text of specified style in document. **: Energy Efficiency Indicators as achieving good quality of life, (Khalil et al 2015)**

Category	Energy efficiency Indicators	
	Achieving good quality of life	Sustaining good quality of life
Energy and CO2	Energy consumption	Same
	Electricity consumption per unit of GDP	Same
	Access to electricity	Renewable energy consumption
		Clean and efficient energy policies
		Climate change action plan
Buildings	Energy consumption of residential buildings	Energy-efficient buildings standards
		Energy-efficient buildings initiatives
		Eco buildings policy
		Number of LEED-certified buildings
Land Use	Green land use policies	Same
	Green spaces per capita	Land use policy
	Population density	Urban sprawl
	Population living in informal settlements	
Waste		Waste recycling and re-use policy
Transport	Size of non-car transport network	Use of non-car transport
	Length of mass transport network	Green transport promotion
	Stock of cars and motorcycles	Urban mass transport policy
	Average commute time from residence to work	
	Congestion reduction policy	

2. ENERGY PERFORMANCE

Energy Performance of Buildings should include a general framework for the calculation of energy performance and building categories together with thermal characteristics of building, air conditioning, ventilation, lighting and appliances aspects considered. These include Active solar systems contribution to domestic water heating based on renewable energy sources, CPH production and District cooling systems. The subsection demonstrates the importance of incorporating an energy performance directive as a Standard in our region such a goal will aid energy savings in large buildings and set regulations to energy efficient designs that are based on Standard calculation methods. The target is to develop standardized tools for the calculation of the energy performance of buildings, with defined system boundaries for the different building categories and different cooling/heating systems. Endeavour is targeted to develop a common procedure for an "energy performance certificate".

Analyses of energy performance would be initiated from power generation pattern in a country, Figure , energy consumption by sector and energy consumption by utility as shown in Figures 3 and 4 for Egypt1, as an example. In commercial buildings, Air conditioning systems can consume as much as 56% of the total energy used in the building. Therefore, it is a challenge to design an optimum HVAC airside system that provides comfort and air quality in the air-conditioned spaces with efficient energy consumption. The conditions of the air to be maintained are dictated by the need for which the conditioned space is intended and comfort of users. Therefore, the air conditioning embraces more than cooling or heating. The comfort air conditioning is defined as "the process of treating air to control simultaneously its temperature, humidity, cleanliness, and distribution to meet the comfort requirements of the occupants of the conditioned space". Air conditioning, therefore, includes the entire heat exchange operation as well as the regulation of velocity, thermal radiation and quality of air, as well as the removal of foreign particles and vapors, Khalil (2013).

It is probable, that one calculation method will not cover all aspects and building categories of the proposed directive. For some applications, more advanced simulation models will have to be used to provide satisfying accuracy. The ongoing and future work on methods for validation and documentation of simulation tools at HBRC and Cairo University Medhat et al (2006) and Khalil (2008), could be valuable in a process of approving models.

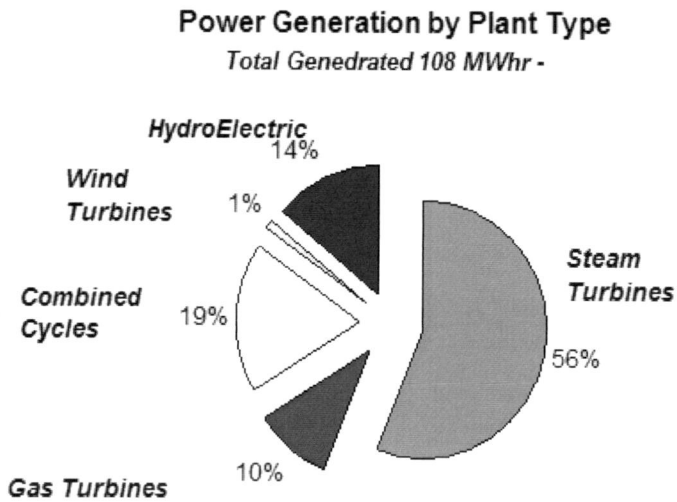

Figure 2: Current Power Generation Technologies in Egypt, Khalil et al (2015)

3. ENERGY EFFICIENCY INDICATORS

Until now, the guidelines and design standards do not provide restricted utilization strategies of the conditioned air in the spaces. Indeed, this situation creates several inefficient systems and consequently expensive energy invoice. In some critical facilities, such as hospitals, HVAC designers face the problem of balancing between the attaining comfort conditions and the efficient energy utilization. The relation between the HVAC system designs and the optimum conditions and optimum energy utilization is still under investigation up today. In recent researches by Khalil (2013), the effect of ventilation design on the comfort and energy utilization is investigated. The effect of the displacement ventilation on humidity gradient in a factory located in the hot and humid region is illustrated, (Kosonen (2002). It was found the strong dependence relation between the correct supplying conditions and comfort. As the world becomes increasingly dependent on electrical appliances and equipment, energy consumption rapidly rises every year. Many programs have been established in various countries to increase end-use equipment energy efficiency. One of the most cost-effective and proven methods for increasing energy efficiency of electrical appliances and equipment is to establish energy efficiency standards and labels, ISO (2009),

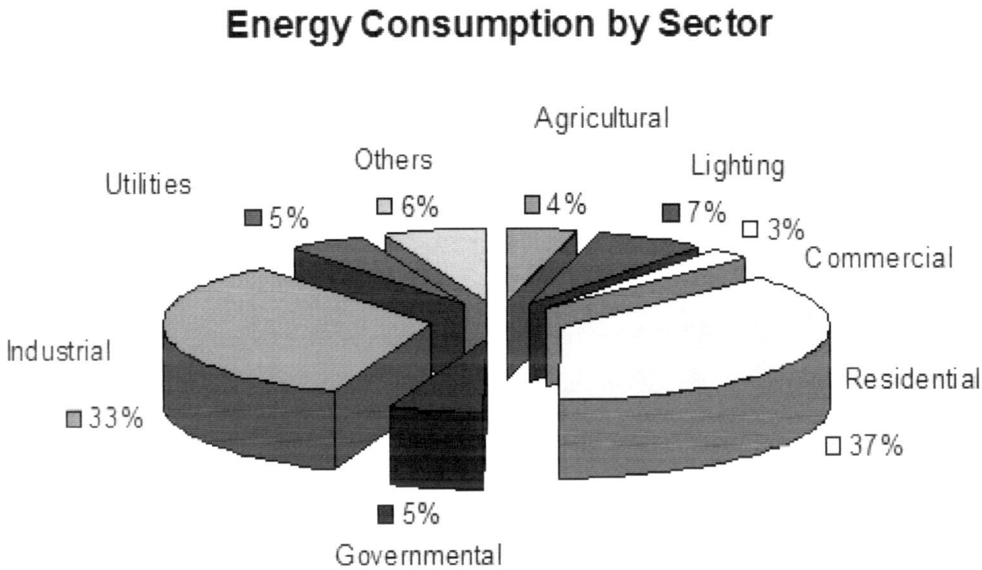

Figure 3: Energy Consumption by Sector in Egypt, Khalil et al (2015)

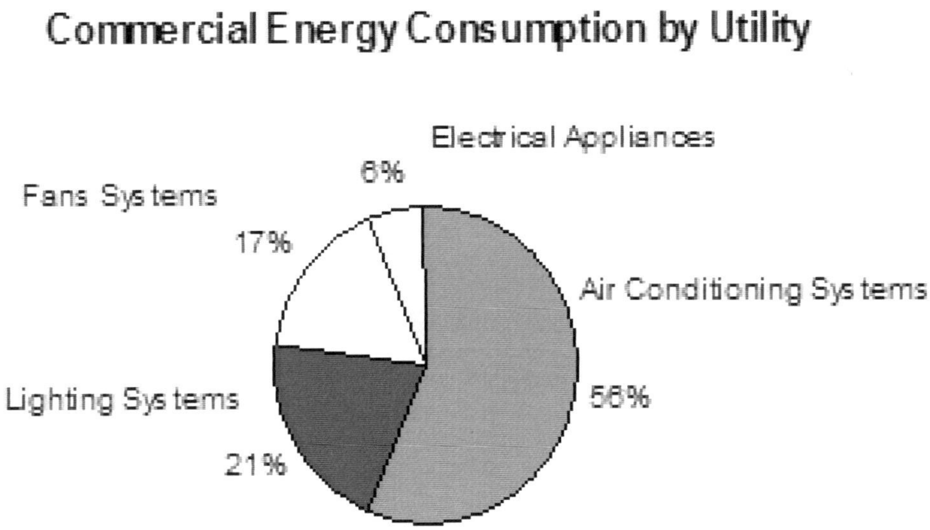

Figure 4: Building Energy Consumptions by Utility in Egypt, Khalil et al (2015)

4. ENERGY EFFICIENCY STANDARDS

Energy efficiency standards are a set of procedures and regulations that prescribe the energy performance of manufactured products, sometimes prohibiting the sale of products less energy-efficient than the minimum standard. The term "standard" commonly encompasses two possible meanings:

1. A well-defined protocol (or laboratory test procedure) by which to obtain a sufficiently accurate estimate of the energy performance of a product in the way it is typically used, or at least a relative ranking of the energy performance compared to other models; and
2. A target limit on energy performance (usually a maximum use or minimum efficiency) formally established by a government-based agency upon a specified test standard.

To achieve such goals of energy efficient design of built environment, the joint working group (ISO JWG163/205) activities aim at the development of a coherent set of standards for the determination of the energy performance of buildings using holistic approach. For this, the JWG joins the expertise of both parent Technical Committees: ISO/TC 163 ("Thermal performance and energy use in the build environment") and ISO/TC 205 ("Building environment design").The scope Covers also: Standardization of the holistic assessment of the energy performance of new and existing buildings as well as building retrofits, in close collaboration with ISO/TC 205 by means of the ISO/TC163/WG4 Joint working group TC 163 & TC 205 Energy performance using holistic approach, including;
- Terms and definitions;
- System boundaries for buildings and technical systems;
- Assessment of the overall energy performance of buildings, taking into account
  - the energy performance of building elements;
  - building related systems (heating, cooling, domestic hot water, ventilation, lighting, system controls, transport, and other energy related systems);
  - indoor and outdoor conditions;
  - local energy production (on site and at district level);
  - (use of) energy sources (including renewable);building commissioning
- Assessment of overall energy efficiency; and
- Means of experimentation

The JWG's Task Group 4, "JWG/AG On the strategy of international standardization on the energy use in buildings", produces a regularly updated strategic view, which includes, VanDijk et al (2011):
- an overview of standards and active, planned or missing work items that are needed for the assessment of the energy performance of buildings using the holistic approach;
- ground rules for the drafting of standards that are to be coordinated by the JWG;
- Proposals on the links with closely related fields of international standardization.

5. ENERGY LABELS

Energy efficiency labels are informative labels affixed to manufactured products indicating a product's energy performance (usually in the form of energy use, efficiency, and/or cost) in order to provide consumers with the data necessary for making informed purchases. Energy labels serve as a complement to energy standards. They provide consumers information that allows those who care to select more efficient models. Labels also allow utility companies and government energy conservation agencies to offer incentive s to consumers to buy the most energy-efficient products. The effectiveness of energy labels is highly dependent on how information is presented to the consumer. Energy efficiency in developing countries plays an important role in achieving global sustainable development. Energy-efficiency improvements can slow the growth in energy consumption, save consumers money and reduce capital expenses for energy infrastructure. Energy consumption is growing rapidly in these countries, yet energy efficiency remains far below levels in developed countries. Energy-efficiency improvements can slow the growth in energy consumption, save consumers money and reduce capital expenses for energy infrastructure. For most developing countries, the foreign exchange needed to finance energy sector expansion is a significant drain on reserves. Additionally, energy efficiency reduces local environmental impacts, such as water and air pollution from power plants, and mitigates greenhouse gas emissions. Standards and labeling programs provide enormous energy savings potential that can direct developing countries towards sustainable energy use. Improved end-use efficiency from Standards and Labeling programs can contribute significantly to developing economies. The main benefits are:

1. Less need to build new power plants. The cost of saving 1 kWh of energy through energy-efficiency programs has proven much less expensive than producing 1 kWh of energy by building a new power plant.
2. Reduced greenhouse gas emissions. Less energy production means less carbon dioxide emissions from power plants. This contributes to environmental benefits such as slowing down environmental pollution and global warming and preserving natural resources and the ecosystem.
3. Improved competitiveness for local manufacturers. Local companies that upgrade the efficiency of their products can compete better with multi-national companies, especially with lower production costs.
4. Higher consumer disposable income. Less spending on electric bills increases consumer purchasing power for other products, which helps local businesses.
5. Increased cash flow in the local economy. With higher disposable income, consumers are more willing to spend, thus, injecting money into the local economy.
6. Improved Trade Balance. Decrease in energy demand will reduce the consumption of indigenous resources (i.e. natural gas and oil), allowing more to be exported (for Lebanon, less to be imported). Increased export earnings (or less import spending) help alleviate trade deficit of Arabian countries.
7. Avoided future energy deficit as power demand rises. Energy exporting countries have become net importers due to dramatic increases in electricity demand. Energy-efficiency programs can help slow down the demand and prevent energy deficit in the future.

In recent years, efforts were concentrated to further develop adequate and necessary standards that help the end user and industry to enhance the energy efficiency. An example of an earlier endeavor is shown here in Figure 5 and relates to ISO 16343. A seven classes energy label is highlighted and in standard colors to set the various levels that should be achieved. Various forms are available. Countries are obliged now to set these levels of EER that designated the classes A….to G. In developing countries the minimum should be larger than 9 and the other upper end of the scale is open to innovation and development , but can be typically 14 or 15 with the inverter technology for room air conditioners.

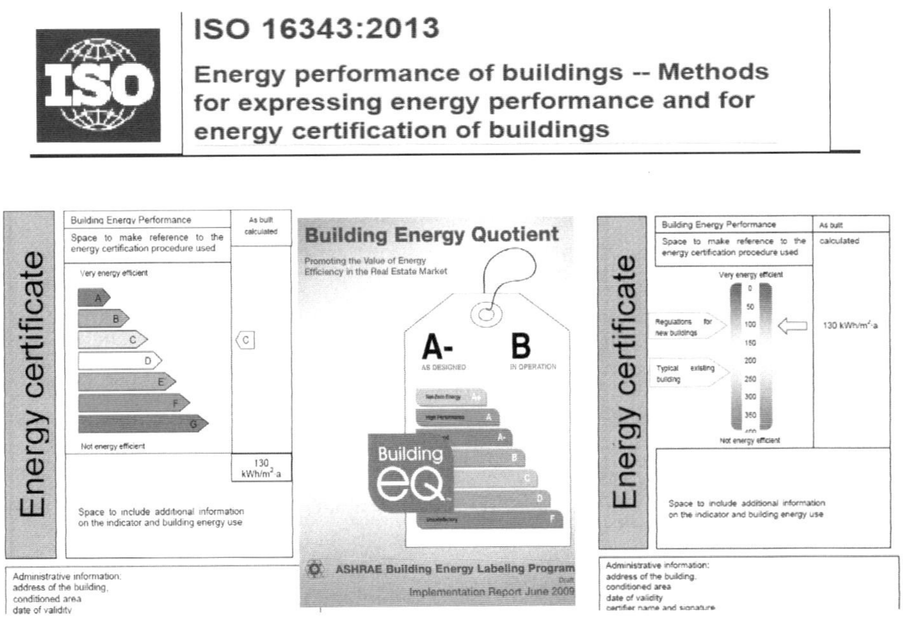

Figure 5: Energy Performance of Buildings ISO Standards 16343

Figure 6 shows the various energy systems interaction in built environment.

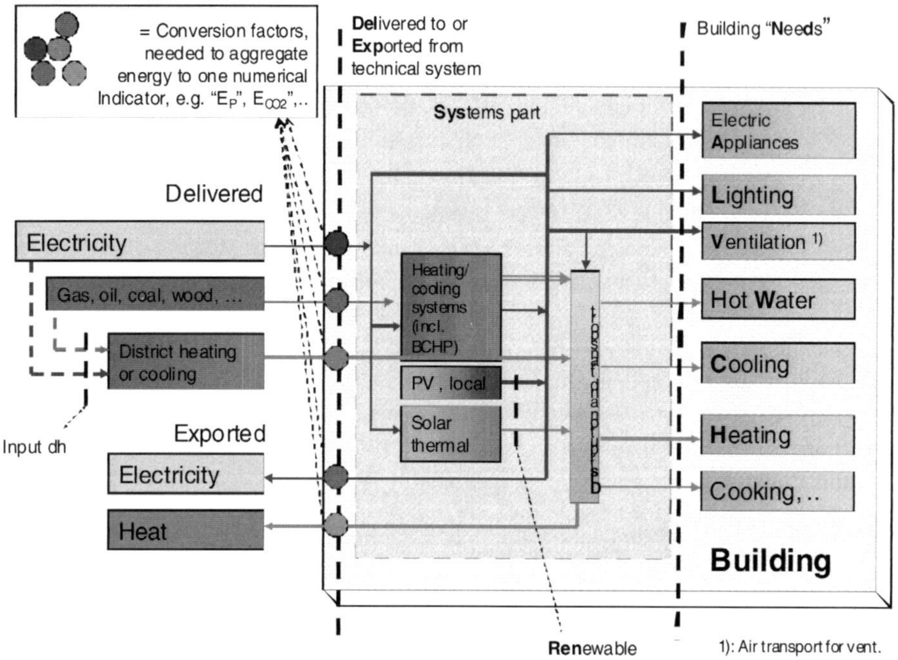

Figure 6: Systems Interaction in Energy Performance Map

The ISO 52000 Series

The design and implementation of public policies on Energy Efficiency may take various forms and require diverse mechanisms and tools, Khalil and Van Dijk (2015) such as:

1. Adequate statistical tools to monitor, analyze, set objectives and forecast energy production, distribution, performance and consumption
2. Public funding of research and development on energy efficient technologies and renewable energies
3. Public-private partnerships for the design and deployment of energy efficient infrastructures and innovative solutions and technologies
4. Policies and regulations to foster energy efficiency and renewable energy covering appliances, buildings, equipment or operation at large, resource utilization and possible related tax or other incentives, as well as labeling, auditing or communication obligations
5. Communication, education and training tools to promote efficient energy management and use at all levels, from companies or public services to consumers
6. Standards based on consensus between stakeholders have been and are being developed at all levels (professional, national, regional and international) to serve as tools to:
7. Support the development and implementation of public policies
8. Assist and motivate stakeholders in improving their own energy efficiency and in increasingly using renewable energies
9. Proposed new standards series may provide a tool box for:

- Terminology, definitions and taxonomy
- Data acquisition, processing and exchange
- Interoperability
- Measurement, analysis and test methods

- Renewable energy, and electrical energy storage
- Design and performance standards and indicators for appliances, equipment, buildings, systems or organizations, and support to the expression of minimum energy performance indicators (MEPs) where applicable
- Labeling and communication standards
- Calculation methods for energy savings, performance and systems
- Management and energy audit standards
- Conformity assessment

Along these lines, the new ISO series 52000 were designed to cover a holistic approach to energy performance in the built environment. The series covers the above aspects. Almost 40 of these standards were published so far.

6. CONCLUDING REMARKS

It can be concluded that it is important to incorporate an energy performance directive as a Standard in our region such a goal will aid energy savings in large buildings and set regulations to energy efficient designs that are based on Standard calculation methods. It is recommended to:

1. Develop standardized tools for the calculation of the energy performance of buildings within urban and rural environments and their relation to providing adequate quality of life.
2. Define system boundaries for the different building categories and different heating/cooling systems
3. Prepare models for expressing requirements on indoor air quality, thermal comfort in winter and summer, visual comfort, etc.
4. Define comparable energy related key values (kWh/m2, kWh per person, kWh per apartment, kWh per produced unit etc.) and to develop a common procedure for an "energy performance certificate"
5. Design, construct and operate a solar decathlon (Building) that can meet the rural and desert requirements and save the diminishing fossil fuel sources.
6. Standardization of the holistic assessment of the energy performance of new and existing buildings as well as building retrofits was carried out and is currently on-going activities, in close collaboration between ISO/TC 163 & ISOTC 205 using holistic approach.

REFERENCES

ISO publications, 2009

Khalil, H.E. and Khalil, E.E., "**Energy Efficiency in the Urban Environment**", Taylor and Francis, CRC Press, USA, 2015

Khalil, E.E. "Energy Performance of Buildings Directive in Egypt: A New Direction, HBRC Journal, Vol. 1, 2005.

Khalil, E.E. "**Air Distribution in Buildings**, Taylor and Francis, CRC Press, USA, 2013

Khalil, E.E. "Air Conditioning and Refrigeration Code For Energy-Efficient Buildings in the Arab World", Journal of Kuwait Society of Engineers, Vol .100, pp. 94-95, September 2008.

Khalil, E.E., and Van Dijk, D., (2015),"The new ISO 52000 series: holistic approach to energy efficiency of buildings", ISO Focus, vol. 23, pp. 12-15, September 2015

Kosonen, R. "Displacement ventilation for room air moisture control in hot and humid climate", Roomvent 2002, pp. 241-244, 2002.

Medhat, A.A. and Khalil, E.E. "Thermal Comfort Meets Human Acclimatization in Egypt" Proceeding of Healthy Building, Vol.2, pp.25, June 2006

Van Dijk, D., and Khalil, E.E, (2011) Future cities – Building on energy efficiency, ISO Focus, pp. 25-27, May 2011

Benefits From Combination of Centralized Ventilation System and Decentralized Conditioning Units

Maciej Danielak, PhD
Member ASHRAE

ABSTRACT HEADING

Each building needs to have regulated temperature as well as needs to be ventilated. These two requierements come from one of most important function of the building: create comfortable indoor climate.

A human being need fresh air to stay healthy and fit. The building need to be thermal conditioned from one side to protect the construction and the mass of the building and because of thermal indoor comfort for a user's. The both needs relate to energy consumption but also with a place (volume) in a building. So, there are both investments cost and consumption costs.

Benefits for both can bring the system witch splits those to needs – hygienic and caloric – and bring a combination between centralized ventilation system and decentralized conditioning system. In this the air exchange is provided by a single bidirectional ventilation unit. The temperature is adjusted on this system by decentralized equipment in the inside of the room and not via the central ventilation unit (air handling unit).

This combines the unbeatable efficiency of decentralized units with the heat recovery capability of a ventilation unit.

The basic system will be presented and compared. The central-decentral system will be described in detail and all the benefits for the building energy consumption and other system component like BMSystem, architecture will be mentioned.

The presentation will present a separately working centralized ventilation unit and decentralized conditioning units with are connected into one system.

INTRODUCTION

Requirements concerning energy efficiency of buildings have become more and more strict. Energy consumption limits for heating, air-condition or ventilation purposes are regulated by law. At the same time, current prices of fossil fuels as well as increasing expectations of consumers pose a great challenge for the designers of new buildings. For greater ventilation air-flows, when heat recovery is required, in most cases central ventilation system equipped with heating and air-conditioning functions is chosen. However, this choice results in many technical problems and to avoid them some other concept complying with hygenic and caloric requirements of a building should be taken into consideration.

A deciding factor in choosing air systems (ventilation centers) equipped with heating and air-condition functions are the installation costs, which are lower as compared to other systems. However, in terms of exploitation costs of a building, combination of caloric and hygenic functions has proved to be an inefficient solution. This paper provides the analysis of significant differences between central and decentral systems as well as their consequences in energy consumption and eventually in building´s exploitation costs.

Maciej Danielak is an Export Sales Director at Kampmann GmbH, Lingen, Germany.

DIFFERENT HVAC SYSTEMS

Heating and cooling systems can be divided into (decentral) and central systems, depending on their location. General definition of decentralised system involves a set of indepentent technical units combined in one coherent system. In terms of heating system, such units include heating equipment connected by pipe system and robotics, while in centralised system there is one main unit regulating the activity of other components. As both technology and life show, it is impossible to find an ideal solution without any flaws. Central system reduces the number of units to minimum, what facilitates service and maintenance as well as optimizes acustic parameters.

The advantages of decentralised system are especially during its utilisation. Individual regulation and independent work of each unit reduce significantly energy consumption of energy. Since its technical equipenent occupies less space in centralised system, it makes the interior arrangements much easier. At the same time the costs of centralised system are lower than decemtralised one. Main functions of HVAC include:

- Hygenic function – providing fresh air in the rooms untilised by people
- Caloric function – piping the excess of heat or providing the rooms with heat definiciency, with the aimto keep the users in thermic balance and comfort

SEPARATION OF HYGIENIC AND CALORIC FUNCTION

As already mentioned, selection of a system and decision regarding its utilisation depends on its compliance with specific hygenic and caloric requirements. These functions are implied by various factors not corresponding to each other. Presence of the users and increased concentration of carbon dioxide (an indicator of the air quality) results in the activation of ventilation mode. As a consequence of the difference between a given temperature and an actual

one, the units will be switched on and both temperatures will be evened out. An optimal solution in terms of investment and exploitation costs is a combination of the two systems and their parallel work. In this case hygenic requirements (regarding ventilation) are realised by the means of centralised system, while all the termdynamic processes take place in decentralised system. Both mechanisms are connected in closed or open system. In the open system there no interference between the mechanisms and the outdoor air is separated from the fresh air.

Decentralised temperature control
Decentralised units only for heating/cooling. Energy transport realised by the water.

Centralised ventilation
The air handling unit only transports air into the building when the sensors also detect a need for outside air

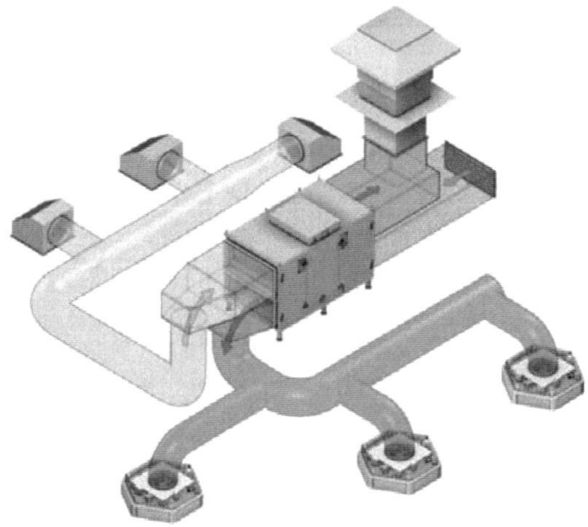

Figure 1 The central-decentral system is a bidirectional ventilation system with efficient heat recovery. In this case the temperature is adjusted to this system by decentralised mechnisms located inside the room and not through the central ventilation unit (air handling unit).

CONTROL OF CENTRAL-DECENTRAL SYSTEM

Combination of centralised and decentralised systems in one building results in significant exploitation efficiency (reduction of current utilisation costs) and simplification of regulating system. The inflow of fresh air and the proper quality of indoor climate is controlled either according to timemodus (eg. office hours) or by means of carbon dioxide detector. Due to the fact that the outside air generates the largests costs, because of the necessity to heat it in winter and cool it in summer, it should be used in an optimal way to create suitable interior conditions. Heating or cooling is realised through the circulation mode, where the relevant option is switched on by the thermostat or as result of the user´s decision (particularly important in terms of cooling process)

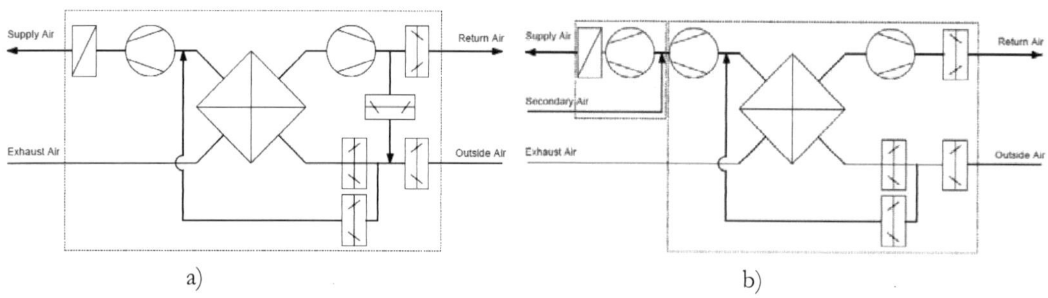

Figure 2 Principle scheme of central (a) and central-decentral (b) system

It is important to underline that separation of hyhenic and carolic functions results in their independent control and does not affect their individual functionalities. It is a great advantage, especially in comparison with complex algorithms applied to control central system. Central system is only responsible for controlling the indoor air quality, while decentral system ensures the energy balance in a room. Both air flows resulting form these two systems do not affect each other (no interference takes places). It happens, because both outdoor air system and air circulation system are open mechanisms. The vicinity of decentral system at the inflow of fresh air results in its proper mainitenance.

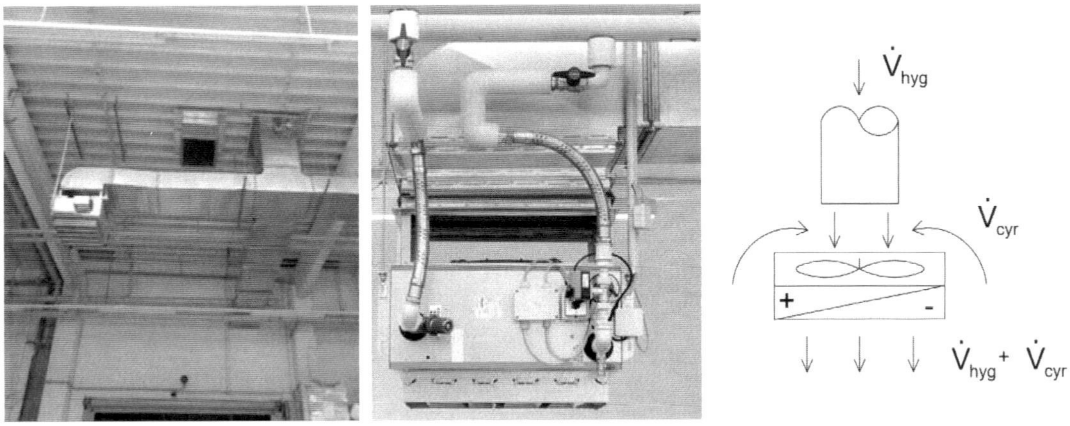

a) b) c)

Figure 3 Decentral unit in central-decentral system

The air handling unit used is employed for pure heat recovery and ventilation. It blows the air out through a duct system around 10 cm above the secondary air (recirculation) fan of an air heater. The ventilation unit runs when ventilation is needed, while the secondary air unit heats or cools the fresh supply air. Only the secondary air unit runs if only heating or cooling is needed.

Secondary air mode avoids excessive ductwork to the heating coil with its associated air resistance, such as bypasses and long air ducts.

When the ventilation air is activated and the room requires no heating/cooling or fresh air provides proper temperature regulation, decentral mechanisms are deactivated. When there is a necessity to provide some heating or cooling, decentral mechanisms are working in circulation mode. Secondary air consists of circulation air and fresh air provided directly through decentral system. It is however ensured by open hydraulic system.

ENERGY SAVING

Integration of central supply system and outdoor heat recovery as well as its decentral heating or cooling up combines the advantages. The heart of integrated system is air handling unit with heat recovery unit with integrated bypass. Its size is calculated based on the amount of fresh air to be provided (hygenic function), which is usually smaller as compared to caloric needs of a given room.

The air, which was initially prepared in heat exchanger is additionalyl heated up or cooled, once it is in a room. Since the airside of decentral units is hydraulically separated from central system, decentral unit is working depending on the needs for mixing or circulating air. The energy savings resulting from the performance of central-decentral instalation can be attributed not only to faster and more exact control system, but also to lower electro consumption of the fans. The analysis involves determining airflows resulting from hygenic needs $\dot{V}_{hyg}$, heating needs $\dot{V}_{heat}$ as well as cooling needs $\dot{V}_{cool}$. For central system the largest airflow in specific time range is selected

$$max\{\dot{V}_{hyg}, \dot{V}_{heat}, \dot{V}_{cool}\} \quad (1)$$

then energy consumption of the fans is calculated. The following relation is applied:

$$\left(\frac{P_1}{P_2}\right) = \left(\frac{\dot{V}_1}{\dot{V}_2}\right)^3 \quad (2)$$

The analysis of decentral system involves calculation of energy consumption of fans in raltion to individual airflows in selected time ranges and summing up the results. In this case, maximum airflow is not selected. The analysis involves the following relations:

Table 1. Air volumes for the dimensioning the HVAC system

$\dot{V}_{hyg}$	$\dot{V}_{heat}$	$\dot{V}_{cool}$
Change of air Timemodus Supply/exhaust air rate	Heat requirement Suppy/return temperature Outdoor/room temperature Heat recovery Heatlimit temperature Dail temperature figuy re	Cooling requirement Suppy/return temperature Outdoor/room temperature Cold recovery Cooling months

Finally, for selected air volumes, nominal power of fans is adjusted. Analogical process takes place in terms of circulation air untis. Below, there is an exemplary simulaton of energy consumption by HVAC fans for the following conditions:

Table 2. Building paramethers

Paramether	Value	unit
Room (Zone) area	700	m^2
Room (Zone) height	3	m

Table 3. Paramethers of the system and building

Ventilation			Heating mode			Cooling mode		
Change of air	0,5	1/h	Caloric requirement	50	W/m^2	Caloric requirement	28	W/m^2
Nominal fresh airflow	1050	m^3/h	Supply	75°C		Supply	6°C	
Timemodus	8,0	h/day	Return	65°C		Return	12°C	
Supply/Exhaust air rate	1,04	-	Room temperature winter	20	°C	Room temperature summer	27	°C
			Outdo tor emperaturewinter	-16	°C	Outd toor emperaturesummer	32	°C
			Heat recovery	75	%	Cold recovery	75	%
			Heatlimit temperature	15	°C	Timemodus	6,0	h/day
						Cooling months	6, 7, 8 ,9	month

The analysis of the exemplary zone (a room) for atmospheric conditions in Berlin clearly show that energy consumption of HVAC fans set up for hygenic (ventilation) and caloric (heating/cooling) needs is over 5-times smaller as compared to central system.

Table 4. Example of electro power of fans in AHU Central System

Supply air, $[m^3/h]$		Qel, kW	Exhaust air, $[m^3/h]$		Qel, kW
100%	113293,2		100%	11329	3,1
90%	10197	2,33	90%	10197	2,26
80%	90641,64		80%	9064	1,59
70%	79311,10		70%	7931	1,06
60%	67980,69		60%	6798	0,67
50%	5665	0,40	50%	5665	0,39
40%	45320,20		40%	4532	0,20
30%	33990,13		30%	3399	0,12
20%	22660,06		20%	2266	0,05
10%	1133	0,03	10%	1133	0,02

Table 5. Example of Energy consumption of fans in AHU Central System

Month	G15, Kd	Tout	Heating		Ventilation		Cooling mode		Sum
			qel1, kWh	qel2, kWh	qel1, kWh	qel2, kWh	qel1, kWh	qel2, kWh	kWh
1	6020,6	13	12	0,8	0,8	0		0	27
2	5271,4	11	11	0,7	0,7	0		0	24
3	4704,8	2	2	0,8	0,8	0		0	5
4	3179,2	2	1	0,8	0,7	0		0	5
5	14814,4	0	0	0,8	0,8	0		0	2
6	5917,4	0	0	0,8	0,7	576	558		1136
7	1819,4	0	0	0,8	0,8	595	577		1173
8	2019,1	0	0	0,8	0,7	576	558		1136
9	12414,7	0	0	0,8	0,8	595	577		1173
10	3089,8	2	1	0,8	0,7	0		0	5
11	4505,0	2	2	0,8	0,8	0		0	5
12	563	1,8	12	12	0,8	0,7	0	0	26
									4714 kWh

Table 6. Example of Energy consumption of fans in AHU Central-decentral System

Supply air, [m³/h]		Qel, kW	Exhaust air, [m³/h]		Qel, kW
100%	10500,8		100%	1010	0,8
90%	945	0,58	90%	909	0,58
80%	8400,41		80%	808	0,41
70%	7350,27		70%	707	0,27
60%	6300,17		60%	606	0,17
50%	525	0,1	50%	505	0,1
40%	4200,05		40%	404	0,05
30%	3150,02		30%	303	0,02
20%	2100,01		20%	202	0,01
10%	105	0,005	10%	101	0,004

Table 7. Exaple of Energy consumption of decentral units

Paramether	Heating	Cooling mode	
Recircilation air	3723	11329	m³/h
Amount of recirculation units	4		
Recirculation/unit	931	2832	m³/h
Electro power	0,1	0,2	kW

Table 8. Example of Energy consumption of fans in central-decentral System

Month	G15, Kd	Tout	Heating qel1, kWh	Ventilation qel1, kWh	Ventilation qel2, kWh	Cooling mode qel1, kWh	Sum kWh
1	602	0,6	60	0,20	0,20	0	60
2	527	1,4	54	0,18	0,18	0	54
3	470	4,8	30	0,20	0,20	0	30
4	317	9,2	29	0,19	0,19	0	29
5	148	14,4	0	0,20	0,20	0	0
6	59	17,4	0	0,19	0,19	144	144
7	18	19,4	0	0,20	0,20	149	149
8	20	19,1	0	0,19	0,19	144	144
9	124	14,7	0	0,20	0,20	149	149
10	308	9,8	29	0,19	0,19	0	29
11	450	5,0	30	0,20	0,20	0	30
12	563	1,8	58	0,19	0,19	0	58
							878 kWh

Although the analysis and juxtaposition are not possibile without discussing detailed assumptions, the following simplified comparison of central and central-decentral system together with nominal parameters acutely illustrates the differences between the two systems.

Central System	Central-decentral system
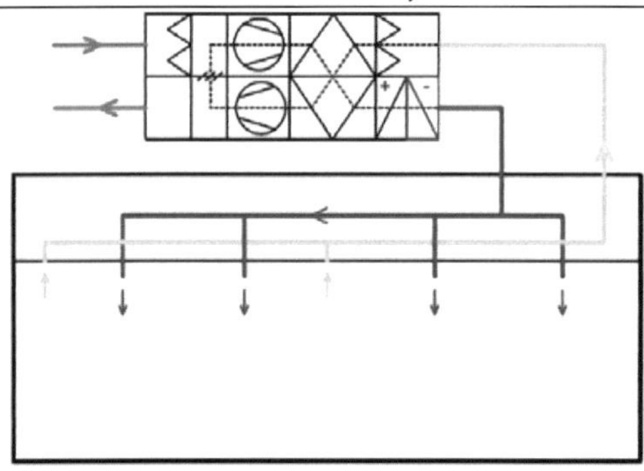	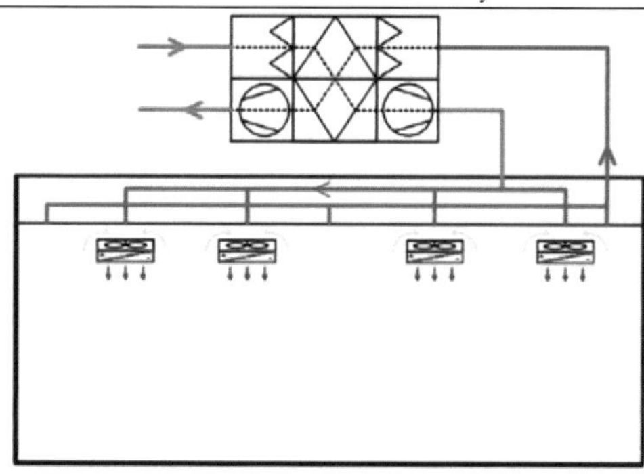
Air handling unit: ON Fresh air volume: 0 m³/h Circulation air volume: 11300 m³/h Cooling capacity: 19,6 kW Power of fans: 6,3 kW	Air handling unit: ON Fresh air volume: 0 m³/h Circulation air volume: 11300 m³/h Cooling capacity: 19,6 kW Power of fans: 0,8 kW
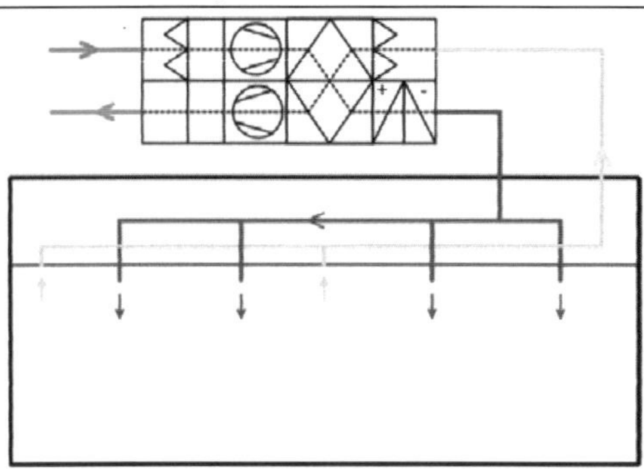	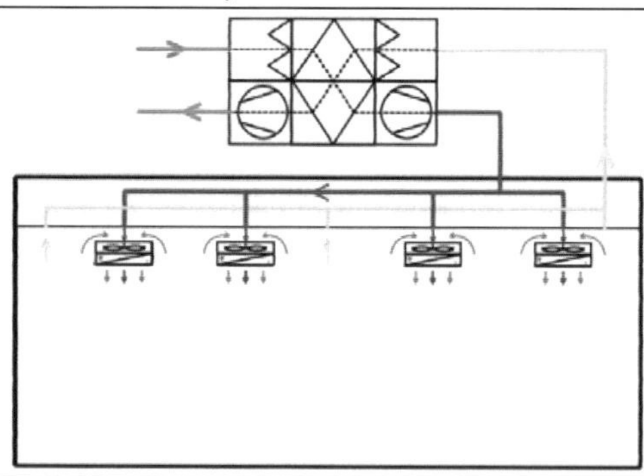

Air handling unit: ON Fresh air volume: 1050 m³/h Circulation air volume: 0 m³/ h Cooling capacity: 0 kW Power of fans: 0,05 kW	Air handling unit: ON Fresh air volume: 1050 m³/h Circulation air volume: 0 m³/ h Cooling capacity: 0 kW Power of fans: 1,65 kW
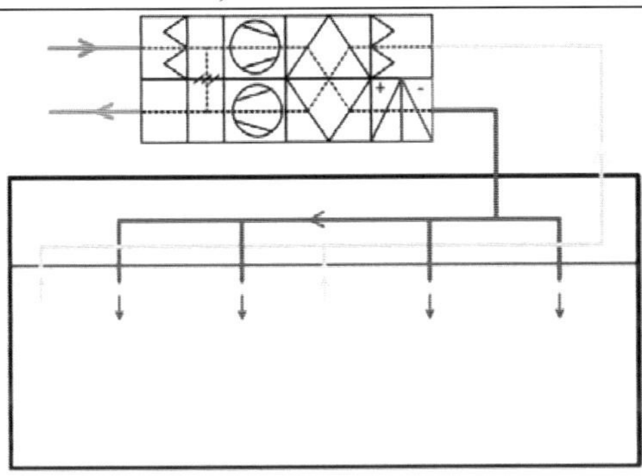	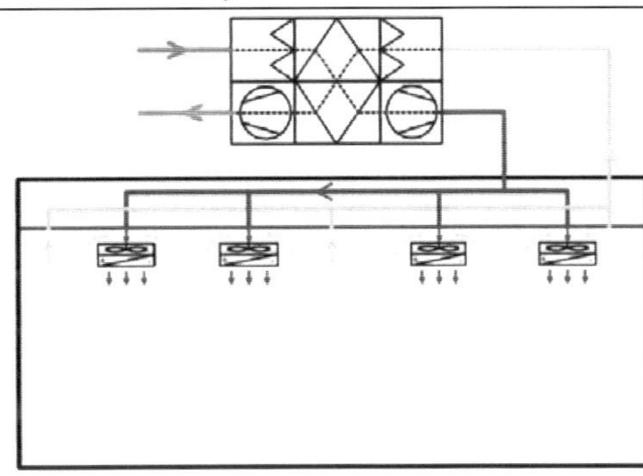
Air handling unit: ON Fresh air volume: 1050 m³/h Circulation air volume: 10250 m³/ h Cooling capacity: 19,6 kW Power of fans: 6,3 kW	Air handling unit: ON Fresh air volume: 1050 m³/h Circulation air volume: 10250 m³/ h Cooling capacity: 19,6 kW Power of fans: 2,4 kW

Figure 4 Comparison of central and central-decentral systems

CONCLUSION

The central-decentral system is a technical solution, which occupies relatively little space and is simple in terms of regulation and controlling. An all-year exploitation of ventilation unit often involves unnecessary air resistance. Air distribution with the use of a cooler in heating mode or with the use of a cooler and a heater or in ventilation mode has proved to be not a sufficiently efficient solution. The comparison of a yearly performance of both systems has showed that in central-decentral system much less energy is consumed by the fans. The main conclusion of the above provided analysis is that the separation of hygienic and caloric funcions in central-decentral system is energy-efficient and environmentally-friendly solution for the ventilation and air conditioning of buildings.

REFERENCES

Hermann Ensink, Lüftung und Temperierung getrennt 2011, TGA
Fachplaner www.kampmann.de

Prediction and Control of Noise and Vibration within a Sport Facility

Ghina Annan, PE **Balsam Nehme, PE**

ABSTRACT

The prediction and control of noise and vibration during sport events has created a tough challenge for sport facilities, as full compliance to minimum standard requirements and city regulations are now mandatory. This dictates designers to achieve suitable reverberation time and utilize desirable while controlling undesirable discrete reflections within occupied spaces and maintaining proper with equally distributed sound energy levels. As such, design aspects shall include all specifications, from room volume and shaping, to the selection and placement of finish materials for walls, floors and ceilings and implementation of reflectors and diffuse elements. Unwanted noise and vibration invade an enclosure by mechanical transmission through solid structure of wood, steel, concrete, or masonry. The noise due to HVAC systems can be transmitted to a room by the air in the ducts or by the metal of the ducts themselves or both, while having the vibration travel great distances through these solid structures with minimal losses. Similarly, water pipes and plumbing fixtures tend to have significant sound carrying abilities. This paper highlights the acoustics, noise and vibration criteria applicable to a sports' facility located in Dubai, and further optimize the acoustic performance of its walls, floors, doors and finishes in occupied and technical spaces to meet the minimum standard requirements. The standard and local regulations referenced in this work are primarily: Dubai regulations, ASHRAE HVAC applications Handbook 200, IEC 60268-16 2003 Speech Transmission Index, BS 8233:2014 and ISO standards.

INTRODUCTION

Fundamentally, sports facilities are not all designed in the same way because of several differences such as but not limited to geometry. Occupants perceive noise as undesirable sound, nonetheless in sport entertainment, this characterization can be confusing. Human auditory system is capable of perceiving the occurrence of a sound; distinguish deviations in frequency, level, and time; identify and localize diverse speech sounds. As such acoustic specialists and designers tend to evaluate what is known by the reverberation time. Practically, the definition of reverberation time is the time required for a sound that is very loud to decay to inaudibility.

In this work, noise and vibration assessment is evaluated through an environmental simulation, day and night noise survey and external mechanical equipment simulation. The environmental simulation involved testing two cases using the IMMI 3d modelling software, such that the first case evaluates the internal sport facility noise effect on external noise sensitive receivers most exposed to the arena breakout noise while second case calculates the break in noise to the arena from external noise. Also, a day noise survey is realized at a number of locations across the site and off the development and a night noise survey include noise measurement at two locations facing the site. Finally, the external mechanical equipment simulation assesses the noise impact of adding air handling units to serve the cooling of the arena onto adjacent existing properties. Results of the assessments must meet the standard and local regulations else noise mitigation measurements are adopted.

DESIGN STRATEGY

The design strategy for this work sports facility takes account of sound insulation in walls, doors, floors and interior

Ghina Annan is a design Mechanical Engineer in the Department of Buildings Sustainability Unit, Dar al Handasah, Beirut, Lebanon. **Balsam Nehme** is the Head of the Department of Buildings Sustainability Unit, Dar al Handasah, Beirut, Lebanon.

finishes. Internal sound insulation should be designed so as to provide a suitable level of acoustic privacy and comfort between noisy areas and occupied areas. The airborne sound insulation figures given in this specification for separating walls represent those measured under laboratory test condition (Rw or STC), then an in situ (DnTw or FSTC) performance requirement to allow for site tolerances and variances in workmanship. The normal gypsum density is of 9.3 kg/m^2 while the sound shield and Fire shield types density is of 12.3 kg/m^2. The accepted variation between laboratory figures and in situ measurement is maximum 5 dB. Table 1 lists the gypsum types properties.

Table 1. Gypsum Types Properties

STC (Rw)	Insulation	Width (mm)	Height (mm)	Number of layers	Stud type/ distance between studs	Gauge of studs
50	80mmRockwool (60kg/m3)	125	4000	1 * 12.5mm each side	C Sound shield/ 600 mm	0.5 mm
55	50mmRockwool (60 kg/m3)	150	4000	2 each side (12.5mm)	C Sound shield/ 600 mm	0.5 mm

All the walls should be plastered with minimum 15mm from both sides. The above construction types are typical only and other wall build ups would be acceptable as long as the outlined Rw or STC are achieved. It is essential that all walls are taken from slab to ceiling in order to achieve the sound insulation of the wall. The construction's performance should be validated and laboratory test certificated attained. The STC adopted ranges between a minimum of 45 to a maximum of 65, depending on the adjacent areas. Table 2 lists the block walls properties.

Table 2. Block Walls Properties

Wall performance (Rw or STC)	Hollow Block (1400 kg/m3)	Solid Block (2000 kg/m3)
50	150	100
55	200	150
60	-	220
65	100 + 30 air void +100	100 + 30 air void +100

As for the doors, doors in sound-isolating partitions should provide sufficient sound isolation performance so that the partition sound isolation is not compromised. The STC of the doors should be as indicated in Table 3.

Table 3. Door And Walls STC

Walls	Doors
50	30
55	35
60	40
65	45

Finally, the floor constructions in occupied areas should achieve in excess of STC 60, IIC 60 and are acceptable in reducing any airborne noise in all cases, except for potentially plant room or special areas spaces.

Sound absorptive materials provide some benefit in terms of reducing sound transfer between adjacent open areas by minimizing the reflected energy that can propagate from space to space. In this project the room acoustic is one of the more important issues to be considered. This discipline regards the development of the aural environment that meets the need and function of a space. The issues include achieving suitable reverberation time, utilizing desirable while controlling undesirable discrete reflections and maintaining proper and equally distributed sound energy levels throughout the spaces. The design aspects include everything from room volume and shaping to the selection and placement of finish materials for walls, floors and ceilings and implementation of reflectors and / or diffuse elements. Table 4 lists the required Reverberation time for various rooms.

Table 4. Room type Reverberation Time

Walls	Reverberation Time (s)
VIP rooms	1 to 1.2
Media room	0.6 to 1
Royal rooms	0.6 to 0.8
TV Studio	0.2 to 0.4
Multipurpose	1.4 to 2
Restaurants	1 to 1.5
VVIP rooms	0.6 to 0.8
TV Studio	0.2 to 0.4
Warm up area	1.4 to 2
Commentators	0.2 to 0.5

Further to the above acoustical separation properties, target reverberation times, as well as general design direction for the acoustical interior finishes are considered. To target the required reverberation times, a study has been performed on the following different types of rooms: Spectator concourse, VIP entrance, VIP lounge, players changing room, Prayer room, open bar, Spectator 's arrival, Chaiman club, Royal toilet, and Spectator's toilet. In each room, the required interior finishes have been designed to achieve the required reverberation time. As an example, Figure 1 illustrates the design adopted for the spectator concourse

Figure 1 Spectator Concorse Interior finish recommendations

SIMULATIONS AND SURVEYS

Environmental simulation

Environmental noise criteria have been extracted from the document Government of Dubai and Municipality of Dubai's "Technical Guidance Number 10 – Requirements for the control of entertainment noise, 2011". The limiting noise criteria taken from this document are summarised in Table 5.

Table 5. Noise criteria applicable for Dubai Arena

Receptor Areas	Allowable Limits for Noise, dBA	
	Daytime (7AM-8PM)	Nightime (8PM-7AM)
Residential Areas with Light Traffic	40-50	30-40
Residential Areas in Downtown	45-55	35-45
Residential Areas with Some Workshops & Commercial or near Highways	50-60	40-50
Commercial Areas & Downtown	55-65	45-55
Industrial Areas (Heavy Industry)	60-70	50-60

The key criteria for the assessment of Dubai Arena music noise breakout are those for Residential Areas with Some Workshops & Commercial or near Highways that state a limit of 50-60dBA during daytime and 40-50dBA during night-time. It is assumed that the majority of the events in the arena finish after night-time (8PM-7AM), therefore the limit of 40-50dBA is applicable.

Internal to External

An environmental model has been built using IMMI 3d modelling software. The internal levels in the arena have been based on measurements taken at typical rock/pop shows. The internal noise spectrum used is indicated in Table 6.

Table 6. Internal noise spectrum of a typical rock/pop event

	63	125	250	500	1K	2K	4K
Internal Noise Spectrum (dB)	114	108	102	99	98	90	86

IMMI utilises the sound propagation method from the international standards ISO 9613-1:1993 and ISO 9613-2:1996. Downwind sound propagation from the Arena to the noise sensitive receivers is assumed for the model. Temperature was assumed to be 25 degrees Celsius and humidity 50%, based on the annual weather averages in the region. The noise propagation has been modelled as breaking out from the arena roof, the Dine and View area glazing façade and the docking bay retractable wall. The lower levels of the building are well buffered by back of house areas and solid constructions.

The proposed roof construction has been used in the break-out model to demonstrate the external environmental noise levels expected during concert events. The predicted sound insulation performance of the roof is provided in Table 7.

Table 7. Sound insulation performance of Dubai Arena roof

	63	125	250	500	1K	2K	4K
Arena roof (dB)	25	24	27	37	47	57	66

The external façade section directly between the bowl and the external shall provide the same acoustic performance as the roof above. An offset between the front of house and expected sound pressure levels at the underside of the roof has been taken from the internal model to determine the predicted noise levels at the roof surface. The loading bay door shall provide a minimum STC or Rw rating of 45dB. Six sensitive receivers have been placed throughout the arena in the most exposed locations to the arena noise break-out. The predicted noise levels from these receivers have been measured to be ranging from 37 to a maximum of 40 dBA, as shown in Figure 2, which is within the allowable limits for noise required.

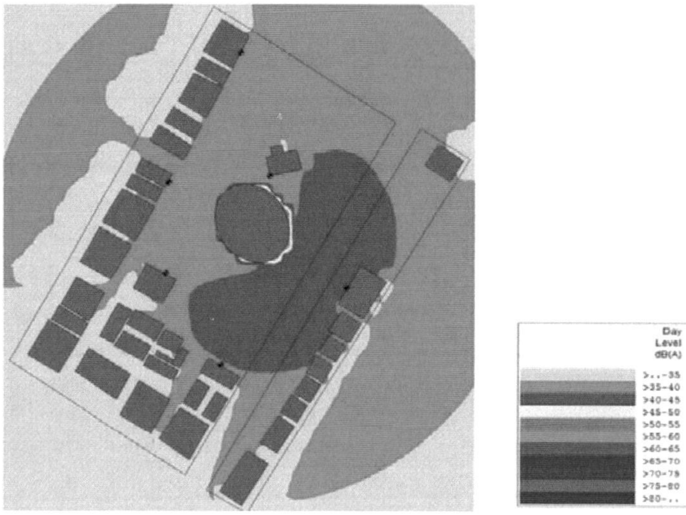

Figure 2 IMMI Noise Mapping

External to Internal Noise Assessment

Break-in calculations have been made to assess the internal background noise levels of the Main Bowl.

Façade information and room details have been compared against the incident sound levels taken from the day survey undertaken on October 31st and November 1st and the night survey and on the 6th and 7th November 2016 after applying a correction due to distance. Tables 8 and 9 indicate the night time background noise survey measured data, and the break in noise calculations in the arena from external noise, respectively.

Table 8. Night Time Background Noise Survey Measured Data

	63	125	250	500	1K	2K	4K	8K
Location 1	57.8	57.5	56.7	52.7	48.1	42.5	35.7	29.5
Location 2	62.4	59.5	54.9	52.8	49.3	47.5	43.2	38.8

Table 9. Break In Noise Calculations In The Arena From External Noise

	63	125	250	500	1K	2K	4K	8K
Break-in Noise Levels	37.6	37.7	24.6	12.9	0.4	0	0	0

Based on the calculations, it is predicted that break-in noise levels of the Main Bowl satisfy the minimum requirements of internal noise levels of NR 35.

Bowl Model

The bowl has been also studied in different sections, including but not limited to: arena roof, arena rear wall, arena side walls, arena lower wall, dine and view wall, external glazing, vip back walls, vvip box, vip premier club. Arena roof area is taken here as an example of the studies performed and it is shown in Figure 3.

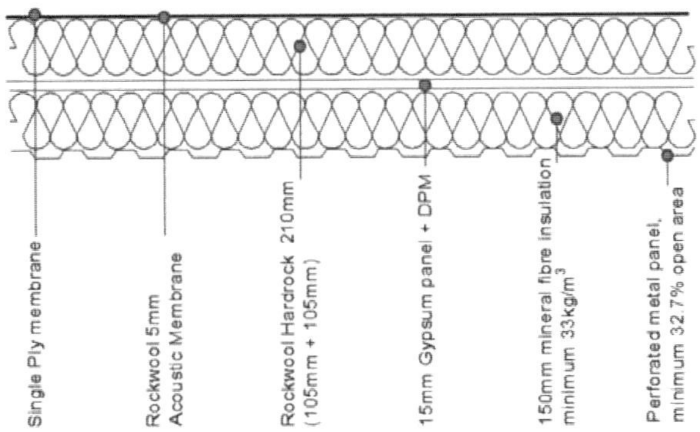

Figure 3 Arena roof requirements

Tables 10 and 11 list the arena roof sound insulation and the arena roof absorption coefficients, respectively.

Table 10. Arena Roof Sound Insulation

	63	125	250	500	1K	2K	4K
Arena Roof (dB)	25	24	27	37	47	57	66

Table 11. Arena Roof Absorption Coefficients

	125	250	500	1K	2K	4K
Arena Absorption Coeff	0.95	0.95	0.95	0.95	0.90	0.80

Additional thickness is required for the lower absorption area to extend the low frequency performance. The middle panel acts as a low frequency resonant panel and further improves the low frequency absorption in the arena. The mid-panel and mass barrier layers are required to ensure the noise-breakout is achieved. A minimum of 32% perforation area shall be provided for the whole roof deck. This should apply to the extents of the lower roof deck.

Surveys

Day Survey

The project will have a number of noise –generating plant such as air handling units on roof, noise from activities, which have the potential to adversely affect the noise-sensitive receivers in the area. Noise surveys were taken to support the planning application for the site. The project may have a number of noise-generating plant which have the potential to adversely affect the receivers in the area. The nearest noise-sensitive receiver is the mosque at approximately 80 meters on north boundary, and also dwellings and offices at 130 meters to 170 meters around the site.

There are two main areas which need to be considered as a result of the noise survey, these are the control of plant noise from the project and the control of the internal ambient noise levels. The relevant criteria for these elements is provided within the Dubai Municipality Requirements for Environmental Noise Limits.

A noise survey was realized on the 31st of October and the 1st of November 2016. This survey includes noise measurements at 9 locations across the site and off the development site at a location representative of the noise source in the area (day 1 between 16:00 and 19:15 at locations 1 to 8, and day 2 between 8:00 and 10:00 at locations 5, 8 and 9). Monitoring location 1 was selected to provide a representative sample of the noise generated by the existing plant power. During the survey there was noise from the construction sites and from an electrical generator on the boundary of the plant power, no noise was perceived from the plant power. Monitoring locations 2 to 6 were selected to provide a representative sample of the noise in the area. During the survey the main noise was generated by the road nearby, the construction sites and the plane routes. Monitoring location 7 was selected to provide a representative sample of the noise from the mosque at Azan and at normal conditions. Monitoring locations 8 and 9 were selected to provide a representative sample of the main noise in the area daytime and nighttime. The measurement position was at 2 meters from the boundary of the road. During the survey, traffic was steady and generally free-flowing. The measured data ranges from 55.5 to a maximum of 75.8 LAeq., according to the location of the sensor, which satisfy the needed requirements.

Night Survey

A noise survey was realized on the 6 and 7th of November 2016. This survey includes noise measurements at a two locations facing the site at a locations representative of the noise level in the area.

Monitoring location 1 was selected to provide a representative sample of the noise at the Residential and Commercial Street considered as the most sensitive residential area at the West site boundary of the site. During the survey there was noise from the street at the North boundary that is directly open to the street, and from the surrounding roads in the Walk city perpendicular to the street from West side. It was noted that the noise from the Walk city roads is mainly due to the type of the road that is stone pavement and have small bumpers, the other main noise was noticed from the frequent Motorcycles and Sports cars in the street.

Monitoring location 2 was selected for the same reason as location 1. During the survey it was noticed that the main noise is from a technical plant at the South side and from the paved roads in the surrounds.

The measured data at the different locations range between 58.9 to a maximum of 60.1 LAeq. The night time background noise survey measured data are indicated in Table 12.

Table 12. Night Time Background Noise Survey Measured Data

	63	125	250	500	1K	2K	4K	8K
Location 1	57.8	57.5	56.7	52.7	48.1	42.5	35.7	29.5
Location 2	62.4	59.5	54.9	52.8	49.3	47.5	43.2	38.8

The model predicts that the noise breaks out from the Arena satisfy the external noise criteria as set by the local authorities. Break-in noise to the Arena from external noise sources was also assessed using noise data from DAR's survey in 2016. It is expected that the Arena will meet the minimum noise requirements of NR35, given the assumed external noise levels and the assumed building envelope sound insulation.

HVAC/MEP Noise Control

Background Noise Criteria (NC)

The background sound level within a space is described in terms of Noise Criteria (NC) level as defined in ASHRAE Guidelines. NC describes the maximum background sound level due to the normal operation of building equipment serving the different areas, typically the air distribution system. Acceptable background noise levels shall be defined based on the intended use of a space, and then the building systems are designed such that this noise does not exceed the limits that are appropriate. The background sound design goals for various spaces are listed in Table 13.

External mechanical equipment simulation

An external sound analysis has been performed to assess the noise impact of adding air handling units to serve the cooling of the arena onto adjacent existing properties such as the mosque and the nearby street. The Air handling units sound level selections are summarized in Table 14.

Table 13. Noise criteria for various spaces

Location	Noise Criteria
Restaurants	40
Royal rooms	25
TV studio	20
Multipurpose	45
VIP rooms	30
Media room	25
VVIP rooms	30
Main arena	40

Table 14. Sound level selections of different air handling unites (dBA)

Frequency (Hz)	63	125	250	500	1000	2000	4000	8000
AHU-MR01-05-01	85	84	82	67	67	62	55	48
AHU-MR01-05-02	85	84	82	67	67	62	55	48
AHU-MR01-05-03	81	80	78	63	63	58	51	44
AHU-MR01-05-04	71	77	71	58	55	54	46	39
A5AHU-MR01-05-05	71	74	70	59	58	58	49	42

The study assumed that the critical receivers are located at 1.65 m above ground level at the nearby street and across the entire façade of the adjacent mosque with a maximum allowable limit of 50-60 dB(A) during daytime and 40-50 dB(A) during nighttime.

The acoustic analysis, conducted using CadnaA software, shows that the noise perceived at the mosque external facade and the nearby street is of a maximum of 37 dB(A) and 32 dB(A), respectively. The model predicts that the noise from the external mechanical equipment satisfy the external noise criteria as set by the local authorities.

CONCLUSION

This paper presented a case study done on a sports facility in an attempt to assess the noise produced. A design for sound insulation in walls, doors, floors and interior finishes were assigned based sports codes and standards. The study was conducted through an environmental simulation, day and night noise survey and external mechanical equipment simulation in order to take into account all noise sources. Results of the assessments meet the needed requirement.

REFERENCES

Dubai regulations
ASHRAE *HVAC applications Handbook* 200
IEC 60268-16 2003 *Speech Transmission Index*
BS 8233:2014 and *ISO standards*

A Four Step Approach for Energy Conservation and Retrofitting Interventions for Residential Buildings

Mohamad Hajj Hassan **Mohamad Abbas** **Bahaa Mneymneh** **Issam Srour**

ABSTRACT

This paper presents a utility data analysis for a residential building and the estimated energy saving results from the use of practical retrofitting energy efficient systems. A four step approach is used to: (1) collect the utility data to estimate an actual base line for energy consumption, (2) conduct an energy audit to determine the existing operating conditions of major energy use equipment, (3) model the building in Autodesk-Revit then export the data into Elite-Chvac (Commercial Heating Ventilation and Air Conditioning) for heat load analysis, and (4) prepare a comprehensive list of energy conservation measures. Several scenarios for possible interventions to the building's cooling, lighting and hot water systems are proposed and evaluated in order to assess the effectiveness of various energy conservation techniques. Based on the results of these simulations, it is concluded that it is possible to reach an overall 44% reduction in energy consumption.

INTRODUCTION

Worldwide energy consumption has risen 85% from 1980 to 2012 (U.S.E.I.A, 2013). The residential sector accounts for 16-50% of that consumed by all sectors, this percentage varies between countries and averages approximately 30% worldwide (Saidur et al., 2007). For this reason, many research studies focused on assessing, modeling, calibrating, and controlling energy use in residential buildings worldwide (Dong et al., 2005, Lopes et al., 2005, Pérez-Lombard et al., 2008, Swan and Ugursal, 2009, Morrissey and Horne, 2011).

In Lebanon, the issues within the electricity sector have been well documented, and since the Lebanese industrial sector is still at its enfancy, the largest portion of the total produced energy is consumed by the residential sector. Houri and Ibrahim- Korfali (2005) surveyed 509 households in Lebanon in an attempt to estimate the average household and per capita energy consumptions, which were found to be 6907 and 1727 kWh respectively.

Increasing power demands and the limited efforts towards improving power generation has led to large economic and environmental costs (Ruble and Nader, 2011). Moreover, little effort has been invested towards decreasing household energy consumption levels in Lebanon. For this reason, this paper presents work targeted at assessing the energy consumption of a residential building, and investigates alternative green, yet cost effective energy systems and solutions. The framework of the study is summarized in Table 1.

Our energy study was carried out on an existing building located in Tripoli, North Lebanon. It is an eleven story residential building, with two basements used for parking and storage. Each story of this building is divided into two apartments, having an area of approximately 140 m^2 each (Figure 2).

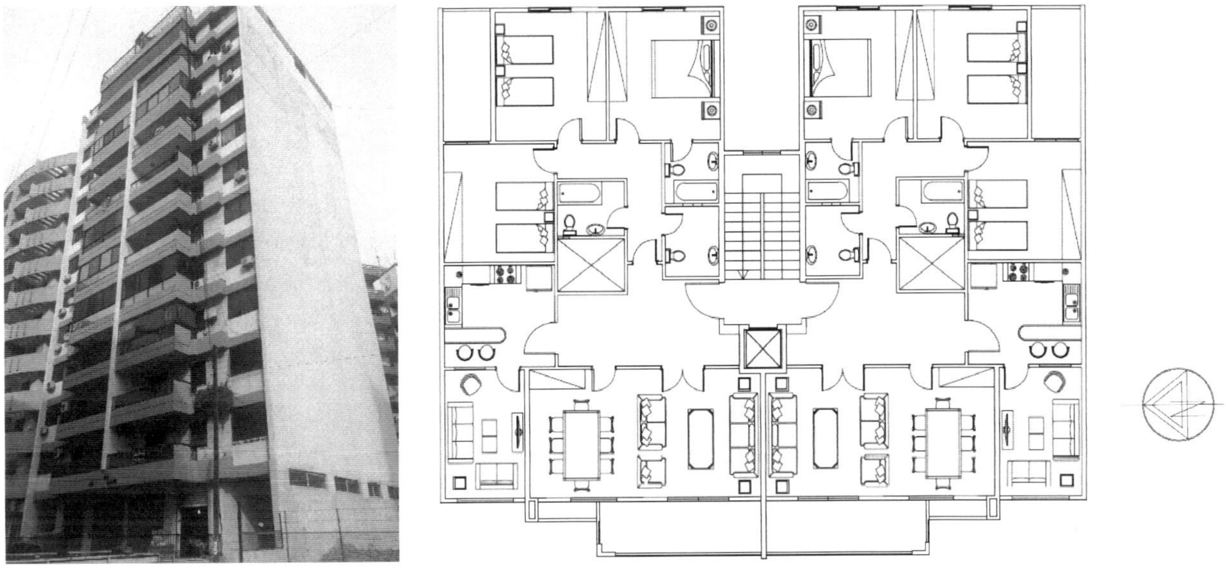

Figure 1 Existing Residential Building in Tripoli and Typical Plan for Two Apartments

Table 1. Summary of the Proposed Framework

Utility Data Analysis	Energy Survey	Energy Modeling	Energy Saving Measures
Utility Rate	Construction Material	Energy Use Breakdown	Energy Conservation Methods
Electrical Energy Use Profile	HVAC System	Heat Load Analysis	HVAC System Retrofit, Temperature Set-points, Heat recovery Systems
Electrical Energy Use per Area	DHW System	Electrical Energy Distribution	Energy Efficient Lighting & Energy Efficient Equipment
Thermal vs. Energy Profile	Lighting Type Density	Baseline for Energy Consumption	DHW System retrofit (Renewable Energy)
--	Equipment Type Density	Adjusted Baseline	LCC for Selected Options
--	Occupancy & Operation		Scenarios of Options

UTILITY DATA ANALYSIS - STEP 1

The electrical utility data of the building over 5 years were collected and plotted as shown in Figure 2a. The actural energy consumption shown represents the average consumption for a typical apartment. The values will be later normalized and adjusted to reflect the energy consumption for the 22 apartments. Subsequently, the outcome of this step is to: (1) determine seasonal effects on the building energy, (2) identify the fuel types (electricity and natural gas), and (3) perform a utility energy use analysis: Thermal vs. Energy profile.

According to the utility bills collected for the years 2011 through 2015, the average electrical utility rate in Lebanon is approximately $0.083/KWh. However, in Tripoli there is no 24 hours operating profile of electrical supply from the main grid, and most households mitigate the shortage through neighborhood generators. In case the bill that is being paid for secondary grid is included, the utility rate will increase to $0.183/KWh. Still this may not be an accurate figure to follow since actual consumption can not be point monitored. Therefore, the initial rate of $0.083/KWh is considered under the assssumption that there is no electrical shedding from the main grid.

After plotting the energy use profile (Figure 2), consumption habits according to the weather conditions are analyzed. First, the correspondent weather data for the city of Tripoli are plotted over the energy profile as shown in Figure 1(b). Noticeably, the energy consumption is closely related to the outdoor dry bulb temperature; consumption increases from July to October and drops in April and December. This is attributed to the use of Air conditioning system during the summer and the fact that the utility bill dates lag by 2 month from the weather data dates.

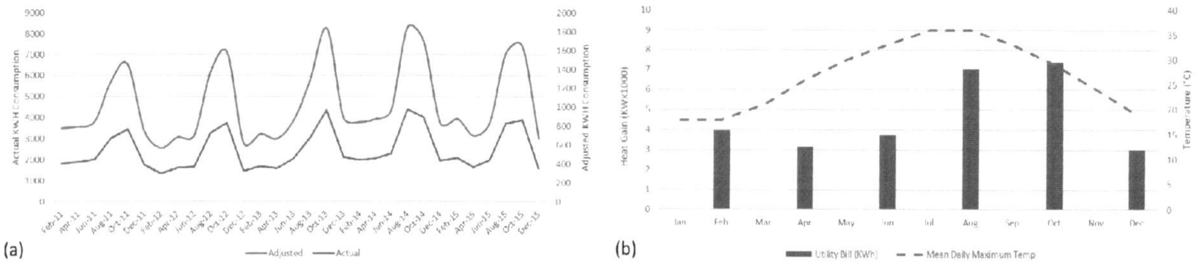

Figure 2 (a) Actual vs. Adjusted Energy Consumption for a Typical Apartment (b) Energy Consumption Relative to the Ourdoor Mean Dry Bulb Temperatures

ENERGY SURVEY - STEP 2

The aim of the energy survey is to characterize in detail the real energy consumption and the thermal environments affecting the apartment thoughout the different periods of the year. The characteristics of the surveyed apartments are described in Table 2. When analysing energy consumption in residential building, several factors are interrelated. Among these two are occupany patterns and operating profiles. Therefore, the intent beihnd this step is to collect as much as possible data for energy modeling inorder to;

1. Determine the existing operating conditions of major energy use equipment (lighting, HVAC systems)
2. Estimate the occupancy, equipment, and lighting (energy use density and hours of operation)
3. Inspect building location, orientation and construction material

The survey includes 4 sections; 1- General Information, 2- Cooling System, 3- Heating System and 4- Domestic Hot Water System. The results of the survey indicate that most of the tenants use split units as a cooling system with capacities for rooms presented in Table 2. For the heating system, most tenants use a combination of portable Gas/ Electric heater. But given that the mean daily temperatures in Tripoli do not drop below 18°C during the winter season, the electrical loads of these heaters will be included as part of the subordinate loads; the primary being the AC system, domestic water heating system and lighting. For heating water, all apartments have 2x100 liters electrical hot water bottlers with a heating element of 1.5 KW each. As for the lighting system, all rooms have incandescent lights with Watt consumption as shown in Table 2. In addition to identifying the exisinting equipment in each apartment, the survey helped in generalizing the average energy use density and hours of operation for the AC, heating and lighting system. The densities for the AC system are depicted as daily and monthly load in Figure 3.

Table 2. Existing Cooling, Lighting and Water Heating Systems

Rooms	Split Unit Capacity	Lighting	Domestic Water Heating
Bedroom 1	12000 BTU/h	Incandescent-200W	
Bedroom 2	12000 BTU/h	Incandescent-200W	100 Liter – 1.5 KW
Bedroom 3	12000 BTU/h	Incandescent-200W	
Kitchen	--	Incandescent-200W	
Living	12000 BTU/h	Incandescent-200W	100 Liter – 1.5 KW
Reception/Dinning	24000 BTU/h	Incandescent-600W	
Entrance/Toilets	--	Incandescent-400W	

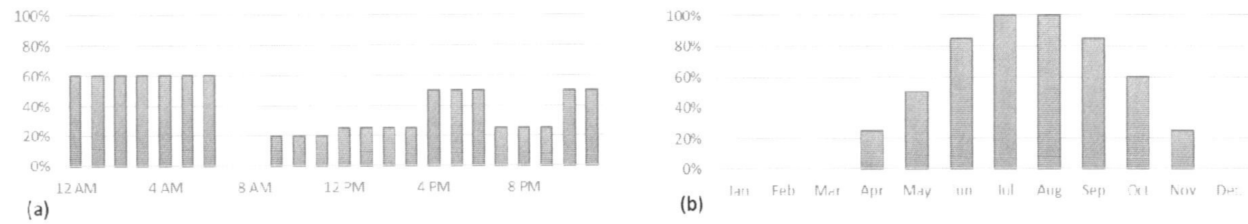

Figure 3 (a) AC System Daily Load Profile (b) AC System Monthly Load Profile

The final part of the energy survey covered the technical features of the building; location, orientation and construction material. Location and orientation were discussed in the introduction. As for construction material, the survey examined the glazing system, exterior walls, floor slab and ceiling slab. All the apartments have single glazing system with aluminum frames. The U-value (a measure of heat transition) for this glazing system is around 5.91W/m2K (ASHRAE, 2017).The U-values for the remaining material were calculated according to the material composition shown in Table 3 and as per the requirements and thermal resistance values set in ASHRAE-Handbook of Fundemantals.

Table 3. Material Composition for Walls and Slabs

Heading Number One	Material Composition	Thermal Resistance (Btu/hr.°F.ft2)
Exterior Wall	Stone cladding, 15cm Hollow Block Plaster	0.48
Floor Slab	2cm Tile 3cm Mortar 5cm Sand 25cm Concrete Plaster	0.3
Ceiling Slab	Ditto	0.254

ENERGY MODELING - STEP 3

The output of the energy survey helped construct an architectural model using Revit; a CAD software used to create 2D and 3D geometry (Autodesk, 2018). The model was then extracted into (Elite-Chvac), a tool used for thermal load calculation and HVAC systems design, via the gbXML interoperability scheme. The load values estimated are studied on an hourly basis for the whole year with an aim to: (1) evaluate building equipment for efficiency, performance and reliability, (2) determine the energy use breakdown, (3) develop a baseline model for building energy use, and (4) calibrate the baseline model using the utility data

The results from the energy survey conducted show that most tenants operate the cooling system at a set point of 19°C. According to ASHRAE (2013), the comfortable indoor temperature can be set at 24°C and 50% RH. Therefore, the intent of the Heat Analysis is to determine:

1. The required cooling capacities if the set point was adjusted to 24°C
2. The actual cooling capacities as it varies from the month of March to the month of November
3. The heat load analysis for the apartments having different exposure; North or South
4. The heat gain breakdown

Figure 4 shows the heat load analysis at set points 19°C and 24°C along with a comparison between the north and the south exposed apartments. The values presented are in KW; each 3.51 KW is equal to 1ton of refrigeration. It is evident the highest cooling demand is during the months of July and August; 5ton for a south exposed apartment at a set point temperature of 19 °C. Moreover, comparing the existing cooling capacities versus the required, a nominal decrease of 25% is perceived. This drop is foreseen for the selected cooling capacities of available split units in the market, hence the trem nominal. However, the actual decrease would range between 12 to 18% depending on the season and a predictable operating profile such as the one shown in Figure 3. The implication of such decrease is presented as a saving profile in Figure 4(a) and is later used to determine the energy savings resulting from setting the indoor temperature at 24°C. Another emenet observation from the figures below is the minor difference of 4% between heat gain values of the north and south exposed apartments. This is asscocited with the balconies that are acting as shading devices and minimizing the effect of solar heat gain for the south exposed apartment.

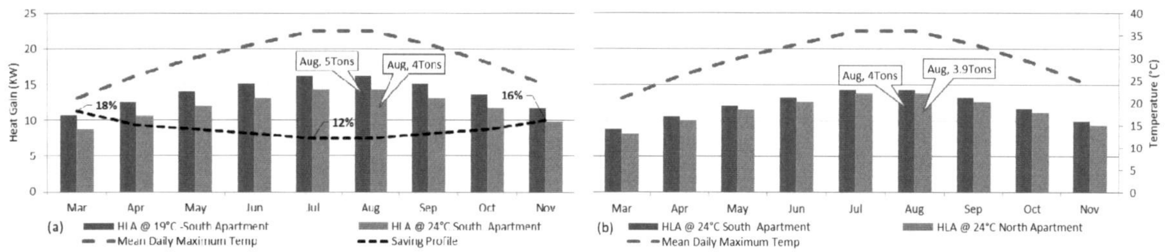

Figure 4 (a) Heat Load Analysis for Indoor Temperatures 19°C & 24°C (b) Heat Load Analysis for North & South Apartments

Heat gain modeling also helps in determining the physical elements in the building that are contributing to the heat gain or loss. Figure 5(a) shows that 37% of the heat gain is from the glazing system, 30% from the walls and 15% from using the lights. These figures give an idea to what sort of energy measures one should follow to reduce the heat gain in a building, but also determines which saving measures to consider and which to dismiss. For example, the 37% heat gain from the single glazing system is caused by its thermal transmittance and solar heat gain coefficient (SHGC) values. Changing the u-value for the existing glass may not be possible without using a double or triple glazing. However, one can change the solar optical properties to reduce the solar heat gain by using green window films. The visible transmittance for the window will drop by around 13% but solar transmittance would decrease by 39%(Hajj-Hassan and Khoury, 2018).

The heat gain break down and the energy use breakdown in Figure 5, infers the use of insulation for walls and changing the glassing system since these are the most two systems consuming energy. But reducing heat gain might not be reflected as energy saving.Therefore, and before going deep into analyzing these options, one should take into consideration the AC system that could be retrofitted for the building. There are 3 choices of AC systems to be used in our building. These are: standard split units, inverter split units and variable refrigerant flow (VRF) system. The drawback out of these three systems is nominal capacities. One can reduce the heat gain to a certain threshold;

however the cooling capacities and energy consumption will remain the same. Therefore, in our analysis we will only study the effect of using LED lights instead of Incandescent lights on these three systems.

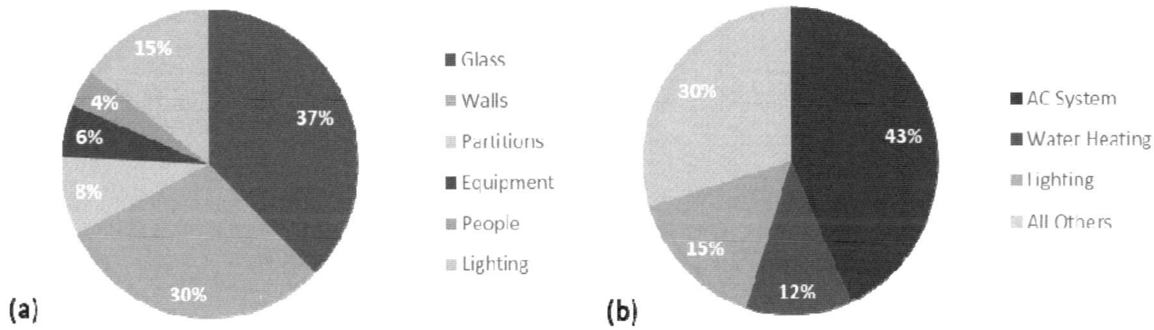

Figure 5 (a) Heat Gain Breakdown for South apartment (b) Energy Use Breakdoen for a Typical Apartment

EVALUATION OF ENERGY SAVING MEASURES - STEP 4

The results of the energy survey and the energy model lead to examine the energy saving potentials of three systems; the AC, lighting, and the DHW system. This analysis will help:

1. Prepare a comprehensive list of energy conservation measures
2. Determine the energy savings due to the various energy conservation measures pertinent to the building using the baseline energy
3. Estimate the initial cost required to implement the energy conservation measures
4. Evaluate the cost-effectiveness of each energy conservation measure using an LCC analysis

AC-System Evaluation:

The Existing AC system (AC-1) is evaluated, and a comparison is made relative to three efficiency opportunities;

AC-1': Keeping the existing split units, but operating them at a set point of 24°C rather than 19°C. The initial Cost of this system is $0, the annual energy consumption is 9,164 KWh and the annual savings is $213.

AC-2: Replacing the standard split unit with an inverter type. This option helps reduce the total required cooling capacity from 6 tons to 4.5 tons. The Initial cost of installing this system is $4,400, the annual energy consumption is 6,337 KWh and the annual savings is $448. The payback period for this system is around 10 years.

AC-3: Installing a variable refrigerant flow system. This option helps further reduce the total required cooling capacity from 6 tons to 4 tons. The Initial cost of installing this system is $4,520, the annual energy consumption is 5,735 KWh and the annual savings is $498. The payback period for this system is around 9 years.

EWH-System Evaluation:

The energy survey showed that all tenants are using 2x100 Liter electric hot water bottles, each with a heating element of 1.5 KW. However, according to ASHRAE Handbook Fundenmatals-2013 a 200 Liter storage tank with a 2KW heating element is sufficient to cover the hot water demand for a single apartment. As for the 22 apartments, a central hot water tank with a storage capacity of 1000 liter and a 72 KW is needed.

According to the calculation performed using the equations presented in ASHRAE-Handbook of HVAC

Applications (Chapter 50- Service Water Heating), two efficiency opportunities were evaluated and compared to the existing water heating system (DHW-1);

DHW-1: The annual energy consumption is 3,147 KWh and the annual energy cost is $261.

DHW-2: Replacing 2x100 Liter tanks with one 200 Liter tank and a heating element of 2 KW. The initial cost of installing this system is $500, the annual energy consumption is 1,899 KWh and the annual savings is $104. The payback period for this system is around 5 years.

DHW-3: Installing a central solar hot water system. The initial cost of installing this system is $1,600 per apartment, the annual energy consumption is 525 KWh and the annual savings is $218. The payback period for this system is around 7 years.

Lighting-System Evaluation:

The energy survey showed that all tenants use incandescent lamps with a total of 2 KW. Two efficiency opportunities were evaluated compared to the existing lighting system (LIG-1);

LIG-1: The initial cost of this system is $20, the annual energy consumption is 4,413 KWh and the annual energy cost is $366.

LIG-2: Replacing the Incandescent lamps with Compact Fluorescent Lamps (CFL). This option will reduce the total KW from 2 to 0.6. The initial cost of installing this system is $100, the annual energy consumption is 1,186 KWh and the annual savings is $268. The payback period for this system is around 4 month.

LIG-3: Replacing the incandescent lamps with light emitting diodes (LED). This option will help reduce the total KW from 2 to 0.4. The initial cost of installing this system is $300, the annual energy consumption is 883 KWh and the annual savings is $293. The payback period for this system is around 1 year.

Scenarios of proposed energy saving measure:

Six energy saving scenarios are propose as shown in Figure 6 and Table 3. The Table shows that Scenario-2 which utilizes AC-1', DHW-2 and LIG-3 costs $800 and reduces energy consumption by 26%. Scenario-4 utilizes AC-2, DHW-2 and LIG-3 costs $5,198 and reduces energy consumption by 37%. Scenario-6 utilizes AC-3, DHW-3 and LIG-3 costs $6,420 and reduces energy consumption by 44%. The cost difference between Scenario-4 and Scenario-6 is $1,222 per apartment and a total of $26,884 for the whole building.

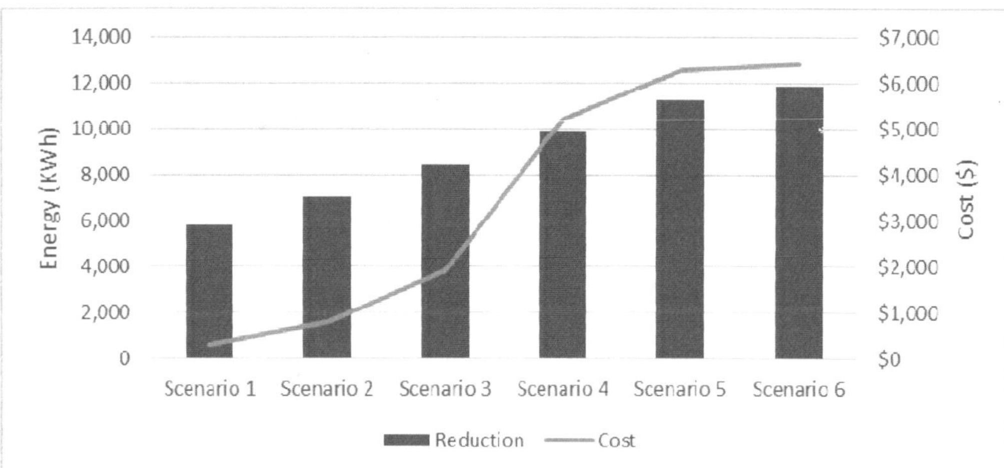

Figure 6 Cost and Energy Implications for Different Scenarios

Table 3. Scenarios of Proposed Energy Saving Measure

	AC System	Water Heating	Lighting	Others
Baseline	AC-1	DHW-1	LIG-1	X-1
KWH	1,173	3,147	4,143	8,041
Scenario-1	AC-1'	DHW-1	LIG-3	X-1
KWH	9,164	3,147	883	8,041
Cost	$0	$0	$300	$0
Scenario-2	AC-1'	DHW-2	LIG-3	X-1
KWH	9,164	1,899	883	8,041
Cost	$0	$500	$300	$0
Scenario-3	AC-1'	DHW-3	LIG-3	X-1
KWH	9,164	525	883	8,041
Cost	$0	$1,600	$300	$0
Scenario-4	AC-2	DHW-2	LIG-3	X-1
KWH	6,337	1,899	883	8,041
Cost	$4,398	$500	$300	$0
Scenario-5	AC-2	DHW-3	LIG-3	X-1
KWH	6,337	525	883	8,041
Cost	$4,398	$1,600	$300	$0
Scenario-6	AC-2	DHW-3	LIG-3	X-1
KWH	5,735	525	883	8,041
Cost$	$4,520	$1,600	$300	$0

CONCLUSION

This paper presented work targeted at evaluating the state of energy consumption in the Lebanese residential sector, and suggesting cost efficient energy systems and solutions, that enhance energy savings. Accordingly, utility data were collected from a residential existing building in Tripoli, aiming to estimate an actual base line for energy consumption. Moreover, the current energy systems used in this building were assessed. In order to propose practical retrofitting energy efficient systems, a model of the building was implemented in Autodesk-Revit, and then the data was exported into Elite-Chvac for heat load analysis. Several energy conservation scenarios were then suggested and evaluated based on their cost and their impact on energy savings. The effectiveness as well as the cost of each scenario is analyzed. The reulst show that a possible of 44% reduction in energy consumption can be reached.

REFERENCES

ASHRAE 2013. *ANSI/ASHRAE Standard 55: Thermal environmental conditions for human occupancy,* Atlanta, American Society of Heating, Refrigerating and Air-Conditioning Engineers.

ASHRAE 2017. *ASHRAE Handbook - Fundamentals,* Atlanta, American Society of Heating Refrigerating and Air-Conditioning Engineers.

AUTODESK. 2018 *Revit vs. AutoCAD* [Online]. Available: https://www.autodesk.com/solutions/revit-vs-autocad [Accessed].

DONG, B., LEE, S. E. & SAPAR, M. H. 2005. A holistic utility bill analysis method for baselining whole commercial building energy consumption in Singapore. *Energy and buildings,* 37, 167-174.

ELITE-CHVAC. 2005. *Commercial HVAC Loads* [Online]. Available: http://www.elitesoft.com/ [Accessed].

HAJJ-HASSAN, M. & KHOURY, H. Energy Modelling and Sustainability: A Review on Current Theories, Practices, and Challenges. 4th International Conference On Building Energy, Environment, 2018.

HOURI, A. & IBRAHIM KORFALI, S. 2005. Residential energy consumption patterns: the case of Lebanon. *International journal of energy research,* 29, 755-766.

LOPES, L., HOKOI, S., MIURA, H. & SHUHEI, K. 2005. Energy efficiency and energy savings in Japanese residential buildings—research methodology and surveyed results. *Energy and buildings,* 37, 698-706.
MORRISSEY, J. & HORNE, R. 2011. Life cycle cost implications of energy efficiency measures in new residential buildings. *Energy and buildings,* 43, 915-924.
PÉREZ-LOMBARD, L., ORTIZ, J. & POUT, C. 2008. A review on buildings energy consumption information. *Energy and buildings,* 40, 394-398.
RUBLE, I. & NADER, P. 2011. Transforming shortcomings into opportunities: Can market incentives solve Lebanon's energy crisis? *Energy policy,* 39, 2467-2474.
SAIDUR, R., MASJUKI, H. & JAMALUDDIN, M. 2007. An application of energy and exergy analysis in residential sector of Malaysia. *Energy Policy,* 35, 1050-1063.
SWAN, L. G. & UGURSAL, V. I. 2009. Modeling of end-use energy consumption in the residential sector: A review of modeling techniques. *Renewable and sustainable energy reviews,* 13, 1819-1835.
U.S.E.I.A 2013. International energy outlook. *Energy information administration (EIA).*

Investigation of Thermal Comfort in a Space Conditioned by Liquid Desiccant Membrane Chilled Ceiling/ Displacement Ventilation System

Racha Seblany, MS

Nesreen Ghaddar, PhD
Member ASHRAE

Kamel Ghali, PhD
Member ASHRAE

Nagham Ismail, PhD

Marco Simonetti, PhD

Fracastoro Giovanni Vincenzo, PhD

Joseph Virgone, PhD

Assaad Zoughaib ,PhD

ABSTRACT

Chilled ceilings combined with displacement ventilation (DV) have gained popularity due to their efficiency in terms of providing thermal comfort and high air quality, with minimal noise and good energy savings. Chilled ceilings can be replaced by liquid desiccant membrane systems integrated with DV system (LDMC-C/DV) to operate at lower ceiling temperatures and to directly dehumidify the indoor environment for improved performance. Thermal comfort is a major concern in any air conditioning system as it is directly related to the health of the occupants. Therefore, the effect of low ceiling temperatures in LDMC-C/DV environments on comfort needs to be investigated. For this purpose, a validated mathematical model of the dehumidifier panel will be integrated with a steady-state DV space model. The integrated model is used to represent the mass and energy balances inside the space and at the two sides of the dehumidifier panel. A parametric study will be carried out under different air supply conditions to predict the air conditions (temperature and relative humidity) in a space conditioned by the LDMC-C/DV system. Accordingly, the thermal comfort of the occupants inside the space is assessed based on the PMV value for different DV supply conditions and inlet desiccant solution temperatures. It was found that at low ceiling temperatures the relative humidty in the space rises but thermal comfort can still be realized under specified DV supply conditions.

INTRODUCTION

The huge urban expansion and the prolonged periods of time that people spend indoors necessitate the adoption of heating, ventilation and air conditioning (HVAC) systems in order to provide the occupants with comfortable, livable indoor conditions (Hao et al. 2007). The growing dependence on HVAC systems in buildings and industries has resulted in a massive increase in energy demand, especially in hot and humid climates (Vakiloroaya at al. 2014). For

Racha Seblany is a graduate student in the Department of Mechanical Engineering, American University of Beirut, Beirut, Lebanon and the corresponding author, email: rys12@mail.aub.edu . Nesreen Ghaddar is a professor in the Department of Mechanical Engineering, American University of Beirut, Beirut, Lebanon. Kamel Ghali is a professor and chair in the Department of Mechanical Engineering, American University of Beirut, Beirut, Lebanon

instance, cooling and ventilation systems contribute to more than 70% of the energy usage in the Gulf countries (El-Dessouky et al. 2004). Therefore, it is of great interest to find lower energy-demanding techniques for air conditioning, with the provision of acceptable air quality and thermal comfort for occupants (Enteria & Mizutani 2011). One of the systems that has been used is the displacement ventilation (DV) system (Bjørn & Nielsen 2002). In a DV system, fresh air is introduced at a temperature slightly cooler than that of the space air. Due to buoyancy, the air rises as it heats up by the internal heat sources and the space is divided into two zones: occupied and contaminated zones. The system provides high indoor air quality, but is constrained by relatively high supply temperature and low air velocity which limit its load removal capacity (Rees & Haves 2001). Consequently, DV systems require an additional cooling mechanism in order to be able to remove larger loads without compromising the air quality and the thermal comfort of the occupants.

Chilled ceilings (CC) can be used with displacement ventilation systems to improve the cooling capacity (Ardehali et al. 2004). In a combined CC-DV system, part of the sensible load is removed by the chilled ceiling, and the other part is removed by the DV system in addition to the latent space load and the pollutants (Novoselac & Srebric 2002). Hence, when appropriately designed, the combined system is able to provide good thermal sensation and high air quality (Ghali at al. 2007). However, one of the disadvantages of the combined CC-DV system is the absence of control over the indoor relative humidity (Fauchoux et al. 2010). It is of great importance to maintain the indoor humidity at a proper level to ensure the well-being of the occupants (Wang et al. 2013). High indoor humidity induces mold growth which might cause breathing problems (Bornehag et al. 2004). On the other hand, low humidity environments can lead to dryness of skin and throats and to irritation of eyes (Sato et al, 2003). To overcome these issues and to avoid condensation at the same time, pre-dehumidification of the supply air is required. One of the used techniques involves cooling the air below its dew point temperature, and then reheating it to the desired supply temperature. However this will result in added energy costs (Wang et al. 2013). Another method for dehumidification of supply air is the use of solid or liquid desiccant techniques. Desiccants are characterized by the ability to absorb moisture due the difference in vapor pressure between the surface of the solution and the adjacent air (Mohammad et al. 2013). But the direct contact between the desiccant and the supply air can be extremely hazardous (Studak & Peterson 1988).

A technique was later developed to directly remove the latent load from the space without any contact between the air and the desiccant by using semi-permeable membranes (Fauchoux et al. 2010). The cool desiccant solution flows over the membrane which is permeable to water vapor and impermeable to liquid desiccant. The flow of air on the other side of the membrane results in heat and mass transfer between the air and the desiccant through the porous membrane. Using membrane-based dehumidification allowed for operation at lower ceiling temperatures (Eldeeb et al. 2013). One of the disadvantages of the system is the condensation risk at the ceiling due to the contradictory effects of the sensible heat loss and the moisture removal by the desiccant. The performance of liquid desiccant membrane chilled ceiling with displacement ventilation (LDMC-C/DV) system was studied in the work of Muslmani et al. (2016). A mathematical model of the system was developed and the energy savings due to decreasing the desiccant solution temperature were found to be significant. However, their model did not take into account the boundary layer that is formed near the ceiling surface. Hout et al. (2017) developed a model for LDMC-C/DV systems that considers the boundary layer zone and integrated it with the two-zone space model of the DV system. They designed charts to determine the ranges of operational parameters at which thermal comfort is ensured and condensation is not likely to occur.

The current work presents a parametric study over a space cooled by a LDMC-C/DV system. The study investigates the air conditions in the occupied zone of the space and the thermal comfort of the occupants in this zone upon varying the ceiling's inlet desiccant solution temperature for different DV supply conditions. Mathematical modeling will be used to achieve this purpose. The steady model developed by Hout et al. (2017) will be used to represent the energy and mass transfer between the desiccant and the surrounding air, and between the two zones of the space model. The air conditions in the lower occupied zone will then be investigated for various operational parameters in order to assess the effect of the ceiling temperature on the thermal sensation of occupants.

LDMC-C/DV SYSTEM DESCRIPTION

A space cooled by liquid desiccant membrane chilled ceiling with displacement ventilation system will be considered in this study. **Figure 1-a** represents a schematic of the system. The DV supply air enters from a grill situated close to the floor level. The air forms a layer near the floor, and warms up due to the heat generated by the occupants and appliances. The contaminated, warm air rises under the effect of buoyancy and is extracted near the ceiling level (Loveday et al. 2002). This results in the formation of upper and lower zones. The lower zone, also known as the occupied zone, is where thermal comfort and good indoor air quality (IAQ) are required for the well-being of the occupants. While the upper zone is a contaminated region where the air conditions are not a concern. The height at which the two zones are separated is known as the stratification height. It is recommended that the stratification height is greater than 1.1 m (3ft 7in) to ensure acceptable IAQ for a typical seated human (ASHRAE 2009). From a physical point of view, the stratification height is the one at which the mass flow rate of the plumes is equivalent to the supply flow rate (Jaluria 1980). At the ceiling level, liquid desiccant solution is introduced into the space. The solution flows over the semi-permeable membrane at a specified inlet temperature and mass flow rate. The flow of the solution and of the air at the two sides of the membrane is assumed to take place in opposite directions (counter-flow configuration). The transfer of heat and moisture between the air and the desiccant through the porous membrane creates a boundary layer near the ceiling. Thus the space is divided into three zones as shown in **Figure 1-b**.

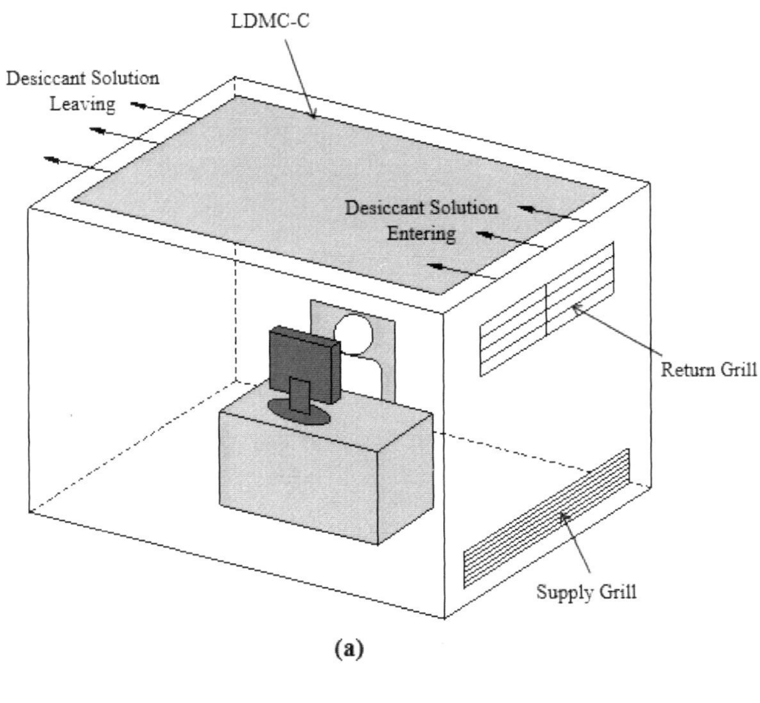

(a)

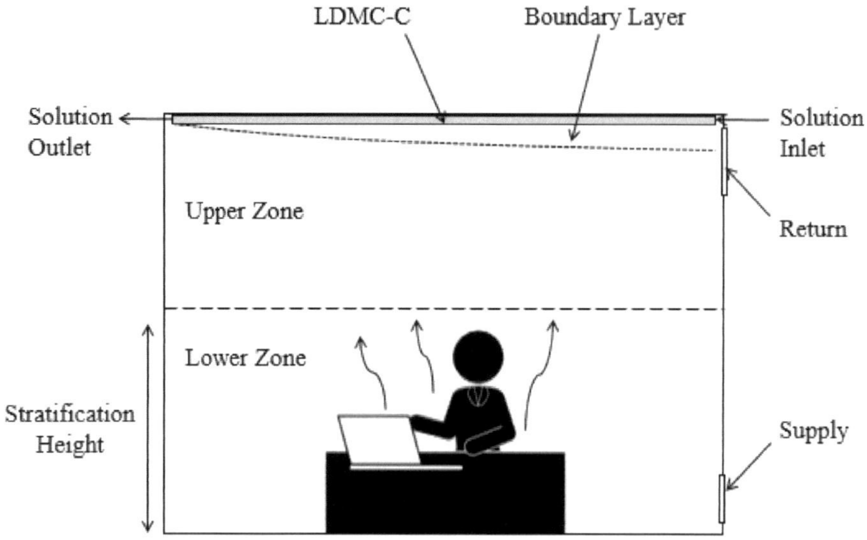

(b)

Figure 1 (a) Schematic of ta space with LDMC-C/DV system and (b) Vertical cross-sectional view of the space with the LDMC-C/DV system.

MATHEMATICAL FORMULATION AND SOLUTION METHODOLOGY

The numerical model developed and validated by Hout et al. (2017) will be used in this paper to describe the steady-state performance of the counter-flow LDMC-C/DV system. In their model, Hout et al. (2017) considered the problem to be unidimensional. They used an axial grid along the membrane width that consists of 100 nodes. Moreover, their equations were based on several assumptions that will be adopted in this work as well. Their mathematical model was basically the integration of two models: The dehumidifier model which represents the heat and mass transfer that take place inside the boundary layer, and the steady-state space model that solves for the temperature and humidity in the occupied and the contaminated zones. The accuracy of the dehumidifier model predictions was investigated experimentally in the work of Hout et al. (2017), and the experimental results were found to be consistent with the numerical data. The equations that represent the energy and mass transfer at the side of the desiccant are respectively:

$$-\dot{m}_{sol}\frac{d}{dx}(Cp_{sol}T_{sol})+U_cW_m(T_{CBL}-T_{sol})+\rho_a h_{fg}U_mW_m(W_{CBL}-W_{sol})=0 \tag{1}$$

$$-\dot{m}_{sol}\frac{d}{dx}(C_{sol})+\rho_a U_mW_m(W_{CBL}-W_{sol})=0 \tag{2}$$

Where $\dot{m}_{sol}$, Cp_{sol}, T_{sol} and W_{sol} are the desiccant solution mass flow rate, specific heat, temperature and humidity ratio respectively. U_c and U_m are the overall heat and mass transfer coefficients. T_{CBL} and W_{CBL} are the respective temperature and humidity ratio of the boundary layer air. Moreover, W_m is the width of the membrane, ρ_a is the density of air and h_{fg} is the latent heat of vaporization of water.

On the air side of the membrane, the heat and mass transfer equations are as follows:

$$-\dot{m}_{CBL}\frac{d}{dx}(Cp_a T_{CBL}) + U_c W_m (T_{sol} - T_{CBL}) + \rho_a h_{fg} U_m W_m (W_{sol} - W_{CBL}) + \frac{\dot{m}_{entrained} Cp_a T_a}{dx} = 0 \qquad (3)$$

$$-\dot{m}_{CBL}\frac{d}{dx}(W_{CBL}) + \rho_a U_m W_m (W_{sol} - W_{CBL}) + \frac{\dot{m}_{entrained} W_a}{dx} = 0 \qquad (4)$$

Where T_a and W_a are the temperature and specific humidity of air in the contaminated zone. $\dot{m}_{entrained}$ represents the mass flow rate that is entrained from the upper zone. Conservation of mass on a control volume of length dx and thickness δ in the boundary layer is applied in order to obtain the entrained mass flow rate which is found to be equal to:

$$\dot{m}_{entrained} = \int_0^{\delta} \frac{d\rho u}{dx} w dx dy \qquad (5)$$

The thickness of the boundary layer is obtained by solving a differential equation that represents the conservation of momentum over a differential element. It is worthy to mention that the velocity of the boundary layer air is based on the profile suggested by Awbi et al. (2003).

The above equations were integrated with a steady space model. The DV space model has been studied in previous literature and is based on the energy and species transfer between the two zones of the space. The integrated LDMC-C/DV model is discretized using the finite volume method and solved iteratively based on the input parameters. These input parameters include the DV supply air conditions (temperature, humidity and mass flow rate), inlet desiccant solution conditions (temperature, concentration and mass flow rate), internal sensible and latent loads as well as the physical and geometrical parameters of the space. The model solves for the: (i) the axial distribution of the temperature and humidity of the air in the boundary layer, (ii) the temperature and concentration of the desiccant solution variations, and (iii) the lumped temperature and humidity of air in each of the upper and lower zones. Convergence is attained once the relative error in predicted parameters does not exceed 10^{-6}. In addition to these parameters, the model predicts the temperature gradient in the space and the stratification height in the room. Furthermore, the predicted mean vote (PMV) that reflects the thermal comfort of the occupants in the occupied zone is obtained based on the correlations of Fanger (1970).

DESCRIPTION OF THE PARAMETRIC STUDY

A parametric study was adopted to assess the steady-state performance of a space conditioned by a LDMC-C/DV system and when there is no risk of condensation. The temperature of the inlet desiccant solution, and thus the ceiling temperature, will be varied under pre-defined DV supply conditions in order to investigate the conditions of air in the occupied zone and consequently predict the thermal comfort of the occupants in this zone.

The space considered in this study is a typical office in Beirut with the following dimensions 5m×5m×3m (16.4ft×16.4ft×9.8ft) and a population of 6 people, each using a personal computer. The internal load is approximately 62 W/m^2 (19.6 Btu/hr.ft^2) consisting of: 70W (239 Btu/hr) sensible heat and 2×10^{-5} kg/s (4.4×10^{-5}

lb/s) of water vapor produced by each occupant, 60W (204.7 Btu/hr) of heat generated by each computer, and 20 W/m² (6.3 Btu/hr.ft²) of heat produced by lighting equipment. The space is conditioned by 100% fresh air supplied by a DV system at a fixed humidity ratio of 0.011 kg/kg. The DV supply air temperature and mass flow rate are varied. Thus, the different DV supply conditions represent the six different scenarios that will be adopted in the parametric study and are shown in Table 1.The inlet desiccant solution is calcium chloride of mass concentration 40% and mass flow rate of 0.05 kg/s (0.11 lb/s). The inlet solution temperature will be varied between 13°C (55.4°F) and 22°C (71.6°F); this range ensures that no risk of salt crystallization is present. Therefore, 60 simulation cases were performed using the different DV scenarios and inlet desiccant conditions.

In the occupied zone, the air conditions (temperature and humidity) were examined while varying the temperature of the inlet desiccant solution for the different scenarios. The thermal comfort of the occupants were investigated as well, represented by the predicted mean vote value obtained by using the correlations of Fanger (1970). Thermal comfort is achieved when the PMV value is within the range of ±0.5.

Table 1. DV Supply Scenarios for the Parametric Study of the LDMC-C/DV System

Scenario	Air supply temperature °C (°F)	Air supply relative humidity %	Air supply mass flow rate kg/s (lb/s)	ACH
1	18 (64.4)	85	0.196 (0.432)	7.8
2	20 (68)	75	0.196 (0.432)	7.8
3	22 (71.6)	66	0.196 (0.432)	7.8
4	18 (64.4)	85	0.1 (0.22)	4
5	20 (68)	75	0.1 (0.22)	4
6	22 (71.6)	66	0.1 (0.22)	4

RESULTS AND DISCUSSION

Thermal comfort inside a conditioned space is directly related to the well-being of the inhabitants, therefore it is important to keep the indoor air properties within the ranges that provide good thermal comfort. The current work represents a parametric study of the thermal comfort within the occupied zone of a LDMC-C/DV environment. Simulations were run by varying the inlet desiccant solution temperature between 13°C (55.4°F) and 22°C (71.6°F) for the suggested scenarios (**Table 1**).

The predicted variation in the temperature and the relative humidity of air within the occupied zone are shown in **Figure 2** and **Figure 3** respectively. As shown in **Figure 2**, the temperature of air in the lower zone undergoes a slow increase of maximum 1.5°C (34.7°F), as the inlet solution temperature increases in all of the scenarios. This is due to the fact that lower solution temperatures lead to bigger temperature gradient between the solution and the air, and consequently to increased load removal capacity of the desiccant. At a fixed supply flow rate, the temperature in the lower zone increases with increasing air supply temperatures. On the other hand, the air temperature is higher at smaller DV supply flow rate and constant supply temperature. It can be seen that the lowest temperatures in the occupied zone were obtained in scenario 1 at low inlet desiccant temperature. The maximum predicted air temperature is 30°C (86°F) and is obtained when the supply temperature is 22°C (71.6°F), the supply flow rate is 0.1 kg/s (0.22 lb/s) (scenario 6) and the desiccant solution is high.

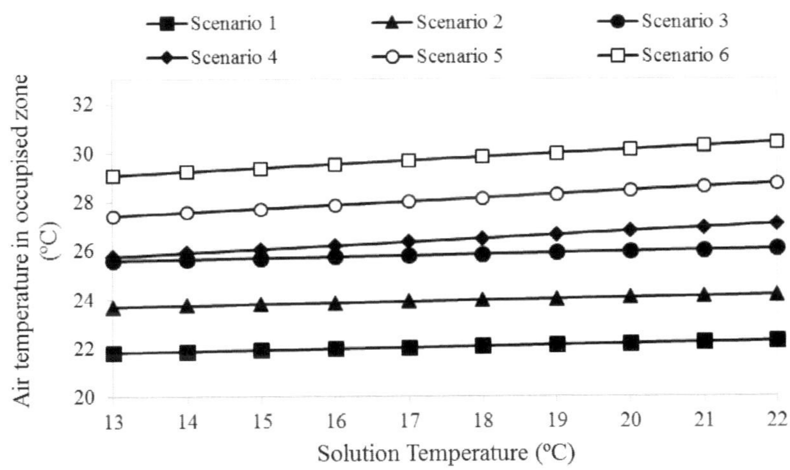

Figure 2: Variation of the temperature of air in the occupied zine as a function of the inlet desiccant temperature for various scenarios.

Figure 3 shows that the relative humidity increases with decreasing the solution temperature, and consequently the ceiling temperature. This is explained by the drop in the air dry bulb temperature that is accompanied by an increase in its specific humidity at lower desiccant temperatures, which lead to an overall rise in the relative humidity of air in this zone. The relative humidity of air is the highest in scenario 1 (around 60%) and is reached at low solution temperature. The relative humidity in the occupied zone decreases with the increase in the supply air relative humidity at a fixed supply flow rate. The same is noticed at lower DV supply flow rate and fixed supply temperature. It is worth mentioning that the rate of change in the both the temperature and the relative humidity is greater at low supply air flow rate. Moreover, it is noted that no condensation was present in any of the simulated cases.

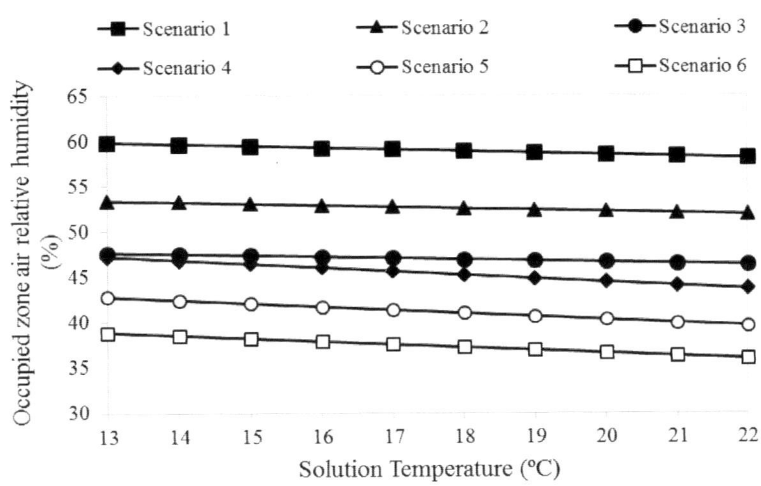

Figure 2: Variation of the temperature of air in the occupied zine as a function of the inlet desiccant temperature for various scenarios.

The aim of the study is to check for the ability of the system to provide thermal comfort at low desiccant temperatures where the relative humidity of air in the space is high. The PMV value reflects the thermal comfort of the occupant and can be used to assess comfort. It is critical that the PMV value remains within the range ±0.5, otherwise the occupants' comfort is jeopardized. **Figure 4** represents the obtained PMV values using the correlations of Fanger (1970) for different inlet desiccant temperatures in the various scenarios. The graph shows that change in the solution temperature has a significant effect on the thermal comfort of the occupants within the space. It can be noticed that the PMV value increases with the increase in the solution temperature because the solution temperature directly affects the ceiling temperature which in turn affects the thermal sensation of the occupants. The figure shows that comfort is achievable at low desiccant temperatures, as in the case of scenario 2 where the solution temperature can drop as low as 13°C (55.4°F) and comfort is guaranteed. In scenario 5, comfort can be obtained when the desiccant temperature is 15°C (59°F) and above. In scenario 1, thermal comfort is reached at high desiccant temperature only (beyond 20°C (68°F)). While in scenarios 3,5 and 6, thermal comfort was not reached at any inlet desiccant temperature.

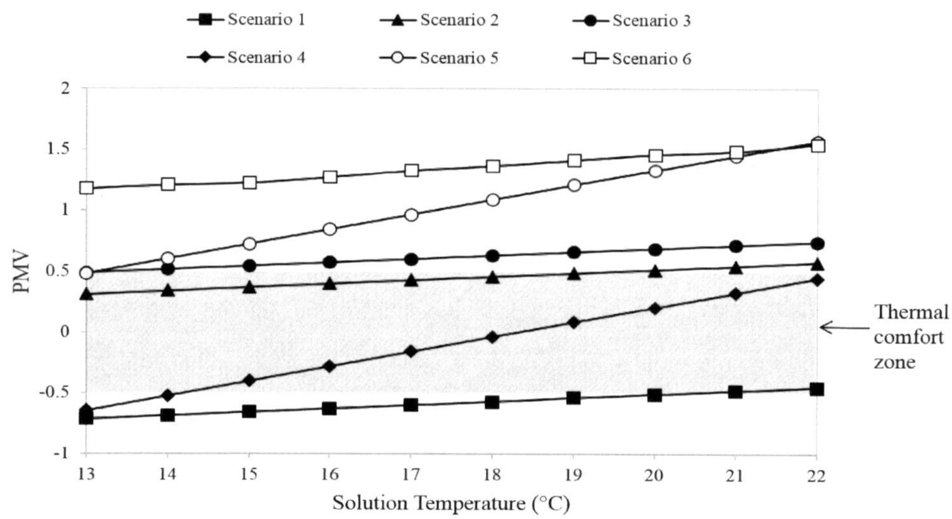

Figure 4: Variation of the PMV value with different inlet desiccant temperatures for various scenarios.

CONCLUSION

A parametric study was carried to investigate thermal comfort in a space conditioned by a combined LDMC-C/DV system when the conditions of air in the occupied zone are changed. A mathematical model that represents the energy and mass transfer at the opposite sides of the dehumidifier was integrated with the DV space model. The simulations were made for different scenarios of the DV system operation while varying the inlet desiccant temperature. It was found that the temperature and humidity in the lower zone change by varying the ceiling temperature that results from varying the inlet solution temperature. At low inlet solution temperatures, the temperature of air in the lower zone seems to become lower while the relative humidity increases. Moreover, the DV supply conditions (temperature, humidity and flow rate) have a significant effect on the air conditions in the occupied zone. It was noted that small changes in the properties of the lower zone air can shift the space in or out of the thermal comfort zone. Furthermore, it was found that at high space relative humidity, which is induced by low desiccant temperatures, thermal comfort is still achievable at specified DV supply conditions.

NOMENCLATURE

DV	Displacement Ventilation
LDMC-C/DV	Liquid Desiccant Membrane Chilled Ceiling with Displacement Ventilation
PMV	Predicted Mean Vote
HVAC	Heating, Ventilation and Air Conditioning
CC	Chilled Ceiling
IAQ	Indoor Air Quality

REFERENCES

Hao, X., Zhang, G., Chen, Y., Zou, S., & Moschandreas, D. J. 2007. A combined system of chilled ceiling, displacement ventilation and desiccant dehumidification. *Building and Environment*, 42(9), 3298-3308.

Vakiloroaya, V., Samali, B., Fakhar, A., & Pishghadam, K. 2014. A review of different strategies for HVAC energy saving. *Energy conversion and management*, 77, 738-754.

El-Dessouky, H., Ettouney, H., & Al-Zeefari, A. 2004. Performance analysis of two-stage evaporative coolers. *Chemical Engineering Journal*, 102(3), 255-266.

Enteria, N., & Mizutani, K. 2011. The role of the thermally activated desiccant cooling technologies in the issue of energy and environment. *Renewable and Sustainable Energy Reviews*, 15(4), 2095-2122.

Bjørn, E., & Nielsen, P. V. 2002. Dispersal of exhaled air and personal exposure in displacement ventilated rooms. *Indoor air*, 12(3), 147-164.

Rees, S. J., & Haves, P. 2001. A nodal model for displacement ventilation and chilled ceiling systems in office spaces. *Building and Environment*, 36(6), 753-762.

Ardehali, M. M., Panah, N. G., & Smith, T. F. 2004. Proof of concept modeling of energy transfer mechanisms for radiant conditioning panels. *Energy conversion and Management*, 45(13-14), 2005-2017.

Novoselac, A., & Srebric, J. 2002. A critical review on the performance and design of combined cooled ceiling and displacement ventilation systems. *Energy and buildings*, 34(5), 497-509.

Ghali, K., Ghaddar, N., & Ayoub, M. 2007. Chilled ceiling and displacement ventilation system for energy savings: a case study. *International journal of energy research*, 31(8), 743-759.

Fauchoux, M., Bansal, M., Talukdar, P., Simonson, C. J., & Torvi, D. 2010. Testing and modelling of a novel ceiling panel for maintaining space relative humidity by moisture transfer. *International Journal of Heat and Mass Transfer*, 53(19-20), 3961-3968.

Wang, N., Zhang, J., & Xia, X. 2013. Desiccant wheel thermal performance modeling for indoor humidity optimal control. Applied energy, 112, 999-1005.

Bornehag, C. G., Sundell, J., Bonini, S. 2004. Dampness in buildings as a risk factor for health effects, EUROEXPO: a multidisciplinary review of the literature (1998–2000) on dampness and mite exposure in buildings and health effects. *Indoor air*, 14(4), 243-257.

Sato, M., Fukayo, S., & Yano, E. 2003. Adverse environmental health effects of ultra-low relative humidity indoor air. *Journal of occupational health*, 45(2), 133-136.

Mohammad, A. T., Mat, S. B., Sulaiman, M. Y., Sopian, K., & Al-Abidi, A. A. 2013. Historical review of liquid desiccant evaporation cooling technology. *Energy and Buildings*, 67, 22-33.

Studak, J. W., & Peterson, J. L. 1988. A preliminary evaluation of alternative liquid desiccants for a hybrid desiccant air conditioner.

Eldeeb, R., Fauchoux, M., & Simonson, C. J. 2013. Applicability of a heat and moisture transfer panel (HAMP) for maintaining space relative humidity in an office building using TRNSYS. *Energy and Buildings*, 66, 338-345.

Muslmani, M., Ghaddar, N., & Ghali, K. 2016. Performance of combined displacement ventilation and cooled ceiling liquid desiccant membrane system in Beirut climate. *Journal of Building Performance Simulation*, 9(6), 648-662.

Hout, M., Ghaddar, N., Ghali, K., Ismail, N., Simonetti, M., Fracastoro, G. V., ... & Zoughaib, A. 2017. Displacement ventilation with cooled liquid desiccant dehumidification membrane at ceiling; modeling and design charts. *Energy*, 139, 1003-1015.

Loveday, D. L., Parsons, K. C., Taki, A. H., & Hodder, S. G. 2002. Displacement ventilation environments with chilled ceilings: thermal comfort design within the context of the BS EN ISO7730 versus adaptive debate. *Energy and Buildings*, 34(6), 573-579.

ASHRAE. 2009. ASHRAE Handbook –Fundamentals, Chapter 9. Atlanta: ASHRAE.

Jaluria, Y. 1980. Natural convection: heat and mass transfer (Vol. 5). Pergamon.

Awbi, H. B. 2003. *Ventilation of buildings*. Taylor & Francis.

Fanger, P. O. 1970. Thermal comfort. Analysis and applications in environmental engineering. *Thermal comfort. Analysis and applications in environmental engineering.*

Interrelationship Between Architectural and Mechanical Aspects of the Building Envelope Design

Hadi Maamoun
Member ASHRAE

ABSTRACT

The building envelope plays an important role in maintaining interior thermal comfort and in minimizing uncontrolled moisture and air movement into a building. A n effective building envelope design works towards minimizing heat transfer and preventing moisture and condensation problems. Because some of these considerations are not always apparent to A rchitects, the interaction between A rchitects and Mechanical engineers is crucial in creating a problem free design. This paper aims to show the relationship bewteen the architectural and mechanical aspects of the envelope design. This paper also details how the integrative design is implemented during the early design phases supported by several building envelope case studies for airports, hotels and institutional facilities. For each case study, a thermal and moisture diffusion analysis is conducted to evaluate the thermal effectiveness of the proposed envelope design and to assess the potential of surface or interstitial condensation within the layers. Different scenarios are also simulated under various climatic conditions during the year and even during a single day to determine the operational behavior of the assembly, the optimal ordering of the assembly layers, the need of some features within the envelope such as vapor barrier and several other critical issues.

INTRODUCTION

The building envelope is the interface between the interior of the building and the outdoor environment.
The selection of building envelope components and the way all layers are ordered and incorporated within the assembly have great impact on the overall thermal performance and moisture management characteristics of the assembly. Other key factors that need to be considered are the outdoor environmental conditions, the interior environment, and the thermal and vapor diffusion resistance properties of the envelope materials.
The objective of this paper is to analyze several case studies for the optimization of the building envelope by addressing Mechanical considerations that are not always apparent to Architects at the early project design stages.
Three different case studies are investigated in this work toward creating a problem free envelope design by having both disciplines, Architecture and Mechanical, to work together closely to address the Mechanical aspects of the design.
For all cases, the envelope performance will be analyzed through a thermal and moisture diffusion analysis to help

Hadi Maamoun, M.E., LEED AP BD+C, Member ASHRAE, is a Senior Mechanical Engineer at Dar Al-Handasah (Shair and Partners), Beirut, Lebanon.

detecting mold growth due to high humidity levels or to detect any surface or interstitial condensation at any interface layer under different critical climatic conditions.

CASE STUDY 1: ROOFING OF CONCOURSE D AT DUBAI INTERNATIONAL AIRPORT

The initial roof composition of the new Concourse D at Dubai Airport as proposed by the Architecture skin designer consisted of the following layers from outside to inside: Kalzip panels, 100mm (4 in.) layer of Rockwool panels, another 100mm (4 in) layer of Rockwool panels, vapor control layer, plaster board, and steel corrugated sheet.

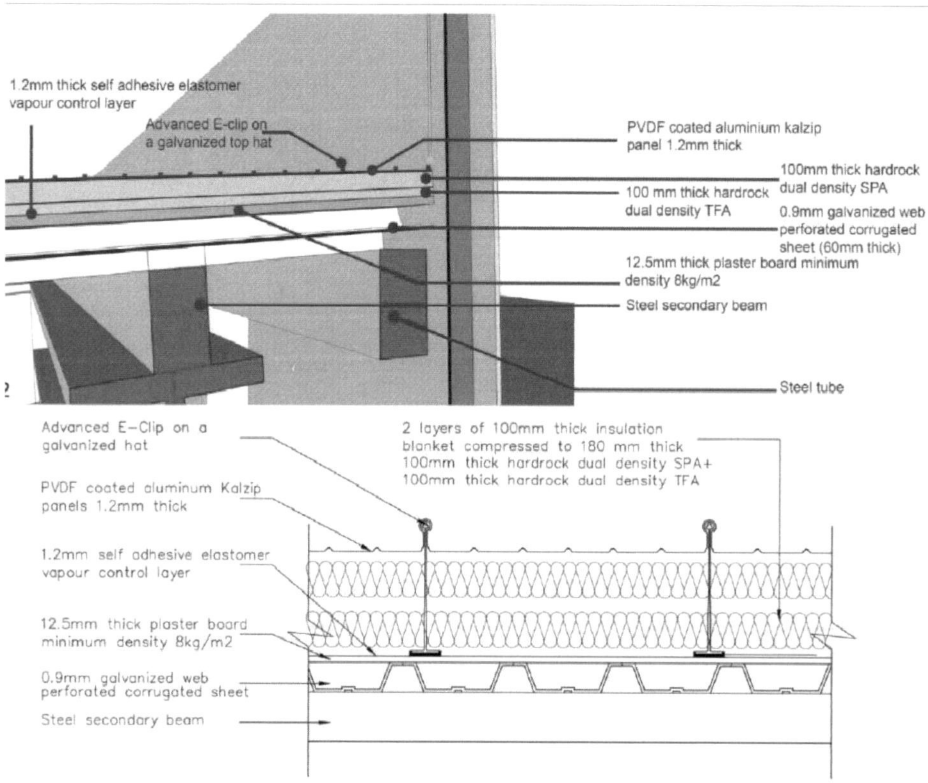

Figure 1 Roof composition proposed by Architect

The thermal and moisture diffusion analysis for the proposed assembly was performed at the peak wet bulb temperature and relative humidity that occurs in Dubai.

No condensation is observed within the layers due to vapor difusion driven by vapor pressure differential.

The results generated from an in house developed spreadsheet showed that the temperature of the vapor control layer is dropping below the dew point of outdoor air infiltrated through the kalzip. This indicates that condensation is likely to occur at the outer side of the vapor control layer due to its temperature dropping below the dew point of the outdoor infiltrated air.

Therefore, providing an insulation layer underneath the elastomer vapor control layer is needed to keep it warmer than the dew point temperature of the infiltrated outdoor air. This agrees with the general recommendation in hot and humid climates that calls for avoiding locating vapor control layers on the cold side of the skin which is towards the internal air conditioned side.

Accordingly, the Mechanical recommendation is to move the vapor control layer one layer outward to become located between the two rockwool panels.

Moreover, to achieve an interface temperature at the vapor control layer greater than the dew point temperature by around 1 to 2 degree Celsius for additional safety, the thickness of the rockwool panel underneath the vapor control

layer to be specified 150mm (6 in.) instead of 100mm (4 in.).
Based on all the above, the optimal roof composition that performs satisfactorily from a Mechanical aspect and at the same time meeting all Architectural considerations is now represented in figure 2 and adopted for the project roofing.

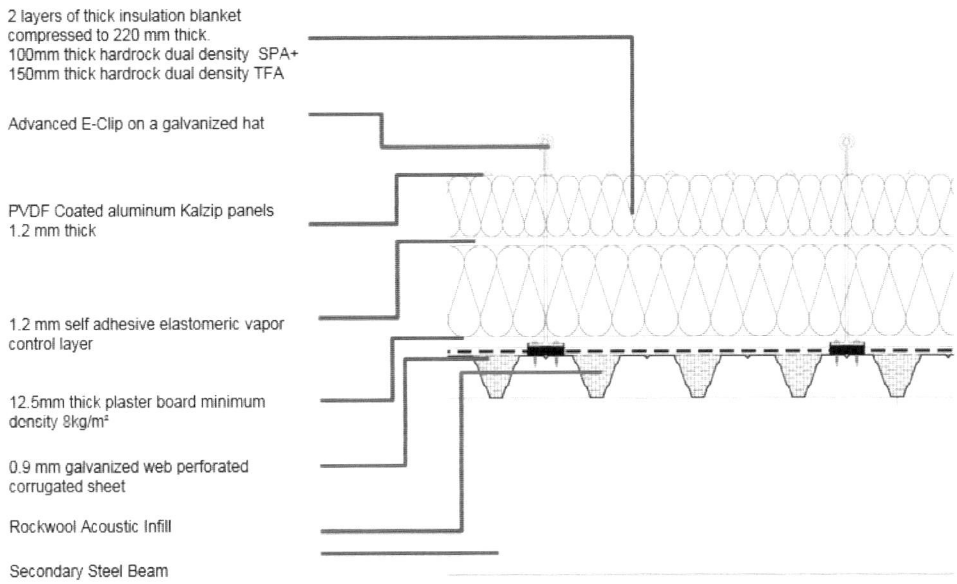

Figure 2 Adjusted roof composition

CASE STUDY 2: EXTENSION TO INTERCONTINENTAL HOTEL, DOHA, QATAR

Wall Composition: Mechanical stone fixation

The initial wall assembly provided by the Architect for the new extension building of the Hotel is shown in figure 3 and it consists of 30 mm (1.2 in.) thick stone mechanically fixed, 30 mm (1.2 in.) air gap, 150 mm (6 in.) thick solid block wall, 50 mm (2 in.) insulation, 100 mm (4 in.) masonry block wall and internal plaster.

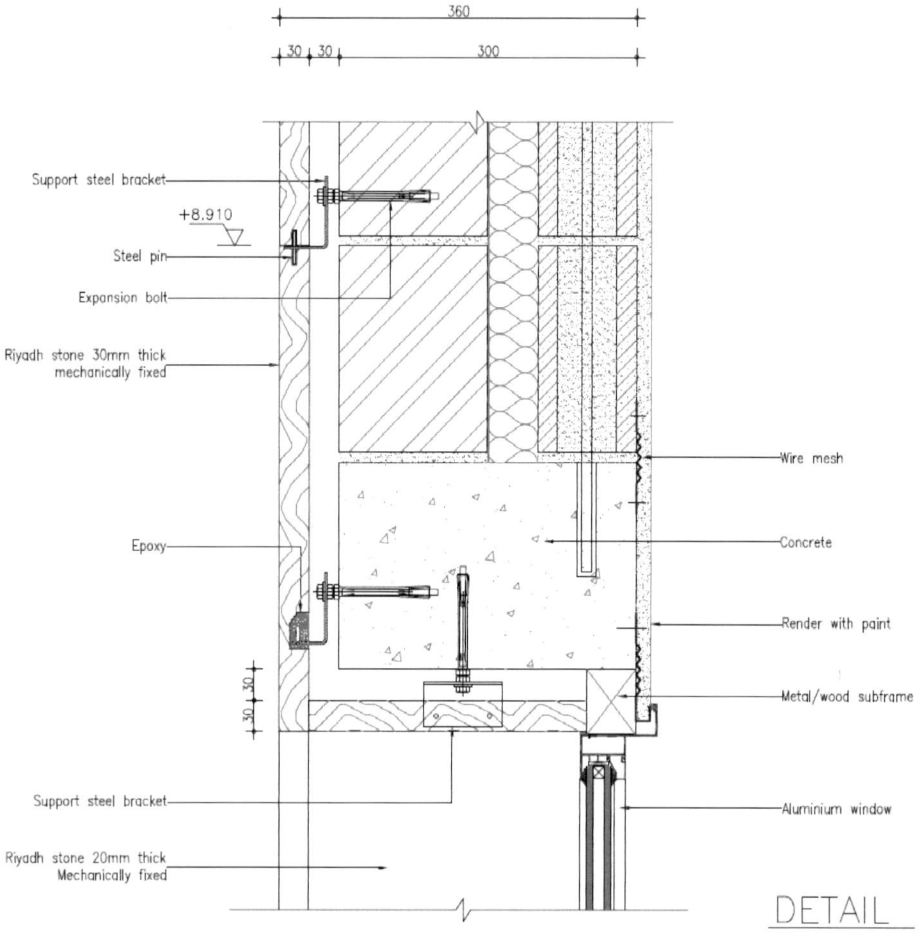

Figure 3 Wall composition proposed by Architect

The overall thermal resitance of the assembly satisfies Ashrae 90.1 recommendation. The thermal and moisture diffusion analysis for the proposed assembly was performed at the peak wet bulb temperature and relative humidity that occurs in Doha.

Project:	Q12046-0100D						
Location:	Qatar	Interior Temp. (°C)	23.0	°C	Interior RH (%)	50	%
Condition:	**Summer**	Exterior Temp. (°C)	35.0	°C	Exterior RH (%)	80	%

Wall Case 1 **Heat Conduction and Vapor Diffusion Properties of Wall**					**Calculated Temperature Drops, Surface Temperature, and Saturation Water Vapor Pressures**						**Calculation of Water Vapor Pressure Drops and Surface Water Vapor Pressures**		
Air Film or Material	**Thermal Resistance [1] (m².K)/W**	**Vapor Permeance[2] ng/(s.m².Pa)**	**Vapor Diffusion Resistance (Pa.m².s)/ng**	**Temperature Drop, °C**	**Interface Temperature, °C**			**Interface Saturation Vapor Pressure[3], Pa**	**Relative Humidity (%)**	**Dew Point Temperature °C**	**Vapor Pressure, Pa Initial Calculation**		**RESULTS**
Inside Air					23.0	296.2	7.941	2810.4	50	12.1		1405	Vapor pressure at Indoor Condition from Psychometric Software
Inside Air Layer	0.12	9200	0.00011	-0.634	23.63	296.8	7.979	2920.0	48	12.1	-2	1407	No Condensation
15mm Plaster	0.02083	532	0.00188	-0.110	23.74	296.9	7.986	2939.4	49	12.4	-32	1439	No Condensation
100mm CMU	0.15	274	0.00365	-0.792	24.54	297.7	8.034	3082.5	49	13.1	-63	1502	No Condensation
50mm Polystyrene	1.7	24.4	0.04098	-8.976	33.51	306.7	8.553	5180.6	43	19.1	-702	2204	No Condensation
150mm Solid Block	0.09	44	0.02273	-0.475	33.99	307.1	8.579	5319.8	49	21.7	-390	2594	No Condensation
30mm Air Gap	0.132	4930	0.00020	-0.697	34.68	307.8	8.618	5530.0	47	21.7	-3	2597	No Condensation
30mm Riyadh Stone mech fixed	0.016	9	0.11111	-0.084	34.77	307.9	8.623	5555.9	81	31.0	-1905	4502	No Condensation
Outside Air Layer	0.044	57000	0.00002	-0.232	35.00	308.2	8.635	5627.8	80	31.0	0	4502	No Condensation
Outside Air					35.00	308.2	8.635	5627.8	80	31.0		4502	
Total	2.27283		0.18068										
	0.43	meets											

(1) & (2) *From "Construction Wall" Sheet*

(3) ***Interface Saturation Vapor Pressure*** values can be found in ***Table 3*** in ***Chapter 1 (Psychrometrics)- 2009 ASHRAE Handbook - Fundamentals (SI)***

Figure 4 In-house spreadsheet used for thermal and vapor diffusion analysis of the upper part of the wall

The interpretation of the results generated from the in-house developed spreadsheet showed that no condensation is occurring due to vapor diffusion through the wall layers.

Condensation due to air infiltration occurs at multiple locations of the external surface of the wall assembly as described below:

- Condensation occurs at the interface between the vertical air gap and the concrete lintel above the window. Recommendation to cover the solid block and the concrete surface with waterproofing paint.
- Recommendation to add flash sill at the intersection between horizontal and vertical stone to drain water.
- A drip edge should be added at the horizontal stone to prevent penetration of drained water.
- Consider slightly protruding the horizontal stone at window sill with respect to window head in order to direct drained water away from vertical stone of external wall below.
- Condensation occurs at the interface between the horizontal air gap and the metal/wood subframe. Recommendation to add flash sill above window to drain horizontal air gap.
- To reduce impact of hot and humid air on masonry wall, it is recommended to consider increasing the thickness of the air gap to 50 mm (2 in.) instead of 30 mm (1.2 in.).

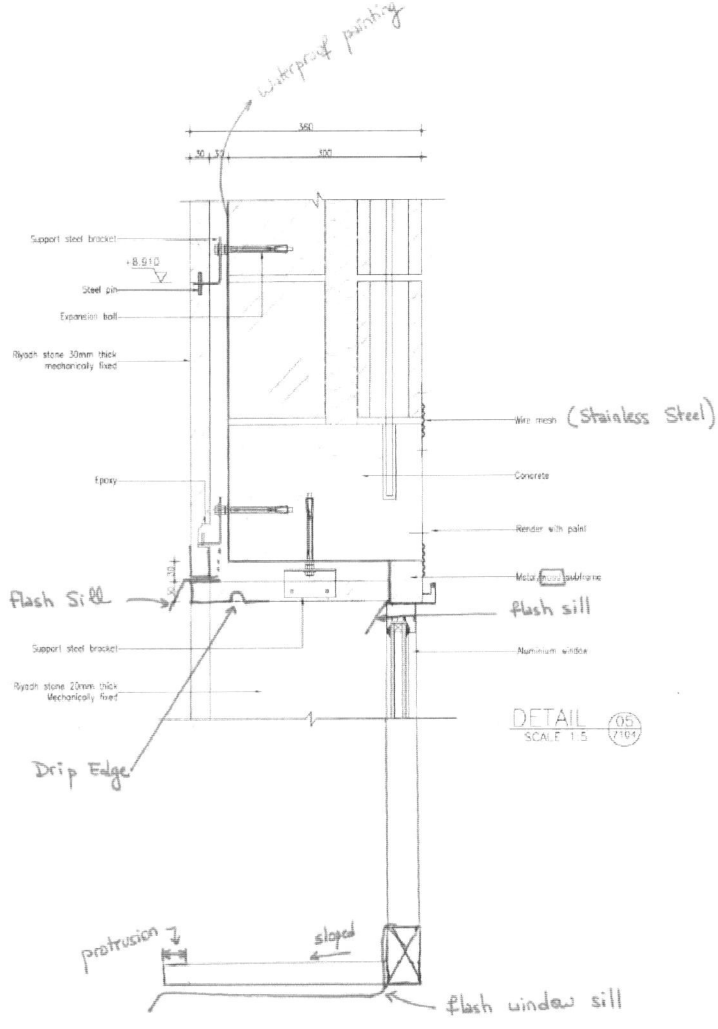

Figure 5 Mechanical comments showing the measures to be applied on the initial wall composition

Roofing Composition

The roof assembly proposed by the building envelope designer consists of the following: 40 mm thick gravel layer, polester fabric filter membrane, 50 mm (2 in.) thick high density rebated edge extruded foam insulation board, water proofing membrane and light weight concrete screed.

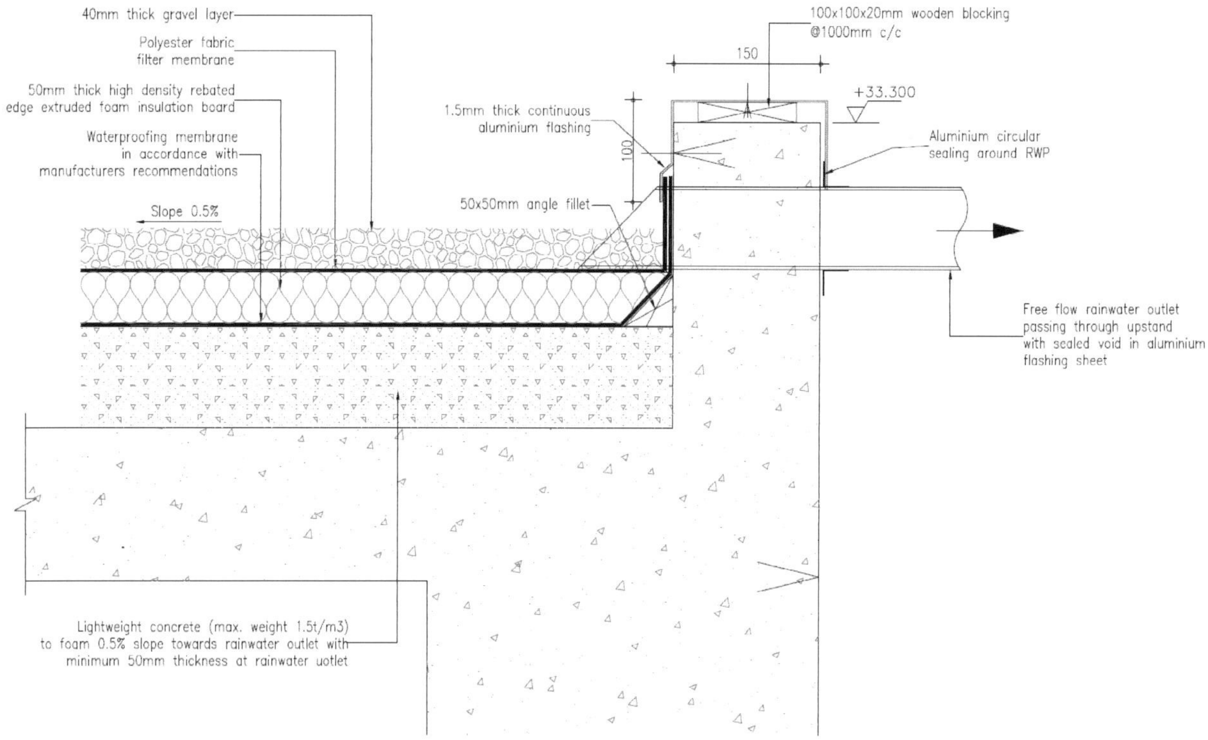

Figure 6 Proposed roofing detail

The thermal and moisture diffusion analysis showed that the overall thermal performance of the roof does not satisfy Ashrae 90.1 recommendation. As such, the extruded polystyrene foam insulation thickness should be increased to 75mm (3 in.).

No condensation is occurring at any layer of the roof provided that the vapor barrier used is a class II barrier at minimum with an allowable permeance to be not lower than 1 perm.

Gravels are energy inefficient as they increase the ability of the roof to store heat and trap moisture. It is therefore recommended to either replace the gravels with white pedestal stones or to provide 100mm (4 in.) thick foam insulation to overcome the heat storage effect of the gravel.

The waterproofing membrane should overlap the metal scupper as shown and the roof flashing to be wrapped underneath the scupper drain.

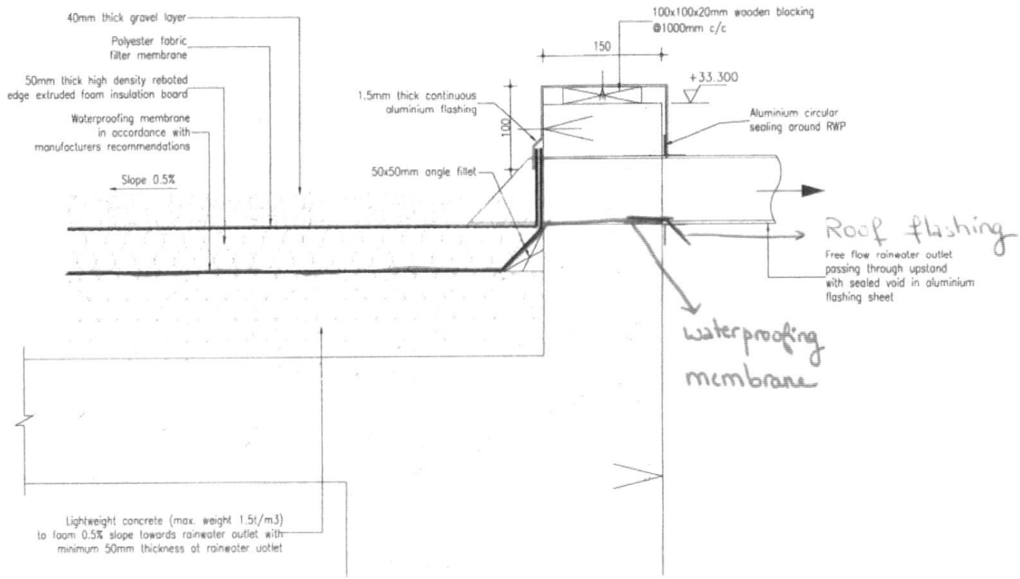

Figure 7 Mechanical comments to be applied on the initial wall composition

CASE STUDY 3: ROOFING DETAIL AT SALAHADDIN UNIVERSITY, ERBIL

The roof assembly proposed by the envelope designer for the new colleges at Salahaddin University consists of the following: 50 mm (2 in.) thick gravel layer on the perimeter of loose laid concrete tiles, geotextile membrane, waterproofing membrane, polyisocyanurate insulation and light weight concrete screed.

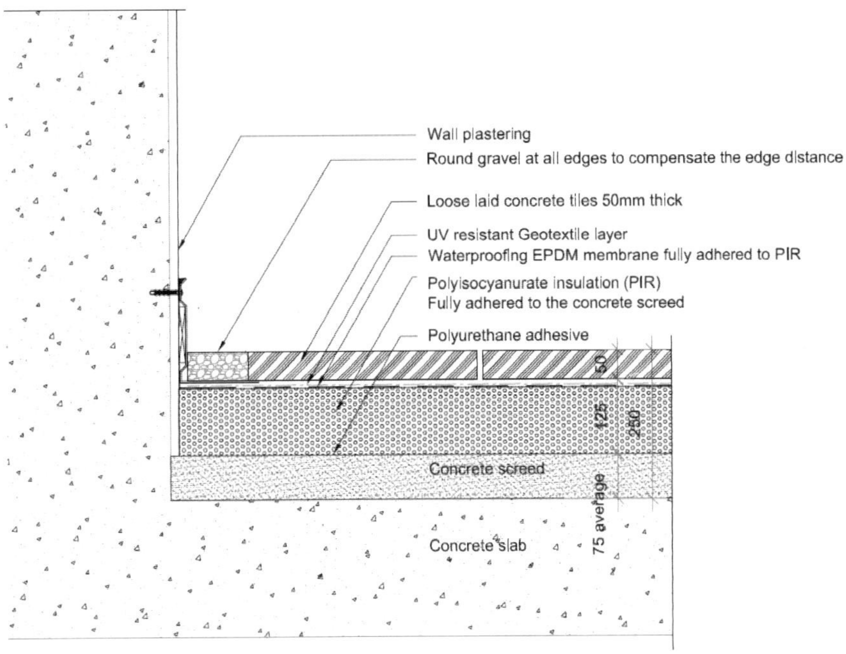

Figure 8 Proposed roofing detail

From a Mechanical perspective, since Erbil has freezing winter conditions, using gravels is not a recommended practice. During the cold season, the watertrapped between the stones will freeze and expand in very irregular shapes causing the damage of the waterproofing membrane and the drainage board. Even though the gravels are used only at the edge of the roof, the recommendation is to avoid using gravels. Making the tiles loose laid is helpful, and in that case, the water shall be trapped between the tiles and water will freeze as a thin continuous layer without damaging the membrane.

To avoid damage at insulation panels joints locations, it is recommended in countries with cold season similar to Erbil to use staggered layers of insulation (2 layers) and not just one layer.

The thermal resistance of some Polyisocyanurate (PIR) insulation panels decreases 20% within few years and is subject to significant loss in cold weather. Mechanical recommendation is to select PIR complying with ASTM C-1303 PIR with LTTR (long term thermal resistance) equal to 15 years.

Architecture specifications to call for both sides faced PIR with UL and FM approvals to resitrict the selection to the best top products in the market.

The selected PIR Insulation should not expand, shrink, and buckle leading to thermal bridges and membrane failure. Accordingly, PIR should be tested according to ASTM-C1289 to ensure dimensional stability is not more than 0.1%.

As for the thermal and moisture diffusion analysis, no condensation is observed at any layer of the proposed assembly.

CONCLUSION

The main objective of this paper is to show through case studies how the integrative design between Architects and Mechanical engineers is crucial in the design of a healthy building envelope that can perform satisfactorily for its intended service life.

The cases analyzed showed that the selection of building materials with appropriate water and thermal transmission characteristics should be part of the skin design process. The analysis focused on highlighting various Mechanical aspects that might not be apparent to Architects and skin designers at the conceptual stages of the project and that might lead to moisture problems in buildings.

ACKNOWLEDGMENTS

I would like to express my gratitude and deep appreciation for the continuous support of the Mechanical Engineering Department at Dar-Al-Handasah and especial thanks to the department Director Mr. Naji Edde and the Head of Department Mr. Jad Laham for their cooperation and assistance.

REFERENCES

Chapter 25, Heat, Air, and Moisture control in building assemblies – Fundamentals – 2013 ASHRAE Handbook – Fundamentals (SI).

Chapter 26, Heat, Air, and Moisture control in building assemblies – Material Properties – 2013 ASHRAE Handbook – Fundamentals (SI).

Ashrae 90.1-2016 standard " Energy Standard for Buildings except Low-Rise Residential Buildings".

The ASHRAE Guide for Buildings in Hot and Humid Climates, Harriman L. & Lstiburek J.

Hygrothermal Engineering Analysis of Walls and Roofs in Hot and Humid Climates

Ghina Annan, ME **Balsam Nehme, ME**

ABSTRACT

Moisture damage in buildings is common, where the consequences can be catastrophic, and the remedial measures expensive. Factors influencing condensation risk include exterior climate conditions, interior environment, thermal and vapor properties of the structure and the heating, ventilation and air conditioning system.

Condensation can occur on building surfaces or can be interstitial, causing serious problems if left unchecked; such as crumbling or soft spots in drywall, decay in wood framing or corrosion of steel, paint peeling, damage to the insulation inside the walls, and mold and mildew problems. Research shows that several human health illness is attributed to damp indoor air quality such as cough, wheeze, asthma and dyspnea.

Controlling moisture is crucial to ensure healthy indoor air quality, occupant comfort and to avoid the deterioration of building structures. Hygrothermal engineering is the combined effect of temperature and humidity transport through the building components, influenced by the envelope material composition and the surrounding climate.

This work is concerned with the assessment of the possibility of condensation in the building structure using WUFI® Pro, a software that performs a transient two-dimensional heat and moisture transport in multi-layered building components using real climatic conditions. Typical walls and roofs used in hot and humid climates, such as typical climates across the United Arab Emirates, are investigated for possible condensation and accordingly building materials are selected with appropriate water and thermal transmission characteristics.

INTRODUCTION

Condensation of walls and roofs is a critical concern for occupants' health and wellbeing especially in Hot and Humid climates as UAE mainly Dubai. The elimination of condensation within walls and roofs is as important in maintaining indoor environmental quality. Condensation may either be observable or concealed, yet both may contribute to dire health consequences as well as building structural damage. The observable type arises as water vapor condenses on an external wall layer and the water droplets are perceived. While the concealed type is when the same process occurs but within the wall section initiating material deterioration and soaking of the wall. The dew point temperature is the state at which air can no longer hold all of the water vapor which is mixed with it, and thus part of this vapor condenses into water. Figure 1 below illustrates an external wall separating a warm interior environment and cold exterior environment, as the water vapor reaches the middle surface whose temperature is below the dew point, condensation becomes inevitable. As a result, the reflection of failures must be part of moisture control calculations based on hydrothermal engineering analysis.

Ghina Annan is a design Mechanical Engineer in the Department of Buildings Sustainability Unit, Dar al Handasah, Beirut, Lebanon. **Balsam Nehme** is the Head of the Department of Buildings Sustainability Unit, Dar al Handasah, Beirut, Lebanon.

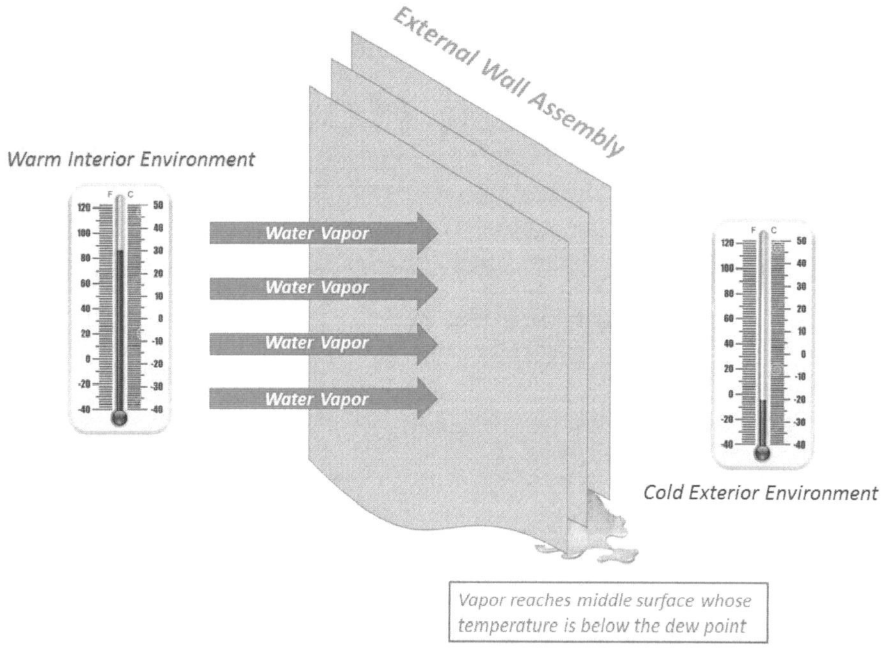

Figure 1 Water vapor Flow across external wall assembly

METHODOLOGY

A major challenge, is to recognize how moisture travels throughout the building envelope system, such as external and walls and roofs, in order to avoid any potential mold growth and to inhibit any prospective structural damage. In this work, a critical hot and humid climate is considered for condensation simulation analysis of multiple external walls and roofs of a residential building located in Dubai, UAE.

The following simulations assess the possibility of condensation in the building structure using a software tool known by WUFI® Pro. WUFI® Pro performs hygrothermal calculations on building component cross-sections, taking into account built-in moisture, driving rain, solar radiation, long-wave radiation, capillary transport, summer condensation. Furthermore, WUFI® Pro consist of a library of material properties of wall, roof, slab compositions and data sets, that are substantial when assigning a material to a wall or roof detail. Thus, each material assigned to the wall or roof detail composition can have the following characteristic properties given by WUFI library:

1. Water vapor diffusion parameters
2. (μ-value, sd-value)
3. Capillary water absorption
4. Reference moisture content
5. Sorption isotherm and moisture retention curve
6. Capillary transport coefficients
7. Bulk density and porosity
8. Thermal and hygric length variations

It is important to note that these data sets have been measured and verified by the moisture research laboratory who is in turn granted accreditation by Germany's national accreditation body DAkkS (Deutsche Akkreditierungsstelle GmbH) in accordance with DIN EN ISO/IEC 17025. (WUFI® Pro)

The simulation Interior design conditions are set to 24 degrees Celsius (75.2 degrees Fahrenheit) and 50% Relative Humidity, while the exterior conditions assumes Dubai weather file, in EnergyPlus (.epw) format, which is acquired from Meteonorm version 7.

Furthermore, In order to comply with Dubai Green building Regulations, all building elements forming the external walls, roofs, must have an average thermal transmittance (U- Value) which does not exceed Roof U= 0.3 W/m2K, External Wall U= 0.57 W/m^2.K. (Green Building Regulations & Specifications)

In this study five construction details are considered as summarized in the table below and shown in figures 2 through 5:

Table 1. Construction Details characteristics

Detail Number	Wall Description	Total U-value (W/m^2.K)	Complies with 'Dubai Green building Regulations'
1	External Wall without insulation	2.475	No
2	External Wall with insulation	0.451	Yes
3	Roof without insulation	1.798	No
4	Roof with insulation	0.212	Yes

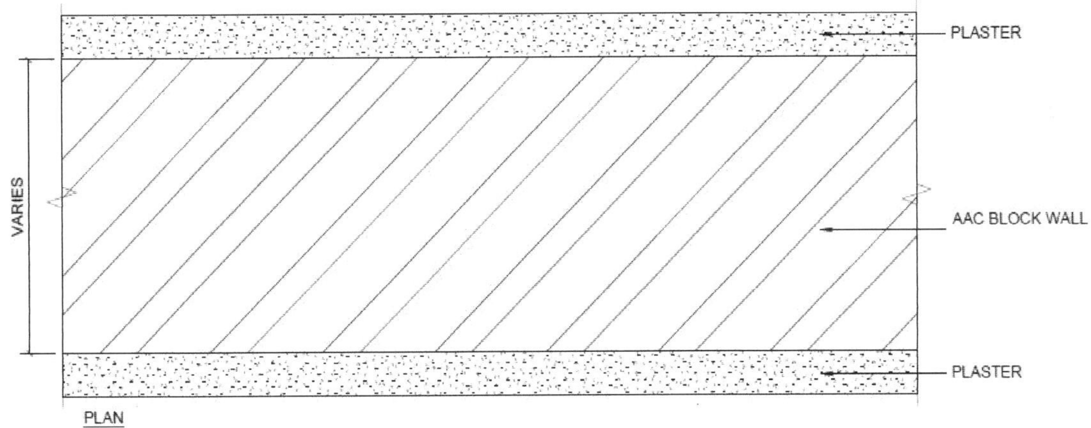

Figure 2 Construction Detail 1 - External Wall without insulation

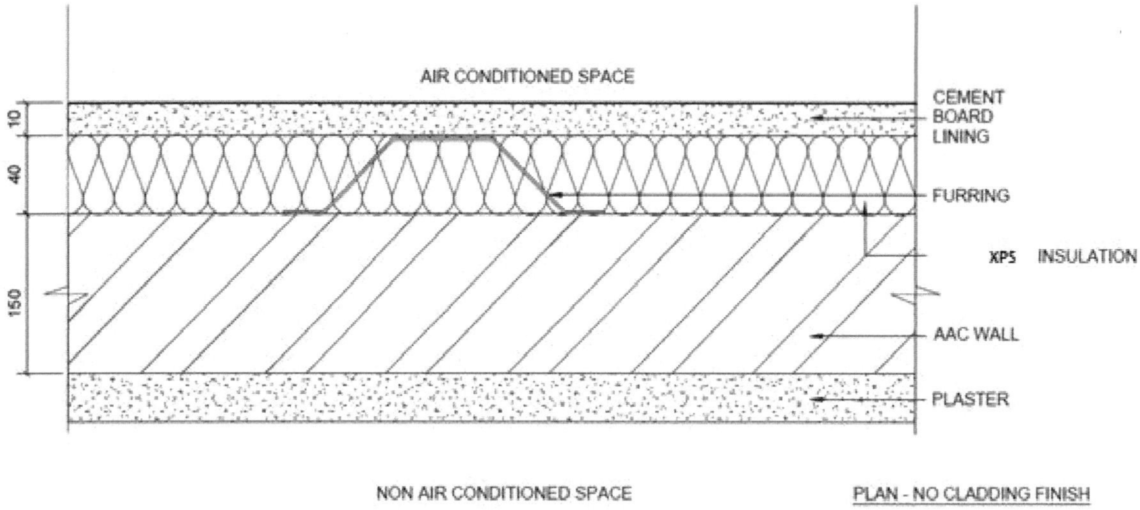

Figure 3 Construction Detail 2 - Construction Detail 2 - External Wall with insulation

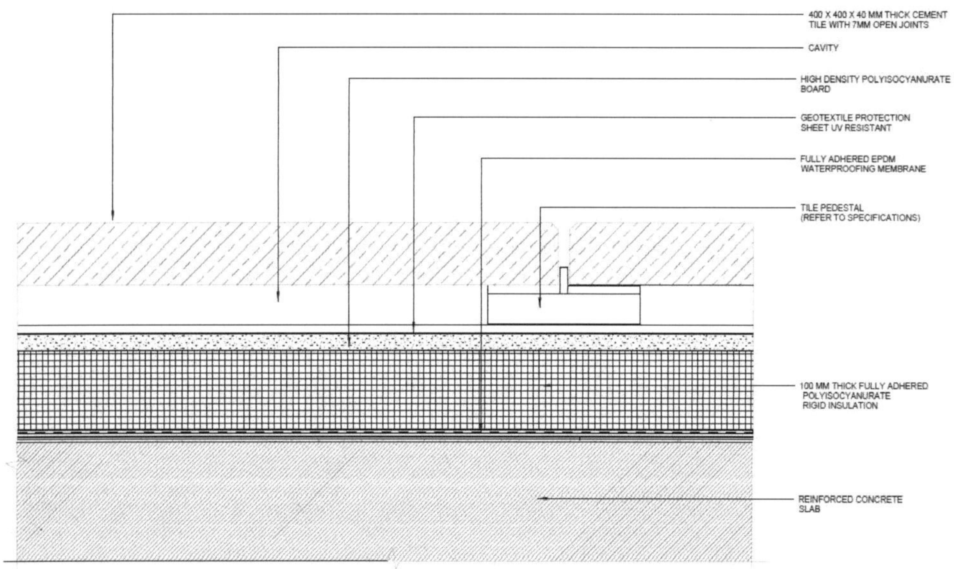

Figure 4 Construction Detail 3 - Roof without insulation

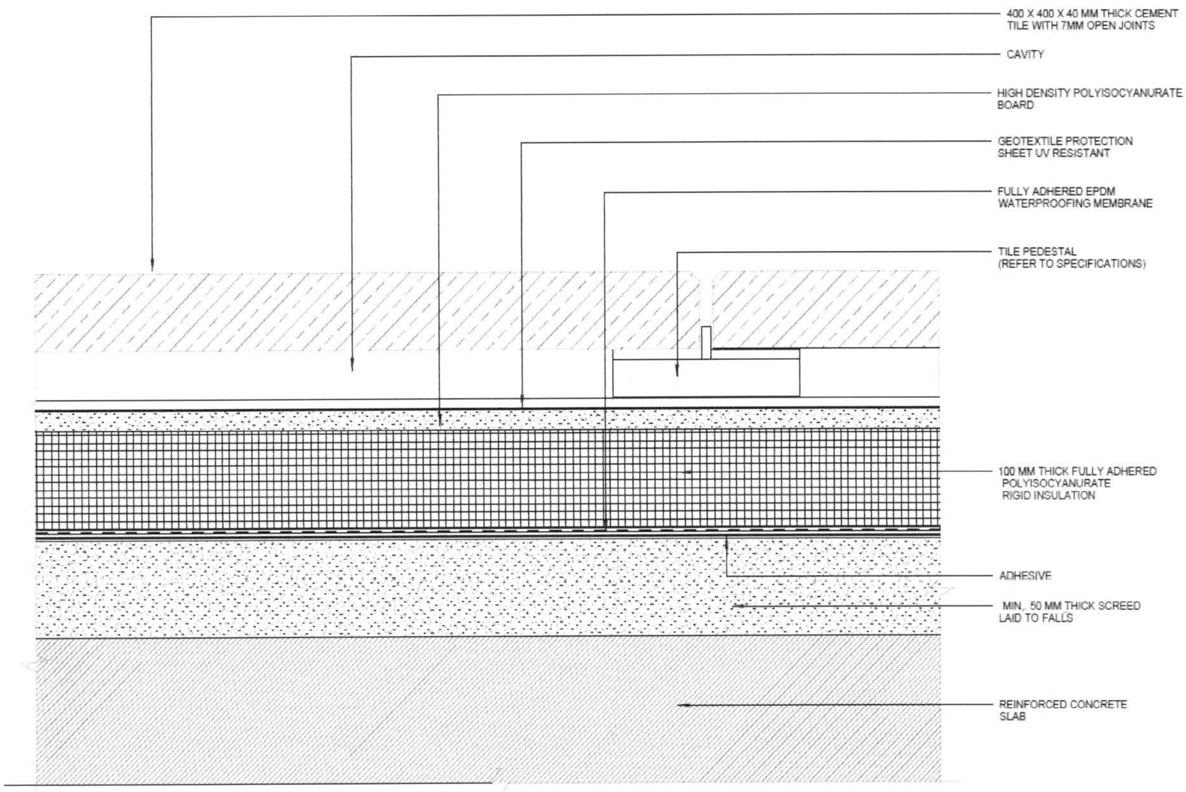

Figure 5 Construction Detail 4 - Roof with insulation

Simulation and Results

The simulation duration of the calculation period is defined to be three years. As such, output graphs showing temperature and relative humidity variation are illustrated below for the four details introduced above in figures 6 through 9. Also, the table below demonstrate a summary of condensation analysis results of the four construction details. The maximum relative humidity and temperature values are indicated:

Table 1. Maximum relative humidity and temperature values reached

Detail Number	Maximum Temperature reached (°C)	Maximum Relative Humidity reached (%)
1	34.0	75
2	26.5	61
3	24.5	75
4	24.0	57

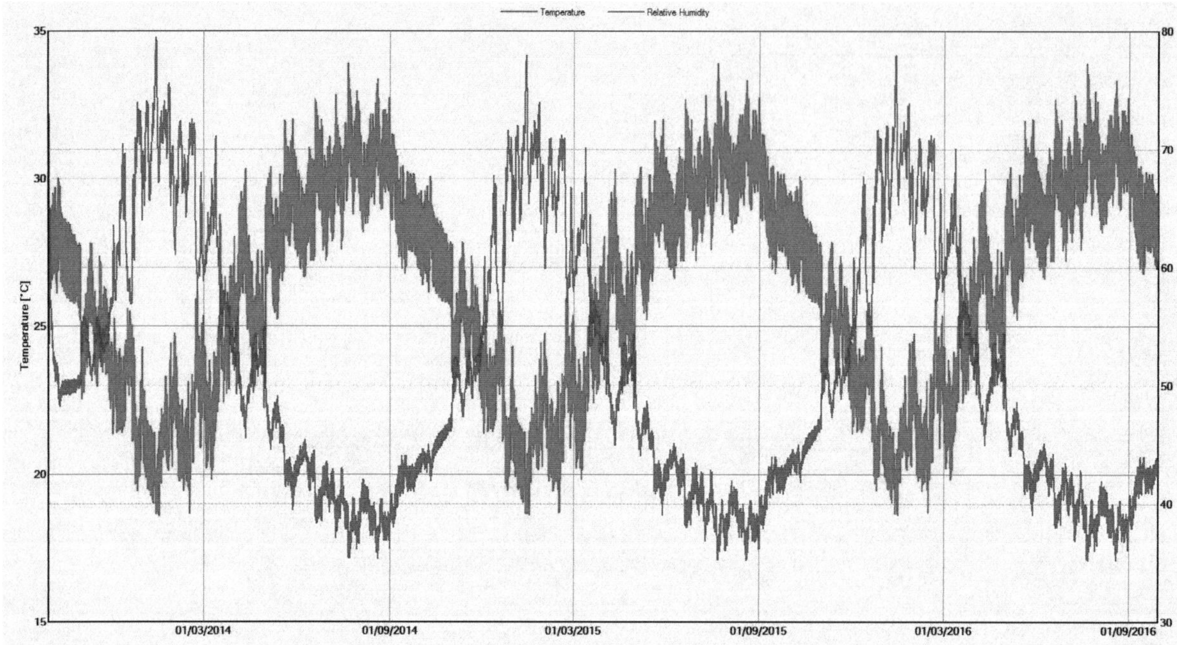

Figure 6 Variation of Temperature and Relative humidity profile of detail 1 throughout 3 years on AAC Block layer

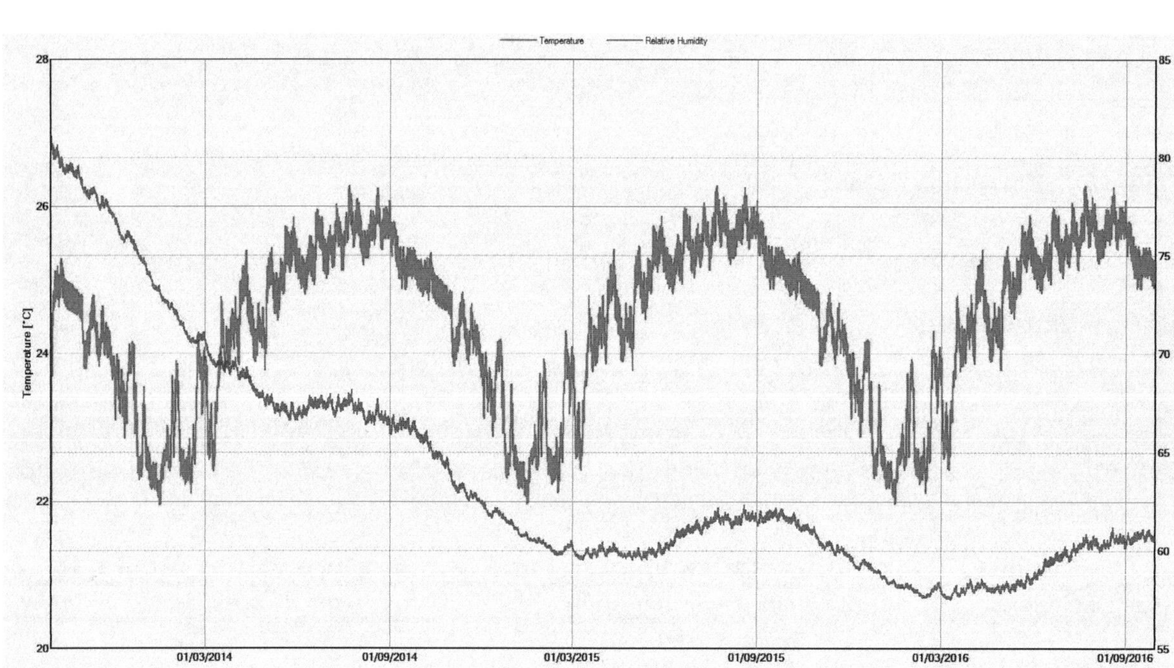

Figure 7 Variation of Temperature and Relative humidity profile of detail 2 throughout 3 years on AAC Block layer

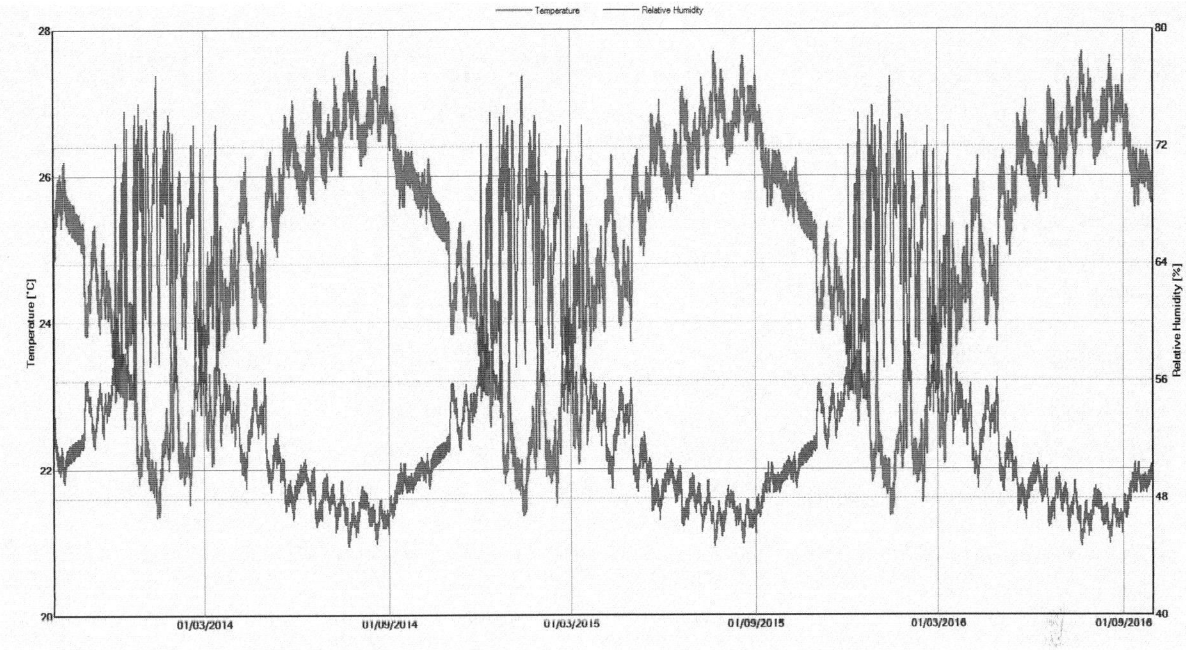

Figure 8 Variation of Temperature and Relative humidity profile of detail 3 throughout 3 years on AAC Block layer

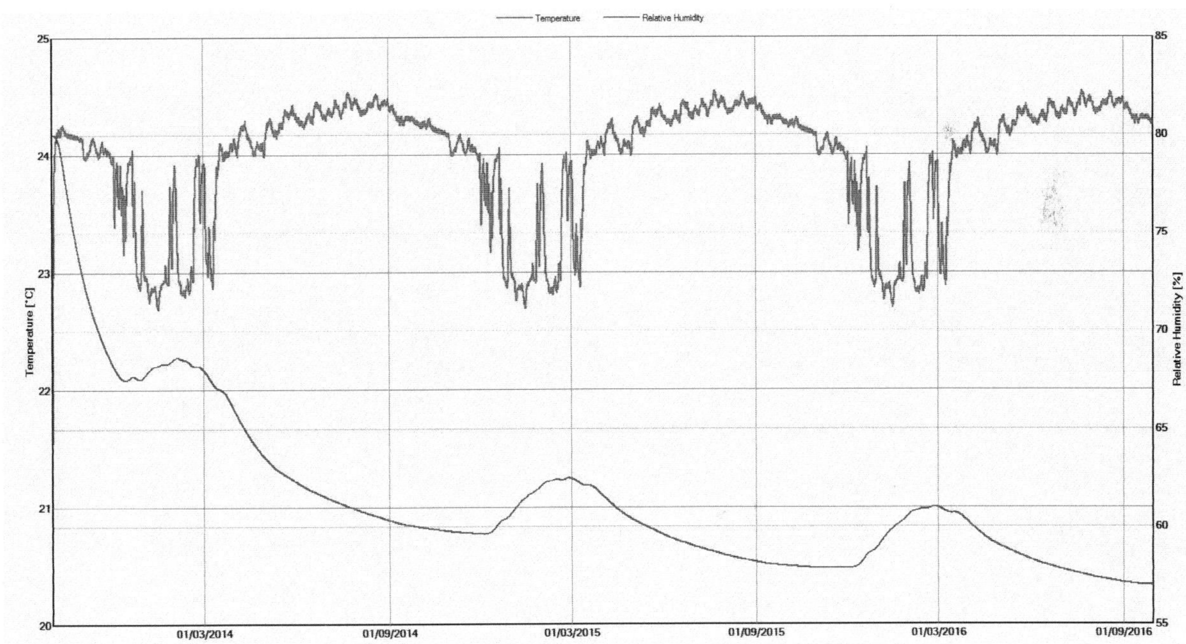

Figure 9 Variation of Temperature and Relative humidity profile of detail 4 throughout 3 years on AAC Block layer

CONCLUSION

An in-depth study was performed to assess the external walls and roof assemblies in hot and humid climates and evaluate whether the presence of insulation and a low total external wall or roof U-value would minimize the opportunity of condensation. It was concluded, sufficient moisture controls, should be implemented before appalling consequences take place. Furthermore, the location of the insultion is important in hygrothermal behavior of the envelope assembly, such that installing the insulation material towards the hot and humid side of the wall is highly effective in this work. Finally, wall and roof assemblies must comply Green Building Regulations & Specifications at minimum.

REFERENCES

https://wufi.de/en/software/wufi-pro/

Green Building Regulations & Specifications, *Government of Dubai, Dubai Electricity & Water Authority, Dubai Municipality.*

Global Trends in Collaborative Infrastructure Development to Foster Economically, Environmentally & Socially Sustainable Built Environment in Regional Economies

Om Taneja, PHD, PE, *Fellow ASHRAE*

ABSTRACT

In the developed or developing economies, the infrastructure systems and their effective and reliable operation is a pre-requisite for delivery of services for HVAC and building systems including associated energy, mobility, water, sanitation, and communication. It is projected that by 2030, more than $50 trillion will need to be invested in infrastructure globally to keep up with GDP, urbanization and climatic changes. Investments of this magnitude can only be delivered through collaboration between cities, regions, and international businesses and investors. Where there are gaps between public funds availability and the needs, Public-Private-Partnerships need to be evaluated and structured to close the gaps. International collaboration common in scientific research, education and finance need to be transposed to capacity building and management of multi-faceted infrastructure projects. Advanced countries can collaborate with regional country counterparts to projects execution, financing and timely delivery. Geo-collaboration enables cities in regional economies to boost their effectiveness by bringing international know-how directly to projects facilitators.

Buildings systems including lighting and HVAC consume nearly 40% of total energy use. To lower their carbon footprint, buildings need to perform better than they had been doing, environmentally, economically and socially. Need in regional economies is for open innovations for self sustainable habitats with minimum pollution levels, optimized energy supplies, adequate security, scalable and adaptable to different situations. A well coordinated collaborative approach is favored to benefit from nexus between smart buildings, HVAC systems, IAQ, environment, power and water infrastructure. There are already collaborative activities, such as regional satellites, regional countries collaboration, National League of Cities, C40 Cities, 100 Resilient Cities, Future-Cities-Collaborative for shared benefits. In the future, the combination of power, water and communications infrastructure with real time data analytics and potential use of Artificial Intelligence will bring us high performance cities and buildings that can also adopt to global weather changes and emergencies.

This paper presents a conceptual framework for development of robust, and resilient infrastructure to suit the needs of regional economies along with fostering skills and job creation. A new approach of using open innovation with a geo-collaborative vision, knowledge, skills, experience and strategies for effective and sustainable delivery of services, goods and policies is discussed. Technologies need to be adapted to serve local needs commensurate with global experiences, to ensure that the right technologies and applications are employed in the right environments, tailored to the specific cultural, social and economic characteristics of the individual city.

INTRODUCTION

It is projected that by 2040, the global population will increase by almost 2 billion people, which is 25% more than today's total population. Continued rural to urban migration will also increase the number of people living in

Om Taneja is currently MD of Associated Consulting Engineers & past Associate Director of GSA Region 2

cities by 46%, pushing global population density up to 61 people per square kilometre from 49 today. (Global Alliance-2016,2017). These demographic patterns will result in increased number of passengers using public transport, more congestion on the road, need for more housing, office buildings, health care and educational buildings and greater demand for power, water, HVAC, sanitation, solid waste and communications services. Construction industry, which is the backbone of developing infrastructure, retrofitting existing buildings, constructing new buildings has an unenviable task of reducing the emission of greenhouse house gases by 50% of current levels (Global Alliance 2016, 2017) while using geo-collaborative efforts in financing, developing and operating efficient infrastructure. Such trends pose challenges but also offer opportunities; either way, they require an adequate response from the industry as a whole.

GROWTH OF BUILDING SECTOR

The global buildings sector is growing at unprecedented rates, and it will continue to do so. Over the next 40 years, the world is expected to build 230 billion square metres in new construction – adding the equivalent of Paris to the planet *every single week*. This rapid growth is not without consequences. While buildings sector energy intensity has improved in recent years, this has not been enough to offset rising energy demand. Buildings-related CO_2 emissions

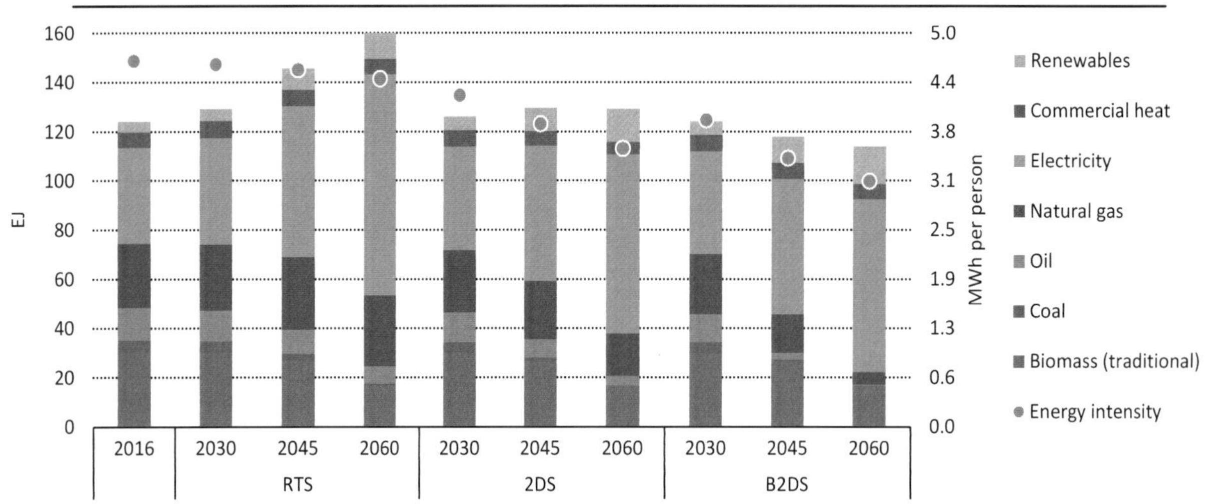

Source: IEA (2017), *Energy Technology Perspectives 2017*, IEA/OECD, Paris, www.iea.org/etp.

Climate Control -RTS - Reference Technology Scenario; 2DS - 2^0 C scenario; B2DS - Beyond 2^0 C Scenario

Figure 1 - Building Sector final energy consumption by scenario and fuel type - 2016-2060

have continued to rise by around 1% per year since 2010, and more than four million deaths each year are attributable to illness from household air pollution. Total spending on energy efficiency in the global buildings sector was less than 9% of the more than USD 4.6 trillion spent on buildings and construction (including renovations) globally. Fortunately, many opportunities exist to deploy energy-efficient and low-carbon solutions for buildings and construction. These solutions will necessitate greater effort to implement strategic policies and market incentives that change the pace and scale of actions in the global buildings market. UN & IEA recommend Net Zero Buildings ((Global Alliance 2016, 2017), which suggests many market incentives and regulatory framework to install high energy efficiency and low-carbon features. This requires prescriptive and performance based international energy codes and standards and a major shift in financing and investments tools that minimize risks and uncertainties. Energy efficient and low-carbon heating and cooling technology investments would reduce final energy demands in buildings by 25 per cent over current levels, the report said. Air conditioning performance is a crucial area to improve. Priority action is recommended for developing policies for energy efficiency and a greater use of renewables, for improving performance of energy efficiency of existing buildings, reduce embodied energy and decarbonize energy sources, achieve net-zero operating emissions, and upgrade adaptation of climate change risks - all of which demand a greater investment in an integrated and collaborative infrastructure amongst different agencies, owners, developers and end users. Smart technologies such as intelligent ventilation systems and connected LEDs allow people to interact with the building and for it to respond to real-time data sensors or occupants' commands. This means that lighting levels, humidity and temperature can be adapted to the preferences of the occupants while at the same time improving building energy performance.

KEY ROLE OF INFRASTRUCTURE & ITS GAPS IN SUPPORT OF THE EFFICIENT BUILDINGS

Infrastructure provides the foundation of our economy and quality of life. Investing in infrastructure, therefore, is essential to supporting healthy, vibrant communities. Many developed countries had invested heavily in infrastructure during and after World wars which is now aging and crumbling. The American Society of Civil Engineers estimates that by 2020, "aging and unreliable" infrastructure will cost American businesses $1.2 trillion (ASCE Foundation 2017) developing and regional economies, because of growth in population, rising wealth and migration from rural to urban areas need to build basic infrastructure for efficient and integrated transportation, wired and wireless communication, safety and security, power and water services to improve the quality of life of people. Furthermore, no matter how well infrastructure is planned, designed or built, there are unanticipated failures due to system and climate related reasons. Therefore, for safety, security and interrupted core services, the systems have to be designed to be either resilient to accommodate mishaps or be able to reconstitute in shortest possible time. Infrastructure needs to be well planned and efficiently operated (World Bank 100 Resilient Cities, Global Alliance 2016). This necessitates that infrastructure should not be over-designed to cover all future potential needs nor should it be undersized that additional supplemenal infrastructure needs to be built at high cost with associated complexities. There are innumerable examples in the industry of infrastrure being built for high future demands that do not occur or for technologies that become outdated and heavy operating burden is incurred. Infrastructure must be designed and billed to users like any other service.

A report from the Economic Policy Institute (Blair Hunter 2017) reiterated the link between infrastructure investment in the U.S., a boost in gross domestic product and the creation of hundreds of thousands of new jobs.[4] For regional economies, there is urgent need to increase GDP and skilled job creations and investing in infrastructure is a favored method. Most regional economies believe that infrastructure is a panacea for creating jobs, learning new skills, improve GDP and possibly become a global power in select areas. However, the reverse also can be true: Poor planning, undue delays in project completion, quality issues, competitive pricing and being able to charge back at reasonable and affordable rates are major setbacks that continue as challenges for the infrastructure sector. For example the buildings' efficiency with hurt due to deficiencies in infrastructure and dire funding needs of a growing economy India, as shown in Figure 2. Urgent need of US $646 Billion over the next five years is required to close the critical infrastructure gaps and India will need investments to the tune of around USD 4.5 trillion till 2040 to develop infrastructure to improve economic growth and community wellbeing (The Economic Times: Global Infrastructure Report).

Geo-Collabortion in infrastructure is the Need of the Day

In an era of Internet, instant communication and Social Media, world is well inter-connected. The international education, research, physical and cyber-security have benefited by mutual cooperation, sharing and providing deeper insights of benefits and pitfalls of different approaches. The development of infrastructure is ripe for a similar or greater collaboration and sharing of data whether in planning, construction, Maintenance, Financing and renewal of infrastructure. The eco-friendly cities would provide world-class facilities with 24-hour power supply and drinking water, mass rapid urban transportation, with bicycle and walking tracks, complete waste and water recycling systems

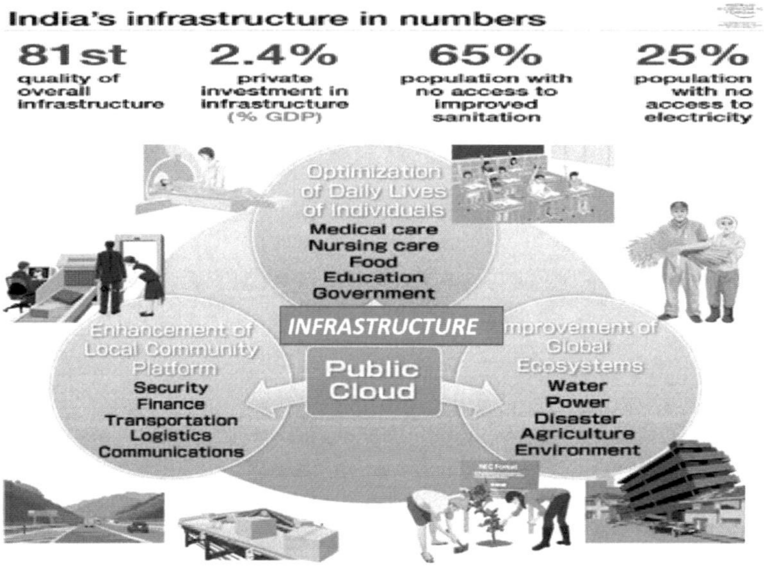

Figure 2 - India's Infrastructure in numbers

for smart grids - digitally managed systems to control energy consumption and smart metering & Accounting, Billing. A well coordinated collaborative approach is favored to benefit from nexus between high performance buildings, HVAC systems, IAQ, environment, power and water infrastructure. There are already collaborative activities, such as regional satellites, regional countries collaboration, National League of Cities, C40 Cities, 100 Resilient Cities, Future-Cities-Collaborative for shared benefits. In the future, the combination of power, water and communications infrastructure with real time data analytics and potential use of Artificial Intelligence will bring us high performance cities and buildings that can also adopt to global weather changes and emergencies.

As the world continues to urbanize, sustainable development challenges will be increasingly concentrated in

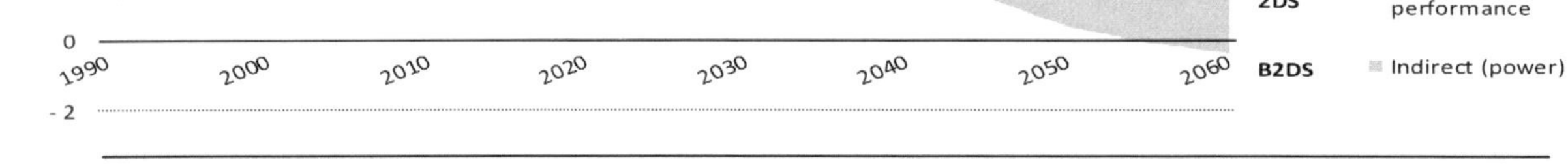

Notes: "Direct emissions reduction" represents a decrease in emissions from reductions in direct fossil fuel consumption in the buildings sector. "Envelope improvements" include measures (including deep energy renovations) that improve the energy intensity of the building envelope. "Technology choice" represents shifts from one type of technology and/or fuel to another (e.g. incandescent lamps to LEDs or gas boilers to electric heat pumps). "Technology performance" represents energy technology efficiency improvements (e.g. higher operational performance for heat pumps). "Indirect (power)" emissions reduction is from improved carbon intensities of power generation, where negative emissions are from carbon capture and storage (CCS) technologies.

Source: IEA (2017), *Energy Technology Perspectives 2017*, IEA/OECD, Paris, www.iea.org/etp

Figure 3 Key contributions to CO_2 emissions reduction in the global buildings sector to 2060

cities, particularly in the lower-middle-income countries where the pace of urbanization is fastest. Integrated policies to improve the lives of both urban and rural dwellers are needed. Many city governments have realised the potential of the collaborative economy to help address the various challenges they face. Beyond individual projects, the term 'sharing city' (Sharing Cities2016) has emerged to describe a more systematic approach to promoting the collaborative economy in cities.

There is a growing acknowledgement and actions promoting that, "institutionalal, economic, social and technical" need to be integrated with core city sub-systems. Technology adoption, while leveraging technologies for envelope improvement, bringing innovations to building HVAC, electrical, lighting and sanitation systems bring direct and indirect reductions in emissions, as projected in the Figure 3. Furthermore, greater resilience and emergency preparedness, resources, outreach and capacity building is needed to enable continuity-of-operations during normal and emergency conditions. 100 Resilient Cities was founded by the World Bank with an ambitious goal: to help cities worldwide build resilience to the growing social, economic, and physical challenges of the 21st century. 100 urban pioneers were selected to spread the resilience movement across the world. 100 is just the beginning. There's a great deal of work ahead to bring this movement to 1,000, then 10,000 cities. Another collaborative Group was Established 11 years ago, known as the C40 Cities, Climate Leadership Group. It includes more than 85 world cities, representing more than 650 million people and one-quarter of the global economy. The world's cities must invest around $375 billion in climate action and low carbon infrastructure over the next four years in order to avoid catastrophic global climate change, the C40 Cities Climate Leadership Group has warned. And in total world cities are likely to need more than $1 trillion in green investment up to 2050, the The World Bank has provided an initial funding of US $4.5 Billion towards these goals with extensive support anticipated from other countries and International Development banks. All risks due to aging and equipment failures, systemic risks due to cross-connectivity and political and regulatory risks needs to be considered and included in any project assessment (UNDP 2012, C40 Cities, Global Alliance 2016; World Bank 100 Resilient Cities 2013).

CHALLENGE- MAKING THE CASE FOR IMPROVED FACILITIES OPERATIONS & MANAGEMENT

80% of the total life cycle costs for infrastructure and buildings occur during operations, while construction cost is 20%. For maintaining asset value, the availability of qualified middle managers, engineers and trained service technicians has become more visible and critical as organizations grapple with energy, costs, climate change issues, finances and customer expectations. Broadly, current challenges in the O&M areas include: limited emphasis on potential efficiencies possible through tighter, better managed operating plans; limited investment in ongoing education and training; gaps in operator's technical expertise and ability; and an overall lack of consistency toward O&M management priorities to successfully harvest the benefits of green technologies and services. Industry has identified skill gaps in technology and operations of infrastructure systems, as shown in Figure 4, that needs to be bridged to plan, design, manage and operate infrastructure in an efficient manner. The required skills base should include Technical, project management, business, behavioral and enterprise Management (UNEP 2011; Global Alliance 2017, Pike Research 2013)

Figure 4: - Growing Skill Gaps of Building Employees in Managing High Performance Buildings

CHALLENGE- LEVERAGE TECHNOLOGY TO MAKE BUILDING SYSTEMS BETTER INTEGRATED, AN "INTELLIGENT"

Data is the gateway for managing infrastructure performance. Technology can help towards mining, analyzing and developing Key Performance Indicators (KPI) and help in reduction in emissions, as shown in Figure 3. Employing Computer Maintenance Management System (CMMS) to track maintenance and customer complaints improves service. . Level of technology introduction can vary from a simple digital controls and spreadsheets to enterprise networked infrastructure monitoring, and diagnostics systems. Integrated design provides evaluation of the cross-component impact of multiple systems as they are changed or upgraded, as shown in Figure 5. All infrastructure and buildings today have physical building systems such as HVAC, lighting, power, elevators, AV equipment co-existing with IT infrastructure. It is conventional for buildings to a have a "Building Controls Management System" and "IT infrastructure" (Adam Elkington 2013, Global alliance 2016).

Challenge to Move Towards The Internet Of Things (IoT)

Most of the modern equipment and devices are embedded with sensors with ability to communicate on different platforms. The resulting information networks promise to, improve business processes, and reduce costs and risks. In most organizations, information travels along familiar routes. Proprietary information is lodged in databases and analyzed in reports and then rises up the management chain. Information also originates externally—gathered from public sources, harvested from the Internet, or purchased from information suppliers. But the predictable pathways of information are changing: the physical world itself is becoming a type of information system. In what's called the Internet of Things, sensors and actuators embedded in physical objects—from roadways to pacemakers—are linked through wired and wireless networks, often using the same Internet Protocol (IP) that connects the Internet. These networks churn out huge volumes of data that flow to computers for analysis. When objects can both sense the environment and communicate, they become tools for understanding detailed operations. When operating environments are monitored continuously for hazards or when objects can take corrective action to avoid damage, risks and costs diminish. Advances in wireless networking technology and the greater standardization of communications protocols make it possible to collect data from these sensors almost anywhere at any time. Massive increases in storage and computing power, some of it available via cloud computing make number crunching possible at very large scale and at declining cost (Global Alliance 2017, Adam Elkington 2012, ACEEE 2017).

CHALLENGE – DEPLOY SPECIFIC TECHNOLOGY FEATURES FOR PERFORMANCE MEASUREMENTS & ACCOUNTABILITY

Use of Building Automation Systems to Monitor & Control Infrastructure Operations

Install Building Automation Systems that are adaptive to changes in pressure, flows, temperatures best suited to building use, occupancy and weather changes and provide surveillance and management of Energy Systems. These systems use a system of sensors, processors, drive units, communication and feedback systems to measure, control, track, perform analytics and report operating data for the infrastructure, store baseline or benchmark data. These systems can alert when resource use is trending upward or downward so that pertinent actions with regards to preventive and corrective measures can be performed.

Need for an integrated, forward thinking approach to NFRASTRUCTURE UPGRADES For CenTral Plant & other Energy Efficiency Measures

It is more effective to use an integrated approach when evaluating central plant systems upgrades and other energy efficiency upgrades and staging operations and maintenance to match the design intent over the life-cycle of a building. It makes all the key issues transparent and easier to address as the project progresses, instead of resolving them after-the-fact. Improving energy efficiency is not just limited to demand side reduction, but also include strategies for a) on-site generation and use of renewable energy sources, photovoltaic cells, geothermal systems or procurement of green power generated from clean renewable sources. How the utilities are procured, any demand penalties during peak cooling periods and on-site generation affect not only overall environmental effect, but also payback analysis, b.)using free energy such as day-lighting and outside air, c) recovering waste energy by using heat generated by equipment for heating and cooling and, d) increasing the efficiency of existing systems such as lighting, envelope and HVAC equipment (ASHRAE 99.1; ASHRAE 189.1 2016) . An integrated approach needs to be used for the entire life-cycle of a central plant, and its auxiliaries. Since most of the central plant operation will occur at off-design conditions, properly applied variable flow pumping can significantly reduce pumping energy costs.

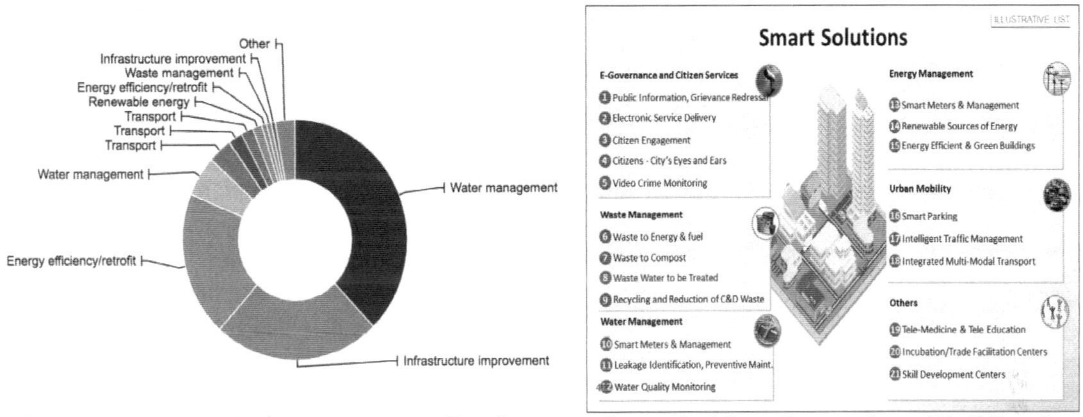

Figure 5. The Low Carbon Investment Landscape & Smart Buildings Integration of Building Systems

Raising Awareness of Occupants & Neighborhood about high performance measures and their Roles and Responsibilities

Tenants are not always made aware of how their use of spaces and equipment affect the energy use and environment and without their active participation and commitments, some of their actions inadvertently negate the benefits of high performance design elements. Encouraging tenants not to use space heaters and fans, and to turn off equipment during off hours, making sure to shutting off lights, power down everything - such as computers, monitors, copiers, kitchen equipment and task lights, can significantly reduce plug loads that affect electrical generation and HVAC systems infrastructure. In U.S. companies alone incur more than $1 billion a year is wasted on electricity for computer monitors that are left on when they shouldn't be. Such activities can change occupant behavior and can help save between 3.5% to 15.2% of energy, particularly with use of smart power strips that automatically shutdown where after an equipment is inactive for a prescribed period.

Life Cycle Environmental and Life Cycle Economic Cost Analysis of Infrastructure Systems

Owners are focusing not only on initial cost, but on the total life-cycle operating cost. The detailed analysis for each of the phases requires substantive analytical focus and is very much a function of local weather, how buildings are used and operational considerations. However, considering the life-span of infrastructure systems of 50 to 100 years, the initial cost of construction is only a fraction of life-cycle cost of operations, distribution, metering, accounting and billings. Therefore, the infrastructure of tomorrow should be required to develop measures for total cost burden, total environmental burden, total energy use burden, and thereafter take a ratio of Environmental Burden per unit cost or Energy use per unit cost. Depending on the priorities, one may be favored over the other – but both are important. The environmental impact assessment also needs to include embedded energy of materials (Seppo Junita et al 2004; Om Taneja, et al 2008). Stages in the life cycle of a material used for chiller plant and its auxiliaries are, Extraction, Manufacture, Transportation, Use as a construction or systems material, Use during occupancy, Operations and Maintenance, Recycle or Disposal

CONCLUSION

Total spending on energy efficiency in the global buildings sector was less than 9% of the more than USD 4.6

trillion spent on buildings and construction, including renovations globally. Fortunately, many opportunities exist to deploy energy-efficient and low-carbon solutions for buildings and construction that require smart infrastructure. By leveraging use of technology, infrastructure can be built to be more scaleable, resilient, and energy and cost-efficient under changing occupancy, weather or other situations. Skilled O&M staff shave between 5% to 20% of energy use by pursuing dynamic adjustments. Priority action is recommended for developing policies for energy efficiency and a greater use of renewables, for improving performance of energy efficiency of existing buildings, reduce embodied energy and decarbonize energy sources, achieve net-zero operating emissions, and upgrade adaptation of climate change risks - all of which demand a greater investment in an integrated and collaborative infrastructure. Most regional economies believe that infrastructure is a panacea for creating jobs, learning new skills, improve GDP and possibly become a global power in select areas. However, the reverse also can be true: Poor planning, undue delays in project completion, quality issues, are major setbacks that continue as challenges for the infrastructure sector. The industry needs pertinent analytical, prescriptive and performance based codes and standards for demand controls and effective operations. The eco-friendly cities would provide world-class facilities with 24-hour power supply and drinking water, mass rapid urban transportation, with bicycle and walking tracks, complete waste and water recycling, systems for smart grids - digitally managed systems to control energy consumption and smart metering & Accounting, Billing. A well coordinated collaborative approach is favored to benefit from nexus between high performance buildings, HVAC systems, IAQ, environment, power and water infrastructure. There are already collaborative activities, such as regional satellites, regional countries collaboration, National League of Cities, C40 Cities, 100 Resilient Cities, Future-Cities-Collaborative for shared benefits. In the future, the combination of power, water and communications infrastructure with real time data analytics and potential use of Artificial Intelligence will bring us high performance cities and buildings that can also adopt to global weather changes and emergencies.

REFERENCES

Adam Elkington ,"New Smart Building Technology to increase Federal Buildings Energy Efficiency", US GSA Press Release, May 14, 2012 (http://www.gsa.gov/portal/category/100731)

The American Council for an Energy Efficient Economy (ACEEE) 2017, "smart Buildings save Energy", A deeper dive into energy efficiency market

Architecture 2030 - Alex Wilson and Allyson Wendt, Architecture 2030, www.architecture2030.org

ASCE Foundation, 2017 infrastructure Report Card-Failure to Act", economic Development Research Group.

ASHRAE/IES Standard 90.1-2016 - Energy Code

ANSI/ASHRAE/USGBC/IES Standard 189.1-2014 - Sustainability Standard

C40 Cities Green Investment Report 2016

Federal Facilities Council of the National Research Council, "Core Competencies for Federal Facilities Asset

Global Alliance for Buildings & Construction, "what is the 4th Industial Revolution", Nicholas Davis, Head of Society and Innovation, Member of the Executive Committee, World Economic Forum Geneva, 2016 & December 2017

National League of Cities, "Shifting Perception of Collaborative Consumption", 2015

Om Taneja and Adam Hinge, "Life Cycle Assessment Measures of Energy Efficient Projects" International Conference on Improving Energy Efficiency in Commercial Buildings, IEECB08, Frankfurt, Germany.

Pike Research, Navigant Consulting, "Smart meters for commercial and industrial markets, May 2013.

Seppo Junita, "The Environmental Impact of an Office Building Throughout its Life Cycle, Dissertation for D. SC degree, Helsinki University of Technology, Espoo, Finland, 2004.

Sharing Cities, "Building Smart Cities Together", www.sharingcities.eu.

Todd Watson, "IBM's 2012 Tech Trends Report: Skills, Skills, and More Skills!" Internet Evolution, Dec 2012

United Nations Development Program (UNDP), "Building Resilience: the importance of prioritizing disaster risk reduction, Hopkins Lecture, University of Canterbury, Aurora Centre, Burnside High School, Christchurch, August 15, 2012

United Nations Environment Programme (UNEP), "Buildings & Climate Change – Status, Challenges & Opportunities", funded by Asia Pro Eco Programme of the European Union, 2007.

World Bank, 100 Resilient Cities 2017

Mathematical Modelling of Hybrid Cooling Vest Integrated with Bioheat Model for Assessing Cooling Effect on Humans in Hot Conditions

Ragheb Raad **Mariam Itani** **Nesreen Ghaddar, PhD** *Member ASHRAE* **Kamel Ghali, PhD** *Member ASHRAE*

ABSTRACT

Workers in hot environments are exposed to increased irritability and to excessive sweating. They are also very likely to lose some of their concentration. This environment will therefore reduce their ability to do physical work and as a result, affect their performance and efficiency. Not to mention that physical work burns additional energy creating extra metabolic heat that adds to the tally. Cooling methods such as cooling vests are therefore necessary to increase the performance and output of the workers. The main objective of this study is to develop a computational model for an evaporative cooling vest and integrate it with a validated bioheat model. Such a vest improves the thermal comfort of the workers, thus allowing them to efficiently perform outdoor physical activities at elevated air temperatures. The vest's model is based on the formulation of the direct evaporative cooling technique. The borders of the channel are respresented by an outer insulated layer and a highly hygroscopic material for the inner layer adjacent to the skin. Ventilation fans located at the lower back are used to blow ambient air into the vest. A mathematical model of the cooling technique is developed and it is coupled with the bioheat model which allows studying the workers' thermal response to the proposed vest. These computations would predict the variation of the torso and mean skin temperature as well as sweat rate and overall thermal comfort during outdoor physical activities and under specific ambient conditions. In this study, the model is tested under a moderate activity level and it is then compared to the case where no cooling is present. The results showed that the proposed system would decrease the torso and mean skin temperatures as well as improve overall thermal comfort of subjects wearing the vest.

INTRODUCTION

The human body has its own mechanism in regulating its temperature, such as sweating and vasodilation. Cooling is achieved through these mechanisms, and the core body temperature is kept within an acceptable range to allow normal functioning of the internal organs. However, in extremely hot environments, it can usually be difficult for the body to maintain a stable core temperature without the aid of external cooling mechanisms. In such environments, outdoor workers may be at risk of heat stress and strains which could lead to health problems (Yi and Chan 2017). The direct consequences of uncompensated heat stress include the loss of efficiency and productivity. Prolonged heat stress exposures lead to long-term effects which are noted by a progressive loss of performance capability, disorganized nervous system activity, heat illnesses and heat injuries. Some of the illnesses caused by heat stress include heat cramps, heat exhaustion, and heat strokes. Similar environments can also lead to loss of concentration, increased irritability and to excessive sweating, hence, decreasing these workers' productivity and efficiency at work (Miller and Bates 2007). Those conditions have shown to increase the workers fatigue level,

Ragheb Raad is a graduate student in the Department of Mechanical Engineering, American University of Beirut, Beirut, Lebanon. **Mariam Itani** is a graduate student in the Department of Mechanical Engineering, American University of Beirut, Beirut, Lebanon. **Nesreen Ghaddar** is a professor in the Department of Mechanical Engineering, American University of Beirut, Beirut, Lebanon. **Kamel Ghali** is a professor and the chairperson of the Department of Mechanical Engineering, American University of Beirut, Beirut, Lebanon.

therefore, forcing them to increase their rest time and to decrease their working time, subsequently, resulting in a significant decrease in their output (Li et al. 2016). Accordingly, cooling vests capable of cooling the individual microclimate around a worker are considered one of the ways that will enhance the workers' resistance to such environments. These vests can be worn while performing the job, thus increasing the subjects' productivity and performance. Many different approaches and methods have been used to design cooling vests. Those approaches include fluid cooling garments (FCGs), phase-change materials (PCMs), and evaporative devices (Yang et al. 2012).

Fluid cooling garments vests are based on the conductive heat exchange between the skin and a circulating cooling fluid (water, air or refrigerant). In the process of microclimate cooling, pressure is the force driving the circulation (Yang et al. 2012). Therefore, thermodynamically speaking, the effective cooling decreases when the inlet temperature of the cooling fluid increases. Similarly, when the flow rate increases, the effective cooling decreases as well, due to the low residence time of fluid. The flow is not present for an enough time in order for the heat exchange to occur (Kayacan and Kurbak 2010). The optimization of the vest performance is achieved by increasing the thermal comfort, maximizing the cooling's efficiency, and minimizing the coolant and power requirements (Flouris and Cheung 2006). In these vests, to guarantee conduction, the selection of optimal inner highly conductive fabric layers is necessary (Cao et al. 2006). However, the configuration of FCGs forces high weights which reduces the energy efficiency of the system (Flouris and Cheung 2006).

On the other hand, PCM vests are equipped with cold packs of phase changing material, which can be any of the following: ice, salts, metals or paraffin waxes. The vest relies on the melting of the packs and the latent heat of sublimation. When changing phases during melting and freezing processes, these materials are capable of storing and releasing large amounts of energy making them an effective way to store latent heat in cooling applications (Mishra et al. 2015). The amount of energy that will be stored depends on the enthalpy of the material in the zone of the phase change (Sarkar and Kothari 2014). PCM cooling vests are simple, functional, and relatively cheap; and they do not require a power supply system (Kenny et al. 2011). This makes them practical for workers in hot environments when physical activity is only for limited periods of time, due to the high carried weight to duration ratio. In PCM cooling, many parameters are studied including the optimal positions of the packs relative to the body's sweat-producing zones, the total insulation, and the amount of heat and moisture transport. This allows for a wide range to reach optimization (Reinertsen et al. 2008). However, PCMs need a cool place to solidify after melting and require a long time to regenerate before they are ready to use. Their added weight hinders moisture transport and they also tend to make the vest bulky affecting the mobility of the wearer, and are sometimes flammable if made from paraffin. Furthermore, the cooling effect of PCMs ends when the whole phase change material melts and their thermal conductivity is very low. This forces extra manufacturing steps and increases in the vest's cost (Mondal 2008).

The evaporative cooling vests consist of an evaporative microenvironment created between the layers of the cooling vest and the skin interacting with these layers. The sweat produced by the human skin is adsorbed into highly hygroscopic materials. Then, through channels, circulating air, extracted from the environment and supplied via ventilation fans, allows for sweat evaporation and thus provides cooling for the human body. This system takes advantage of the relatively significant latent heat of evaporation of water which is seven times larger than the latent heat of melting ice (Johnson 2013) and at least ten times larger than that of common phase-change materials (Yazdi and Sheikhzadeh 2014). This highlights the increased cooling capacity of evaporative cooling vests with respect to PCMs or FCGs. Moreover, evaporative cooling vests can operate for relatively long durations and do not require the use of expensive coolants.

In this study, an evaporative cooling technique is integrated with a bioheat model in order to check its effect on human comfort during activity and compare it to the case where no cooling is present. The integrated model will predict the torso and mean skin temperatures and the temperature and humidity of the microclimate air adjacent to the back as well as overall thermal comfort.

DESCRIPTION OF THE PROPOSED SYSTEM

Evaporative cooling vests have been applied mainly on the human torso (Chinevere et al. 2008; Sun and Jasper 2015; Xu and Gonzalez 2011) since the areas of the torso, like the lower back, the upper back and the abdomen contribute the most to the thermal comfort of the human body, much more than the limbs and extremities as the former possess more sensitive thermal sensors (Arens et al. 2006; Guéritée et al. 2015; Nakamura et al. 2008; Nakamura et al. 2013). The proposed evaporative cooling method will subsequently focus on thermally cooling the back of the human body.

The evaporative cooling vest has an outer insulative layer in order to avoid the heat exchange between the microclimate air and the hot ambient air. In addition, the vest has an inner layer that will be in contact with the skin of the human torso. This inner layer is made of a hygroscopic sheet in order to absorb the sweat that the back secretes. A side view of the skin layer, the hygroscopic sheet and the outer vest layer is shown in Figure 1(a). Ventilation fans located at the lower back extract ambient air of certain temperature and humidity and supply it to the vest. The supplied air flows through a rectangular channel of thickness δ, as shown in Figure 1(a), where sweat evaporation occurs. Figure 1(b) shows a back view of the vest including the ventilation fans. The length, L, and width, W, of the vest are equivalent to those of the back segment.

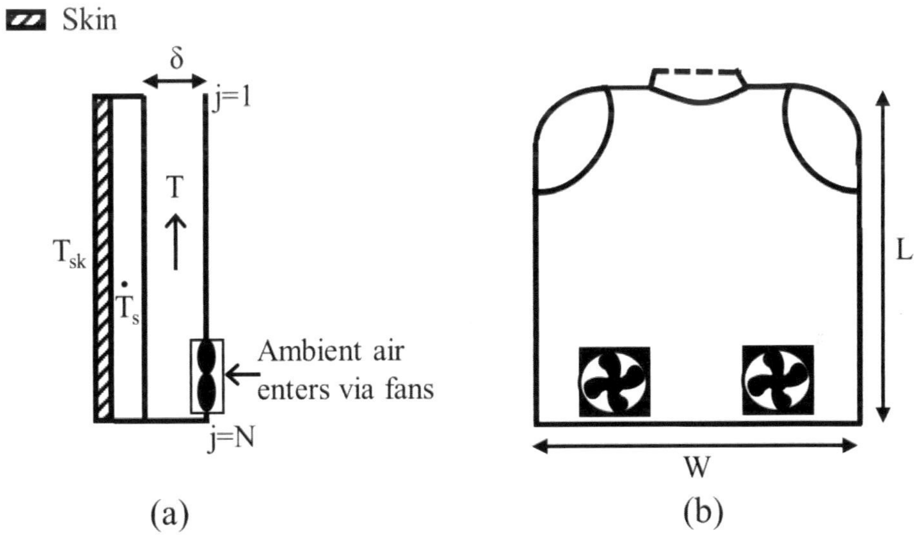

Figure 1 Schematic of the cooling vest showing (a) a side view of the skin layer, the hygroscopic sheet and the channel model and (b) a back view of the vest including the ventilation fans.

MATHEMATICAL METHODOLOGY

A computational model of the evaporative cooling vest is developed in this study and then integared with the validated bioheat model of Karaki et al. (2013) and the comfort model of Zhang et al. (2004). The integrated model will be used to predict human thermal responses and overall comfort as well as the variation of the temperature and humidity ratio of the microclimate air in the vest channel.

Evaporative Cooling Vest Model

The evaporative cooling vest model assumes that the air flow inside the channel, shown in Figure 1(a), is laminar, since the velocity of the microclimate air is below 1.5m/s (4.92 ft/s), which results with a Reynolds number of about 1800. The outside layer of the vest is insulated in order to avoid the heat exchange between the inside air and the outside air. This would therefore maximize the capacity of moisture extraction by the inside air. The hygroscopic sheet, which is chosen to be a cotton sheet, is assumed to be of a relatively high thermal conductivity, leading to small temperature variations across its length. This is why a constant temperature boundary condition is assumed for the microclimate air. Therefore, the Nusselt number with such boundary conditions (constant temperature at inner layer and insulation at the outer layer) is equal to 4.86. Since this is a process of water evaporation into air, the Lewis number is assumed to be 1. Therefore, the mass transfer coefficient h_m is found as follows:

$$h_m = \frac{h_w}{\rho C_p} \tag{1}$$

By psychrometry, the enthalpy h of moist air is found as follows:

$$h = Cp_a * T + h_{fg} * X_w \tag{2}$$

where T is the temperature of air in degrees Celsius and X_w is the humidity ratio (ASHRAE 2005). As ambient air is introduced into the channel, the humidity ratio will increase. So the change in enthalpy can be calculated using the following relationship:

$$\frac{\partial h}{\partial x} = Cp_a * \frac{\partial T}{\partial x} + h_{fg} * \frac{\partial X_w}{\partial x} \tag{3}$$

where $\frac{\partial X_w}{\partial x}$ is such that the net flow of moisture in the channel is equal to the moisture exchange with the wet sheet at the back and represented as follows:

$$\frac{\partial X_w}{\partial x} = \frac{h_m}{v\delta}(X_w^* - X_w) \tag{4}$$

where X_w^* is the humidity ratio of saturated air in the channel, such that:

$$X_w^* = 0.62198 \frac{P_{sat}}{P - P_{sat}} \tag{5}$$

where P is the atmospheric pressure and P_{sat} calculated at the the air temperature:

$$P_{sat}(T) = e^{\left(\frac{-5.8002206\times10^3}{T + 1.3914993 - 4.8640239\times10^{-2}\cdot T + 4.1764768\times10^{-5}\cdot T^2 - 1.4452093\times10^{-8}\cdot T^3 + 6.5459673\cdot\log(T)}\right)} \tag{6}$$

Equations (3) and (4) then lead to the following heat balance inside the channel:

$$\rho C_p v\delta \frac{\partial T}{\partial x} = h_w (T_s - T) \tag{7}$$

An energy balance is then performed on the hygroscopic sheet by including the conduction of the sheet with the skin, the convection with the air and the mass transfer term, as follows:

$$\frac{k_e}{\delta_e}(T_{sk} - T_s) + h_w(T - T_s) + \rho h_m h_{fg}(X_w - X_w^*) = 0 \qquad (8)$$

where T_{sk} is the temperature of the skin. k_e and δ_e are the effective back layer thermal conductivity and thickness respectively, such that:

$$\delta_e = \delta_{sk} + \delta_s \qquad (9)$$

$$k_e = \frac{\delta_e}{\frac{\delta_{sk}}{k_{sk}} + \frac{\delta_s}{k_s}} \qquad (10)$$

Discretized Equations of Evaporative Cooling Vest Model

The equations that represent the heat and mass transfer processes inside the vest are then discretized using the finite difference method over N elements, as shown in Figure 1(a). The temperature of the hygroscopic sheet was assumed to be slightly variable, therefore, the balance on the sheet was solved at each element.

For the air heat balance, the discretization process yields the following equations:

$$\rho C_p v\delta \frac{T_{(j)} - T_{(j-1)}}{\Delta z} = h_w(T_{S(j)} - T_{(j)}) \qquad (11)$$

$$T_{(j)} = \frac{h_w \Delta z T_{s(j)} + \rho C_p v\delta T_{(j-1)}}{h_w \Delta z + \rho C_p v\delta} \qquad (12)$$

with j varying from 1 to N.

The discretized form of the mass balance in the channel is:

$$\frac{X_{w(j)} - X_{w(j+1)}}{\Delta z} = \frac{h_m}{v\delta}(X^*_{w(j)} - X_{w(j)}) \qquad (13)$$

$$X_{w(j)} = \frac{\Delta z h_m X^*_{w(j)} + v\delta X_{w(j+1)}}{\Delta z h_m + v\delta} \qquad (14)$$

The temperature of the hygroscopic sheet is found as follows:

$$T_{s(j)} = \frac{\frac{k_e}{\delta_e} T_{sk} + h_w T_{(j)} + \rho h_m h_{fg}(X_{w(j)} - X^*_{w(j)})}{\frac{k_e}{\delta_e} + h_w} \qquad (15)$$

Integration with the Bioheat and Comfort Models

The transient multi-node bioheat model of Karaki et al. (2013) can predict segmental and mean skin temperatures, core body temperature as well as sweat rate under different environmental conditions and metabolic rates or activity levels. The microclimate air conditions of the back segment, predicted by the evaporative cooling vest model, will be considered as the ambient conditions of the back segment to predict the corresponding skin temperature and sweat rate at each time step. Moreover, it is important to mention that the maximum amount of moisture that the microclimate air can extract from the cotton sheet is equal to the sweat rate at that time step. The sweat rate therefore acts as a constraint that is added to the model. Overall thermal comfort is predicted using the model of Zhang et al. (2004) which is based on a scale from -4 (very uncomfortable) to +4 (very comfortable). The integration between the models permits assessing the effect of the proposed system on human thermal responses and overall comfort.

In order to utilize the integrated model, different parameters need to be provided as inputs to be able to solve for the desired outputs. The input parameters include the metabolic rate of the human, the ambient conditions (temperature and relative humidity), the initial thermal state of the human body, in addition to the evaporative cooling vest dimensions and microclimate air mass flow rate. The evaporative cooling vest model uses mass and energy balance equations as well as the back skin temperature and sweat rate to predict at the new time step the variation of the microclimate air temperature and humidity ratio in the channel. The channel is discretized with a step equivalent to 0.002m (0.079in). Then, the bioheat model updates the skin temperatures and sweat rate by using the new microclimate air conditions as the ambient conditions for the back segment. Once the segmental and mean skin temperatures and core temperatures as well as their rate of change are found, the overall thermal comfort can be predicted. A time-step of 0.02s is used to solve for the transient variation of the different parametrs.

CASE STUDIES

Two case studies were adopted in this study, where a human was assumed to be perform activity with a metabolic rate of 4 MET, which is equivalent to cycling at low velocities, for a duration of 30 minutes. The ambient conditions were assumed to be at 40°C (104°F) and 25% relative humidity. The first case study considers a human where no cooling vest is worn and the second case study represent a human performing activity while wearing the evaporative cooling vest. The dimensions and properties of the evaporative cooling vest layers are provided in Table 1. The evaporative cooling vest resulted with a microclimate air velocity in the channel of 0.50m/s (1.64 ft/s) using a channel thickness of 0.01m (0.39 in).

Table 1. Parameter Values Used in the Computations

Symbol	Parameter	Value	Unit
L	Length of the channel	0.60 (1.97)	m (ft)
W	Width of the channel	0.45 (1.48)	m (ft)
h_{fg}	Latent heat of evaporation of water	2501 (1075)	kJ/kg (btu/lb)
δ_s	Thickness of cotton sheet	1.00 (0.04)	mm (in)
k_s	Thermal conductivity of cotton sheet	0.46 (3.19)	W/m.K (BTU.in /hr.ft^2.F)
δ_{sk}	Thickness of the skin	1.00 (0.04)	mm (in)
k_{sk}	Thermal conductivity of the skin	0.3 (2.1)	W/m.K (BTU.in /hr.ft^2.F)
k_w	Thermal conductivity of wet air	0.027 (0.189)	W/m.K (BTU.in/hr.ft^2.F)

RESULTS AND DISCUSSION

The torso skin temperature variation with time is plotted in Figure 2 for the two simulated cases; with and without the evaporative cooling vest. As seen from the figure, the torso skin temperature in all the conditions started from the same initial value of 36.33°C (97.39°F). After that, the torso skin temperature started to increase due to the heat generated under the moderate activity level in the no cooling case. However, the torso skin temperature decreased with time when using the cooling vest. At the end of the simulation time, the torso skin temperature in the no cooling case reached a maximum value of 36.60°C (97.88°F), while that in the cooling case reached a minimum value of 35.81°C (96.46°F). Thus, the evaporative cooling vest reduced the torso skin temperature by a maximum of 0.79°C (1.42°F) from the no cooling case.

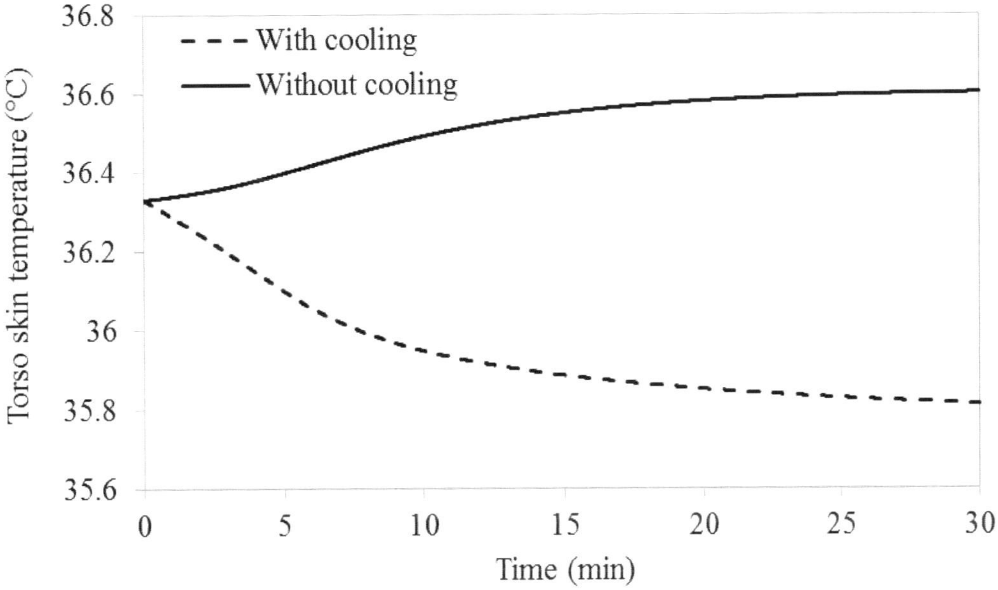

Figure 2 Plot showing the variation of the torso skin temperature as a function of time for the case where the evaporative cooling technique is present and for the case where it is not.

The variations of the temperature and humidity ratio of the microclimate air adjacent to the back at the end of the 30-min exercise period, corresponding to the cooling vest case, are shown in Figure 3. It can be seen that the inlet conditions of the channel are the same as the ambient air conditions of 40°C (104°F) and 0.0116 kg/kg (0.0116 lb/lb). As the air flow in the channel, it absorbs water from the hygroscopic cotton sheet, which in turn absorbed the sweat secreted by the back segment. Thus, the humidity ratio of the microclimate air increases reaching a maximum value of 0.0158 kg/kg (0.0158 lb/lb), before leaving the channel at the top of the back segment. Moreover, the temperature of the microclimate air decreases, as it flows in the channel, reaching a minimum value of 36.48°C (97.66°F).

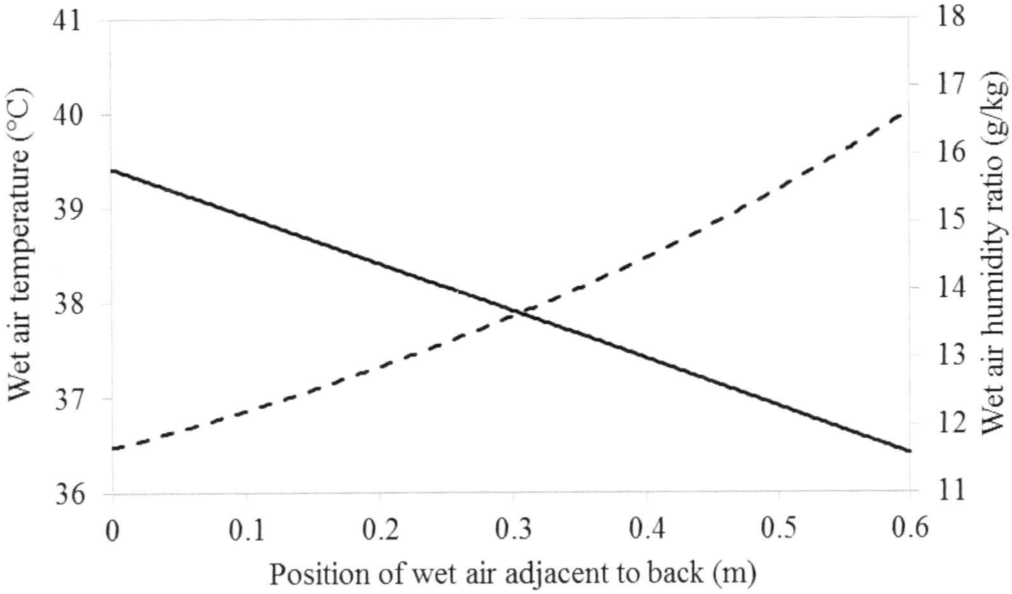

Figure 3 Plot showing the variation of the temperature and humidity ratio of the microclimate air adjacent to the back in the case of cooling, at the end of the 30-min exercise period.

The time-averaged mean skin temperature and overall thermal comfort for the no cooling and cooling cases are shown in Table 2. It can be noticed that the evaporative cooling vets was able to reduce the mean skin temperature by 0.23°C (0.41°F) on average. This reduction in mean skin temperature allowed for improved overall thermal comfort levels. The time-averaged overall thermal comfort was improved by about 8.5%, when the human subject is cooled via the evaporative cooling vest. It should be noted that the change in body weight due to sweating, during the 30-min exercise duration and at the simulated conditions, was less than 1%, indicating that no considerable dehydration was present to reduce activity performance and productivity (Casa et al., 2000; Stand 2009).

Table 2. Time-averaged mean skin temperature and overall thermal comfort

Case	Time-averaged mean skin temperature in °C (°F)	Time-averaged overall thermal comfort
Without cooling	36.63 (97.93)	-0.894
With cooling	36.40 (97.52)	-0.818

CONCLUSION

In this study, an evaporative cooling vest was integrated with a bioheat and comfort model to predict its cooling performance and compare it with the case where no cooling is undergone. The model was used to predict the torso and mean skin temperatures and temperature and humidity ratio of the microclimate air adjacent to the back. The results showed that such the proposed evaporative cooling vest improved the overall thermal comfort due to the decrease in the torso and mean skin temperatures. In future work, parametric studies need to be done on the geometry of the system, on the inlet flow rate, and on the ambient conditions in order to reach best design of the cooling vest and suggest which design can be used under which ambient conditions.

NOMENCLATURE

ρ = density of air
C_p = constant volume specific heat
v = velocity of air in the channel
δ = thickness of the channel
h_w = convection heat transfer coefficient of wet air
Δz = length of discretization element

Subscripts

s = cotton sheet
sk = skin
w = wet air

REFERENCES

Arens, E., Zhang, H., and C. Huizenga. 2006. Partial- and whole-body thermal sensation and comfort - part II: Non-uniform environmental conditions. Journal of Thermal Biology 31 (1-2): 60-6.

ASHRAE. 2005. ASHRAE Handbook—Fundamentals. Atlanta: ASHRAE.

Cao, H., Branson, D.H., Peksoz, S., Nam, J., and C.A. Farr. 2006. Fabric selection for a liquid cooling garment. Textile Research Journal 76 (7): 587-95.

Casa, D.J., Armstrong, L.E., Hillman, S.K., Montain, S.J., Reiff, R.V., Rich, B.S., Roberts, W.O. and J.A. Stone. 2000. National athletic trainers' association position statement: Fluid replacement for athletes. Journal of Athletic Training 35 (2): 212-24.

Chinevere, T. D., Cadarette, B.S., Goodman, D.A., Ely, B.R., Cheuvront, S.R., and M.N. Sawka. 2008. Efficacy of body ventilation system for reducing strain in warm and hot climates. European Journal of Applied Physiology 103 (3): 307-14.

Flouris, A. D., and S.S. Cheung. 2006. Design and control optimization of microclimate liquid cooling systems underneath protective clothing. Annals of Biomedical Engineering 34 (3): 359-72.

Guéritée, J., House, J.R., Redortier, B., and M.J. Tipton. 2015;2014;. The determinants of thermal comfort in cool water. Scandinavian Journal of Medicine & Science in Sports 25 (5): e459-71.

Johnson, J. K. 2013. Evaluation of four portable cooling vests for workers wearing gas extraction coveralls in hot environments.ProQuest Dissertations Publishing.

Karaki, W., Ghaddar, N., Ghali, K., Kuklane, K., Holmer, I., and L. Vanggaard. 2013. Human thermal response with improved AVA modeling of the digits. International Journal of Thermal Sciences 67 : 41-52.

Kayacan, Ö., and A. Kurbak. 2010. Effect of garment design on liquid cooling garments. Textile Research Journal 80 (14): 1442-55.

Kenny, G.P., Schissler, A.R., Stapleton, J., Piamonte, M., Binder, K., Lynn, A., Lan, C.Q., and S.G. Hardcastle. 2011. Ice cooling vest on tolerance for exercise under uncompensable heat stress. Journal of Occupational and Environmental Hygiene 8 (8): 484-91.

Li, X., Chow, K., Zhu, Y., and Y. Lin. 2016. Evaluating the impacts of high-temperature outdoor working environments on construction labor productivity in china: A case study of rebar workers. Building and Environment 95 : 42-52.

Miller, V. S., and G.P. Bates. 2007. The thermal work limit is a simple reliable heat index for the protection of workers in thermally stressful environments. Annals of Occupational Hygiene 51 (6): 553-61.

Mishra, A., Shukla, A., and A. Sharma. 2015. Latent heat storage through phase change materials. Resonance 20 (6): 532-41.

Mondal, S. 2008. Phase change materials for smart textiles – an overview. Applied Thermal Engineering 28 (11): 1536-50.

Nakamura, M., Yoda, T., Crawshaw, L.I., Kasuga, M., Uchida, Y., Tokizawa, K., Nagashima, K., and K. Kanosue. 2013. Relative importance of different surface regions for thermal comfort in humans. European Journal of Applied Physiology 113 (1): 63-76.

Nakamura, M., Yoda, T., Crawshaw, L.I., Yasuhara, S., Saito, Y., Kasuga, M., Nagashima, K., and K. Kanosue. 2008. Regional differences in temperature sensation and thermal comfort in humans. Journal of Applied Physiology 105 (6): 1897-906.

Reinertsen, R.E., Færevik, H., Holbø, K., Nesbakken, R., Reitan, J., Røyset, A. and M., Suong Le Thi. 2008. Optimizing the performance of phase-change materials in personal protective clothing systems. International Journal of Occupational Safety and Ergonomics 14 (1): 43-53.

Sarkar, S., and V.K. Kothari. 2014. Coaling garments-A review. Indian Journal of Fibre & Textile Research 39 (4): 450-8.

Stand, A.P. 2009. Exercise and fluid replacement. Medicine and science in sports and exercise, 39 (2): 377-390.

Sun, Y., and W.J. Jasper. 2015. Numerical modeling of heat and moisture transfer in a wearable convective cooling system for human comfort. Building and Environment 93 (2): 50-62.

Xu, X., and J. Gonzalez. 2011. Determination of the cooling capacity for body ventilation system. European Journal of Applied Physiology 111 (12): 3155-60.

Yang, Y., Stapleton, J., Diagne, B., Kenny, G., and C.Q. Lan. 2012. Man-portable personal cooling garment based on vacuum desiccant cooling. Applied Thermal Engineering 47: 18-24.

Yazdi, M.M., and M. Sheikhzadeh. 2014. Personal cooling garments: A review. The Journal of the Textile Institute 105 (12): 1231-50.

Yi, W., and A. Chan. 2017. Effects of heat stress on construction labor productivity in Hong Kong: A case study of rebar workers. International Journal of Environmental Research and Public Health 14 (9).

Zhang, H., Huizenga, C., Arens, E., and D. Wang. 2004. Thermal sensation and comfort in transient non-uniform thermal environments. European Journal of Applied Physiology 92 (6): 728-33.

Effect of Inter-segmental Ventilation on the Segmental Heat Losses by Means of Electric Circuit Analogy

Nagham Ismail, PhD

Nesreen Ghaddar, PhD
Member ASHRAE

Kamel Ghali, PhD
Member ASHRAE

ABSTRACT

The flow characteristics and heat transfer in the presence of inter-connection between clothed segments are studied for the walking clothed upper human body part in cross wind. The aim of the work is to develop a computationally effective model to estimate the clothing ventilation and heat transfer of the inter-connected upper human body part. The model is based on the analogy between (i) the air flow in the microclimate air layer and an electric circuit composed of resistance and inductance elements, and between (ii) the heat losses from the human body and the thermal electric circuit resistance. The two developed model are coupled allowing the estimation of segmental heat losses generated at constant skin temperature.

The coupled ventilation circuit and heat circuit models predicted both the segmental clothing ventilation rate and heat loss for different walking speeds and external wind. The developed model of the segmental ventilation and heat transfer from the clothed human segment were validated by conducting experiments on a walking thermal manikin at constant skin temperature using tracer gas method placed in an environmental chamber under windy conditions. The segmental ventilation and heat losses measured from the thermal manikin experiment were compared to the results of the ventilation-heat circuit model. Good agreement was observed between the ventilation-heat circuit model and the experimental data with a maximum error of 16 % which is still within the experiment standard deviation range. Results show that the segmental heat losses for the arm and the trunk attain its minimum for the lowest relative velocity and its maximum for the highest relative velocity. Evidently, when the relative velocity increases, the trunk and arm ventilation increase leading to an increase in segmental heat losses. The main reason is that the penetrated air is coming from the ambient condition. It is slightly heated in the fabric nodes and enter to the microclimate air layer leading to significant heat losses from the heated skin.

INTRODUCTION

Energy exchange between the human body and its thermal environment has been the focus of many researches because of its influence in human thermal comfort (Dear et al. 1997). However, estimating the energy exchange was always challenging since it is significantly influenced by the clothing types and characteristics (Gomez et al. 2006). Indeed, the flow of heat is evidently related to the flow of air from the human skin to the environment since the diffusion in fabric is basically neglected. Therefore, the ventilation induced by relative wind or walking condition resulting in air penetration through the fabric to microclimate air trapped between skin and fabric is the most influencing parameter in heat exchange (Wan et al. 2007, Ghaddar et al. 2008). The microclimate ventilation between a specific garment and the environment includes three mechanisms: air exchange through the fabric with the environment, air exchange through garment apertures with the environment, and air exchange between local body parts' microclimates (Ke et al. 2014). The air exchange through clothing has been thoroughly studied in the literature (Birnbaum et al. 1978, Lotens 1993) while the air exchange through open apertures has been modeled mathematically and estimated empirically by recent researches (Ghaddar et al. 2010[a], Ghaddar et al. 2010[b]).

Nagham Ismail is a lecturer at Lebanese University, Department of Industrial Engineering, Saida,

Lebanon. Nesreen Ghaddar and Kamel Ghali are professors at the American University of Beirut, Beirut, Lebanon

The air exchange between local body parts has been rarely investigated. The main reason is that adding the inter-connection will add complexity to the ventilation model. Nevertheless, accounting for the inter-segmental ventilation is important since it constitutes a real physical boundary condition in the prediction of segmental ventilation (Ismail et al. 2016[a], Ismail et al. 2016[b]). The accurate estimation of segmental ventilation rate will lead to accurate estimation of segmental heat losses as ventilation and heat losses are related. The importance of the accurate estimation of segmental heat losses between the human body and the environment is highly considered at the upper human body since it is considered as the most influential clothed segment in bringing thermal comfort (Ghaddar et al. 2010[a]). Furthermore, knowing how much is the air exchange between segments and its direction could be also used in the design of clothing. Particularly, the human body zone at the interconnection where the inter-segmental ventilation occurs possesses a highly humid microclimate air layer especially in the Axilla zone (under arm pit zone). This will generate an accumulation of sweat and a local discomfort. Thus, the ventilation of this zone is highly needed to improve the heat lost from the warm skin. Clothing companies have noticed the critical need of the ventilation in this specific zone; therefore, they tried to provide ventilation channels under the arm pits and promoting these ventilation channels in improving sensation in warm temperatures or during high physical activity (Cotter et al. 2010). From these examples, it is clear that it is important to study the inter-segmental ventilation that can lead to the migration of the accumulated sweat in the Axilla zone and hence enhance the local thermal comfort. Another important reason to study the inter-segmental ventilation is to predict the contaminant distribution migration that finds its way inside the microclimate air layer of the trunk. Such particles could deposit on the chest skin and produce dermal contamination.

Although some research studies attempt to estimate the inter-segmental ventilation, however, this was accompanied with complexity that restrict the solution to the case of sedentary human body (Ismail et al. 2016[a], Ismail et al. 2016[b]). Extending this model for the walking condition will require transient 3-D CFD simulations. Therefore, it is of interest to use the simplified model developed by Ismail et al. (Ismail et al. 2016[c]) in estimating the segmental and inter- segmental ventilation. The simplified model is based on an electric circuit analogy that transforms the 2-D cylinder model of the conventional ventilation model to a lumped model using electric circuit analogy for the air flow composed of a resistance-inductance circuit. The electric resistance described the resistance to the air flow while the electric inductance described the resistance to the change in air flow rate when oscillation occurs in walking condition. This simplified model will be easily connected to a thermal resistance network to account for the associated segmental heat losses (Ismail et al. 2017).

Therefore, the aim of this study is to develop a simplified and comprehensive model that predicts accurate clothing ventilation and segmental heat losses based on the analogy between the air flow in the microclimate air layer and an electric circuit. The model takes into account the inter-connection between the segments of the clothed human upper part driven by difference of pressure in the microclimate air of the trunk and the upper arms. In order to accurately predict the segmental heat losses, the segmental ventilation rates are determined while taking into account the inter- connection and are used to correct for the clothing dry resistances in a thermal resistance network model. The coupled ventilation-heat electric circuit model will be validated experimentally by estimating the segmental heat losses based on a constant skin temperature mode of the walking clothed manikin system.

METHODOLOGY AND PHYSICAL CONFIGURATION

The research methodology consists of developing a simplified approach based on thermal resistance network model that connects all the human body parts. Since the segmental heat losses prediction is basically related to the segmental ventilation, a simplified ventilation model based on an electric circuit analogy is developed. The simplified ventilation model is taking into account the inter-segmental ventilation between segments. Therefore, the methodology begins by describing the physical configuration of the upper human body part. Then, the heat resistance network is described followed by a description of the ventilation electric circuit. The coupling between the inter- connected heat and ventilation model is validated by conducting experiments on a thermal manikin at different wind conditions and body motion for a constant skin temperature.

The upper human body consists of the clothed segments representing the clothed arm (lower and upper), and the clothed trunk that could be divided into chest and shoulder. The shoulder segment is where the inter-segmental ventilation occurs. Figure 1 represents the physical configuration of the upper human body part consisting of one large segment representing the two arms and the trunk above the shoulder level, and three independent cylinders extending downward from the large cylinder. This physical configuration was adopted by Ismail et al. (2016[a], 2016[b]) after studying the flow past two cylinders of different diameters for sedentary case. Accordingly, it is found that the flow is consid
to the flow around a single bluff body above the shoulder level and similar to the flow around independent cyli

the shoulder level. This is due to the gap separating the large cylinder representing the trunk and the small cylinder representing the arm. The physical model is divided into ten annular zones LF, BF, UF, UB, TF, TB, SFa, SBa, SFt, SBt associated respectively with the lower human clothed arm in front and back sides, the upper clothed arm in front and back sides, the clothed trunk in front and back sides, the clothed shoulder part connected to the arm in front and back sides, and the clothed shoulder part connected to the trunk in front and back sides. Each annular cylinder consists of a porous cylinder surrounding a heated cylinder representing the clothed arm. The fabric covered cylinder has a radius r and height H. The air layer trapped between the concentric fabric and skin cylinders is of thickness Y and it separates the inner solid cylinder maintained at the skin temperature Tskin and the outer porous cylinder represented by an isotropic fabric layer of permeability α and thickness ef.

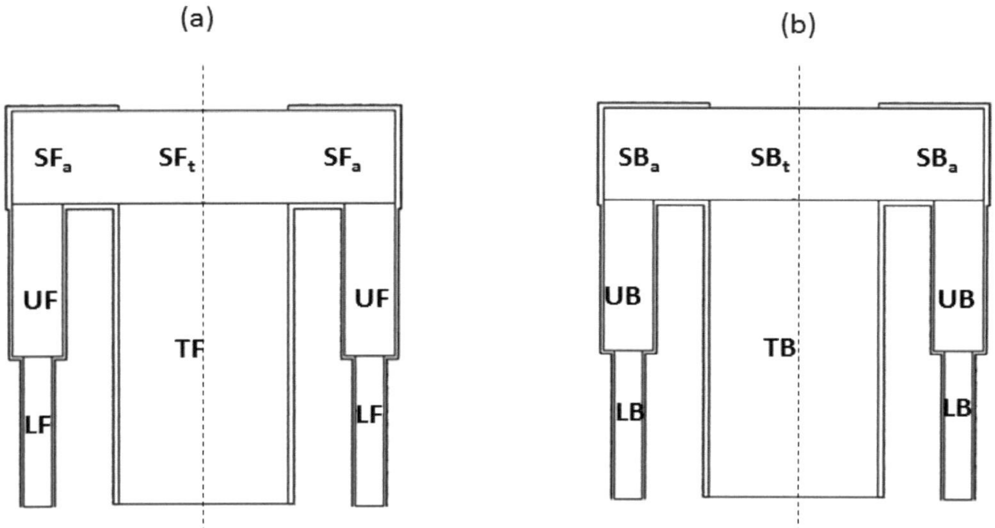

Figure 1 Schematic of the physical configuration showing (a) front view (b) back view of the upper clothed body

ELECTRIC CIRCUIT ANALOGY OF THE HEAT MODEL FROM THE UPPER HUMAN BODY PART

This section represents the heat resistance model adopted to find the segmental heat losses of the upper human body part at constant skin temperature of 34°C. It is noteworthy that in real life scenario, the human body skin temperature is not constant because the human thermal response depends on the environmental conditions, the activity level, and the clothing properties in a way to keep a constant core temperature. However, in this study, the skin temperature is held constant in such a way that the model could be validated using a thermal manikin in a constant skin mode. Once the model is validated, it could be easily integrated with a bio-heat model that predicts the human thermal response as done recently by Ismail et al. (2017). In order to predict the heat and moisture transfer from the human body, Lotens (1991) developed a steady-state heat resistance network model that describes all the mechanisms of heat and moisture transfer from the skin of the clothed segment to the environment. The heat network includes the radiative heat exchange, the convective resistances due to the air motion inside the microclimate air layer and outside in the environment, the dynamic resistance which includes the dry resistance of the clothing corrected for radial ventilation of renewal air (Lotens 1991). The open aperture of the clothed segment induces a presence of additional heat exchange path between environment and microclimate air due to ventilation through opening. This heat resistance network has been used in different studies to find the heat losses at different conditions. Ismail et al. (2017) coupled the Loten's model to the electric circuit analogy of the ventilation model in order to find the heat losses from the independent clothed segment subjected to windy condition and experiencing walking mechanism. However, the coupling assumes an independent clothed segment concept and ignores the heat losses due to the flow rate from the clothed body segments. In this section, we will take the example of the clothed trunk and extend it to all other segments. The trunk segment consists of two zones as shown in Figure 1: the front zone (TF) and the back zone (TB). The air flowing from TF to TB and vice versa leads to a convective heat transfer. Furthermore, air could pass from the trunk to the shoulder part (TF-SF and TB-SB). This flow of air could also create a convective heat transfer. Figure 2 (a-b) illustrates respectively the heat transfer mechanism from each clothed body segment and the heat network associated with the clothed trunk. It is shown that the temperatures of the trunk in front and back sides (TTF, TTB) are co the temperature of the shoulder in front and back sides (TSF, TSB).

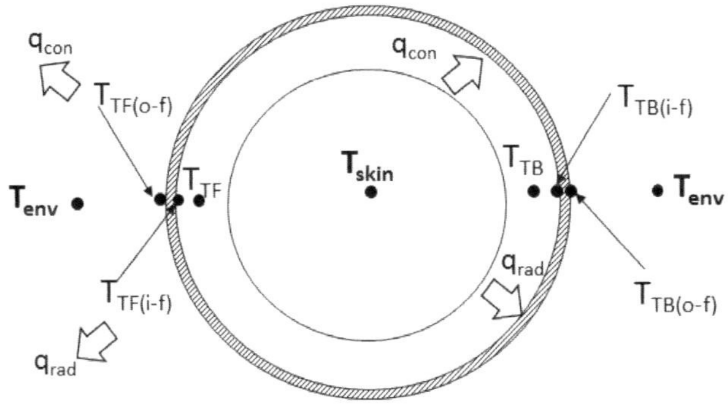

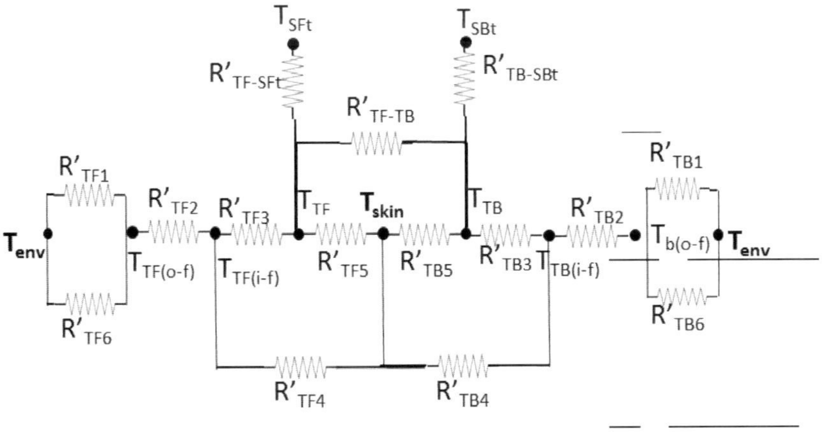

Figure 2: Schematic of (a) different mechanisms of heat transfer from the independent segment representing the trunk (b) thermal resistance network

In the same manner, the thermal resistance of Figure 3 illustrates the heat transfer in the shoulder zone connected to the trunk and connected to the arm. In fact, the inter-connection between the clothed segment microclimates allows air to be exchanged. The flow of air between the segments permits the convective heat exchange. Therefore, as shown in Figure 3(a), there is a heat exchange between the trunk and the shoulder connected to it. In order to relate the temperature of the SFt and SBt zones, an equivalent resistance representing the heat exchange due to inter-segmental ventilation is used and it is given by:

$$R'_{int\,er-segm} = R'_{SFt-SFa} + R'_{SFa-SBa} + R'_{SBa-SBt}$$

All the thermal resistances are summarized in Table 1.

Table 1: Thermal resistance equations corresponding to the thermal network that represents the heat transfer from the clothed trunk

Resistance	Definition	Equation	Similar resistance equation
R'_{TF1}	Convective resistance from the outer fabric to the environment of the front side of the trunk	$\frac{1}{a.V_w^{\,b}.(\pi r_{TF} H_{TF})} \times \frac{r_{TF}}{(r_{TF}+Y_{TF})}$ where $a = 4.8 * (2.r_{TF})^{-0.33}$ and $b = 0.5$	R'_{TB1} R'_{SFt1}, R'_{SBt1} R'_{SFta}, R'_{SBta} R'_{UF1}, R'_{UB1} R'_{LF1}, R'_{LB1}
R'_{TF2}	Fabric dry resistance corrected by the dynamic resistance for ventilation of the front side of the trunk	$\frac{1}{(\frac{1}{R_d} + \rho Q_{r,TF} \times cp_a)} \times \frac{r_{TF}}{(r_{TF}+Y_{TF})}$	R'_{TB2} R'_{SFt2}, R'_{SBt2} R'_{SFa2}, R'_{SBa2} R'_{UF2}, R'_{UB2} R'_{LF2}, R'_{LB2}
R'_{TF3}	Convective resistance from the inner fabric to microclimate air of the front side of the trunk	$\frac{1}{h_{i-a}} \times \frac{r_{TF}}{(r_{TF}+Y_{TF})}$ where $h_{i-a} = 9.6$	R'_{TB3} R'_{SFt3}, R'_{SBt3} R'_{SFa3}, R'_{SBa3} R'_{UF3}, R'_{UB3} R'_{LF3}, R'_{LB3}
R'_{TF4}	Radiative resistance to the skin of the front side of the trunk	$\frac{1}{h_r} \times \frac{r_{TF}}{(r_{TF}+Y_{TF})}$ where $h_r = 3.7$	R'_{TB4} R'_{SFt4}, R'_{SBt4} R'_{SFa4}, R'_{SBa4} R'_{UF4}, R'_{UB4} R'_{LF4}, R'_{LB4}
R'_{TF5}	Convective resistance from the skin to microclimate air of the front side of the trunk	$\frac{1}{h_{s-a}} \times \frac{r_{TF}}{(r_{TF}+Y_{TF})}$ where $h_{s-a} = 10.5$	R'_{TB5} R'_{SFt5}, R'_{SBt5} R'_{SFa5}, R'_{SBa5} R'_{UF5}, R'_{UB5} R'_{LF5}, R'_{LB5}
R'_{TF6}	Radiative resistance to the skin of the front side of the trunk	$\frac{1}{h_r} \times \frac{r_{TF}}{(r_{TF}+Y_{TF})}$ where $h_r = 3.7$	R'_{TB6} R'_{SFt6}, R'_{SBt6} R'_{SFa6}, R'_{SBa6} R'_{UF6}, R'_{UB6} R'_{LF6}, R'_{LB6}
R'_{SFt7}	Convective to the environment through the opening of the front side of the shoulder	$\frac{1}{(\rho Q_{x,SFt} \times cp_a)} \times \frac{r_{SFt}}{(r_{SFt}+Y_{SFt})}$	R'_{SBt7} R'_{LF7}, R'_{LB7}
R'_{TF-TB}	Convective resistance from the front to the back microclimate air of the trunk	$\frac{1}{\rho Q_{\theta,TF} \times cp_a}$	$R'_{SFa-SBt}$ $R'_{SFa-SBa}$ $R'_{SBa-SBt}$ R'_{UF-UB} R'_{LF-LB}
R'_{TF-SFt}	Convective resistance from the trunk front to the shoulder front zone microclimate air	$\frac{1}{\rho Q_{x,TF} \times cp_a}$	R'_{TB-SBt} R'_{UF-SFa}, R'_{UB-SBa} R'_{LF-UF}, R'_{LB-UB}

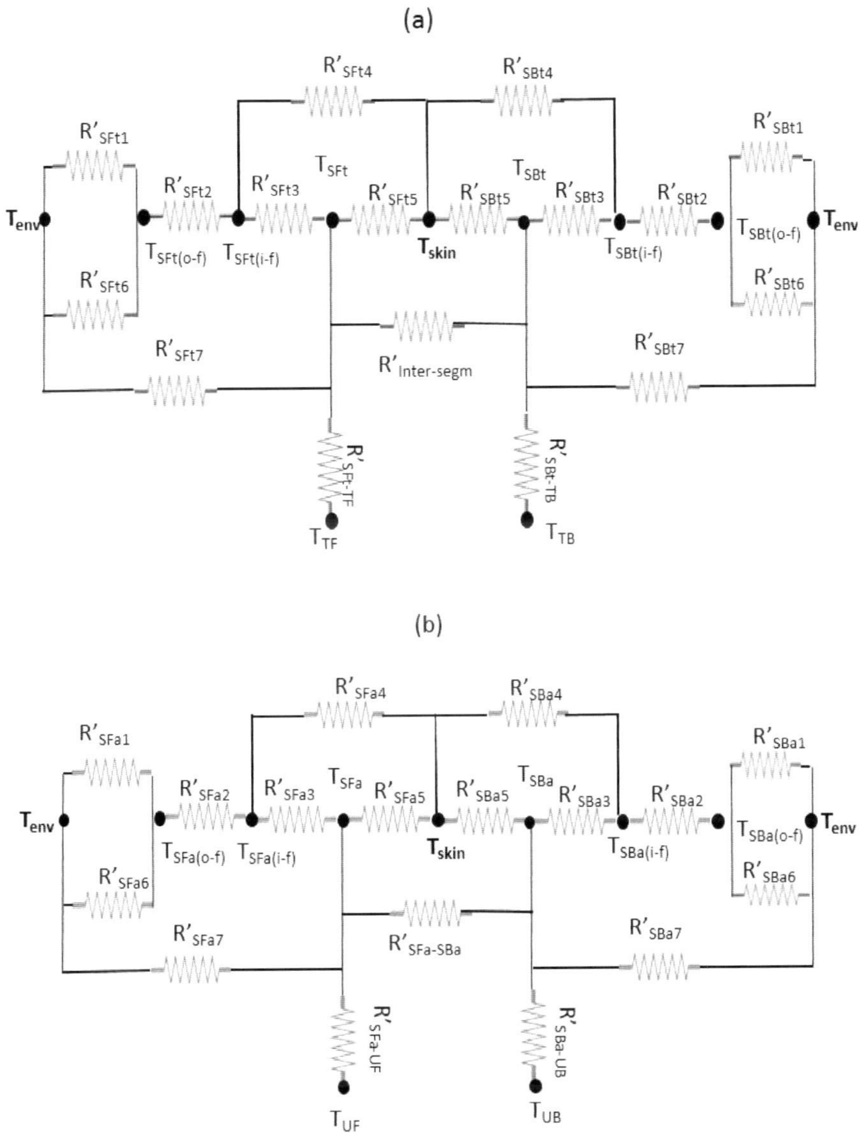

Figure 3: Schematic of thermal resistance network of (a) shoulder zone connected to the trunk and (b) shoulder zone connected to the arm

ESTIMATION OF THE DYNAMIC RESISTANCE FROM THE INTER-CONNECTED VENTILATION MODEL

The dynamic resistance includes the dry resistance of the clothing corrected for radial ventilation of renewal air. Therefore, in order to estimate the dynamic resistance, the simplified ventilation model based on the electric circuit analogy is developed. The ventilation model takes into account the inter-segmental ventilation and requires revising the segmental electric circuit of Ismail et al.(2016[c]) . If we take the example of the lower arm zone (LF, LB), the inter- connection between the lower arm and the upper arm adds a supplementary axial flow rate that passes from the lower arm to the upper microclimates in the front and back sides (LF-UF, LB-UB) or vice versa. By adopting the same methodology, the flow rate between the lower and upper arm is represented by a resistance-inductance circuit. The upper arms and the trunk are connected to the shoulder zone on both front and back sides exchanging some air flow rate. The upper clothed arm is connected to the portion of the shoulder exchanging air with it (UF-SFa, UB-SBa) while the clothed trunk is connected to the other portion of the shoulder exchanging some air with it (TF-SFt, and TB-SBt). These exchanges are represented by a resistance-inductance circuit between the upper arm and the shoulder pressures and by a resistance circuit between the trunk and the shoulder since the trunk is not experiencing any oscillation.The inter-segmental ventilation is defined as the net air flow between the clothed human body parts. This air flow could occur from the trunk to the arm and vice versa.Therefore, the electric circuit representing the walking human body is composed of the electric circuits characterizing each clothed segment connected between each other. This electric circuit is shown in Figure 4. All the resistances and inductances representing the ventilation used are presented in Table 2.

EXPERIMENTAL METHODOLOGY

The aim of this experiment is to measure the heat losses from the human body at different wind and walking conditions in order to validate the integrated ventilation and heat model. The experiments were conducted using 20- zone "Newton" thermal manikin (Measurement Technology Northwest, Seattle Washington, USA) equipped with a walking mechanism (Burke et al. 2011). The surface temperature or heating power of each body segment of the manikin could be measured and controlled individually. Wire sensors are embedded on the thermal zones to afford the requested skin temperature with a standard deviation less than 0.01 °C. The heat flux generated to provide the skin temperature were recorded during experiments at each body part by ThermDAC® software. The manikin is clothed with a long sleeved cotton shirt and trousers where the dimensions are illustrated in Table 3. The air permeability and the dry resistance of the shirt and pants were measured experimentally by SDL MO2IA air permeability device and the sweating guarded hotplate (Model 306-200/400) respectively. These properties are given in Table 4. Two large fans with a diameter of 0.5 m were installed vertically on a stand with controllable wheels to adjust the distance between the manikin and the fans stand for the desired velocity at the front side of the manikin. An additional single fan is used whenever we want a trunk relative velocity that is different than the clothed arm. In fact, when considering the walking condition, the trunk is moving in the real case scenario; however, the thermal manikin trunk stays at the same place while the clothed arm is oscillating. The wind speed was measured by the manikin wind speed sensor which is an air velocity transducer (model 8475-06) at 3-min intervals at different locations and a mean value of wind speed is reported for each test scenario. The manikin and the fans were placed in a controlled environmental chamber. The chamber air temperature was measured by two manikin sensors at different heights (0.9 m and 1.6 m) and the average was about 25 °C ± 0.2 °C. The relative humidity was measured at 1.6 m height and it was about 40 ± 3 %. The experiment is conducted on a thermal manikin using a constant temperature mode at different wind and walking conditions. The thermal manikin was turned on at constant segmental skin temperature of 34 °C. The walking and windy conditions are set and the heat losses are given by the Thermdac software of the thermal manikin. Figure 5 shows the experimental set up with the walking thermal manikin.

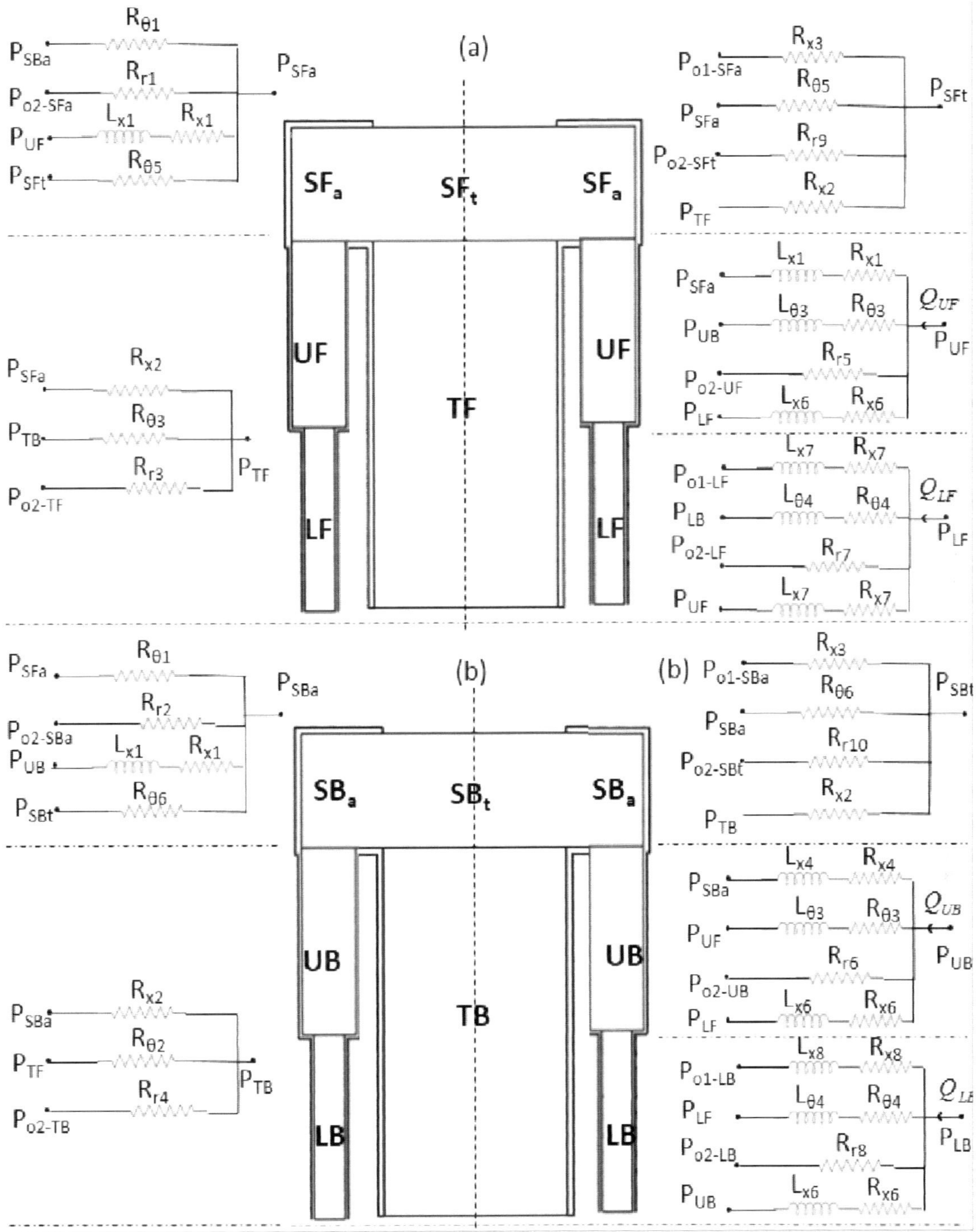

Figure 4: Network representing the analogy of the air inside the microclimate air layer of clothed upper human body for (a) the front side and (b) the back side

Table 2: Definition of the resistances and inductances in each direction

Direction	Symbol	Equation	Symbol	Equation
Axial	R_{x1}	$\frac{12\mu H_{UA}}{\bar{Y}^3_{UA}(\pi r_{UA})}$	L_{x1}	$\frac{\rho H_{UA}}{\bar{Y}_{UA}(\pi r_{UA})}$
	R_{x2}	$\frac{12\mu H_{TF}}{\bar{Y}^3_{TF}(\pi r_{TF})}$	L_{x4}	$\frac{\rho H_{UB}}{\bar{Y}_{UB}(\pi r_{TF})}$
	R_{x3}	$\frac{12\mu H_{SFt}}{\bar{Y}^3_{SF}(\pi r_{SFt})}$	L_{x6}	$\frac{\rho H_{UF}}{\bar{Y}_{UF}(\pi r_{UF})}$
	R_{x4}	$\frac{12\mu H_{UB}}{\bar{Y}^3_{UB}(\pi r_{UB})}$	L_{x7}	$\frac{\rho H_{LF}}{\bar{Y}_{LF}(\pi r_{LF})}$
	R_{x5}	$\frac{12\mu H_{TB}}{\bar{Y}^3_{TB}(\pi r_{TB})}$	L_{x8}	$\frac{\rho H_{LB}}{Y_{LB}(\pi r_{LB})}$
	R_{x6}	$\frac{12\mu H_{UF}}{\bar{Y}^3_{UF}(\pi r_{UF})}$	R_{x9}	$\frac{12\mu H_{UF}}{\bar{Y}^3_{UF}(\pi r_{UF})}$
	R_{x7}	$\frac{12\mu H_{LF}}{\bar{Y}^3_{LF}(\pi r_{LF})}$	R_{x10}	$\frac{12\mu H_{SBt}}{\bar{Y}^3_{SF}(\pi r_{SBt})}$
	R_{x8}	$\frac{12\mu H_{LB}}{\bar{Y}^3_{LB}(\pi r_{LB})}$		
Angular	$R_{\theta 1}$	$\frac{12\mu(\pi r_{SFa})}{\bar{Y}^3_{SFa} H_{SFa}}$	$L_{\theta 3}$	$\frac{\rho(\pi r_{UF})}{\bar{Y}_{UF} H_{UF}}$
	$R_{\theta 2}$	$\frac{12\mu(\pi r_{TF})}{\bar{Y}^3_{TF} H_{TF}}$	$L_{\theta 4}$	$\frac{\rho(\pi r_{LF})}{\bar{Y}_{LF} H_{LF}}$
	$R_{\theta 3}$	$\frac{12\mu(\pi r_{UF})}{\bar{Y}^3_{UF} H_{UF}}$	$R_{\theta 5}$	$\frac{12\mu(5\pi r_{SFa}/12)}{\bar{Y}^3_{SFa} H_{SFa}}$
	$R_{\theta 4}$	$\frac{12\mu(\pi r_{LF})}{\bar{Y}^3_{LF} H_{LF}}$	$R_{\theta 6}$	$\frac{12\mu(5\pi r_{SBa}/12)}{\bar{Y}^3_{SBa} H_{SBa}}$
Radial	R_{r1}	$\frac{\Delta P_m}{\alpha\pi r_{SFa} H_{SFa}}$	R_{r5}	$\frac{\Delta P_m}{\alpha\pi r_{UF} H_{UF}}$
	R_{r2}	$\frac{\Delta P_m}{\alpha\pi r_{SBa} H_{SBa}}$	R_{r6}	$\frac{\Delta P_m}{\alpha\pi r_{UB} H_{UB}}$
	R_{r3}	$\frac{\Delta P_m}{\alpha\pi r_{TF} H_{TF}}$	R_{r7}	$\frac{\Delta P_m}{\alpha\pi r_{LF} H_{LF}}$
	R_{r4}	$\frac{\Delta P_m}{\alpha\pi r_{TB} H_{TB}}$	R_{r8}	$\frac{\Delta P_m}{\alpha\pi r_{LB} H_{LB}}$
	R_{r9}	$\frac{\Delta P_m}{\alpha\pi r_{SFt} H_{SFt}}$	R_{r10}	$\frac{\Delta P_m}{\alpha\pi r_{SBt} H_{SBt}}$

Table 3. Basic dimensions (circumference) of the upper part of the thermal manikin and the medium size of the jacket

	Bust (*cm*)	Waist (*cm*)	Hip (*cm*)	Neck line (*cm*)	Cuff (*cm*)	Height (*cm*)
Thermal manikin size	93	74	93	35	11	65
Shirt size	120	110	116	38	14	65

Table 4: Fabric properties of the thermal manikin ensemble

	Material	Air permeability (m/s)	Dry resistance (K·m2/W)
Shirt	100 % cotton	1.38	0.046
Pants	75 % cotton, 25% polyester	0.25	0.041

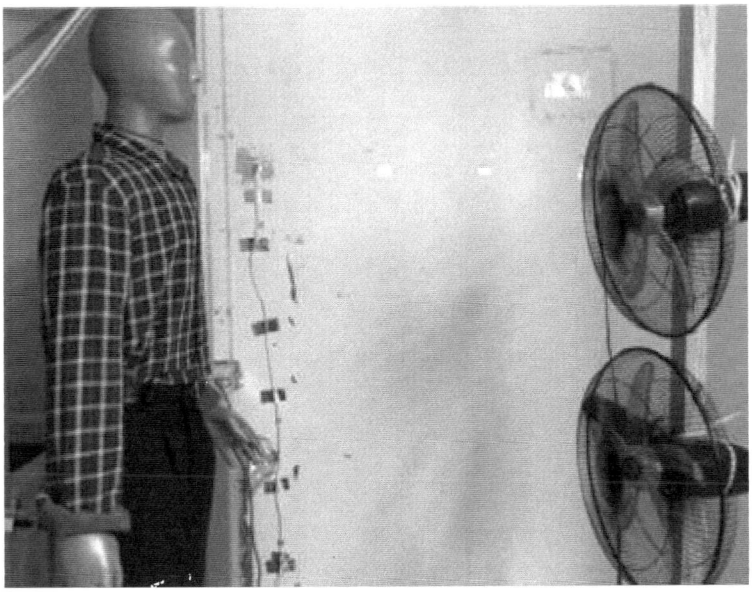

Figure 5: Figure showing the walking manikin in the face of the fans

MODEL VALIDATION WITH EXPERIMENT

The thermal manikin was run using the constant skin temperature of 34°C. The ambient temperature of the environmental chamber is 25.5 °C ± 0.2 °C while the relative humidity was about 60 ± 3 %. The experiments were repeated at different conditions of wind speeds (V_w < 0.11 m/s, V_w=1 m/s and V_w=1.2 m/s). Furthermore, different walking speeds are investigated: no walking (V_{walk}= 0 m/s); walking at 0.8 m/s corresponding to a DSPM of 38 and a frequency of about 50 *rpm*; walking at 1.25 m/s corresponding to a DSPM of 60 and a frequency of about 80 *rpm*. The inputs to the integrated simplified model are the clothing air permeability, the geometrical parameters of the clothed arm an oscillation parameters for the arm, the walking speed for the trunk, the ambient temperature and wind speeds. In the integrated simplified model, the height, radius, and air gap of eight zones are presented in Table 5.

Table 5: The geometrical inputs of the air gap size, radius, and height of the eight zones.

	LF/LB	UF/UB	TF/TB	SF/SB
Y (cm)	0.5	0.8	4	4
R (cm)	3	5	12	17
H (cm)	30	25	40	10

The oscillating parameters depend on the walking speed at a typical angle of oscillation $\phi_{max} = 20°$.

Table 6 presents the segmental heat losses for the arm and trunk. Good agreement is seen between the segmental heat losses recorded experimentally by the thermal manikin at constant skin and environmental temperatures of 34 °C and 25.5 °C respectively and the predicted values from the coupled ventilation and thermal network model with a maximum relative error of 8 %. The segmental heat losses for the arm and the trunk attain its minimum for the lowest relative velocity and its maximum for the highest relative velocity. Evidently, when the relative velocity increases, the trunk and arm ventilation increase leading to an increase in segmental heat losses. The main reason is that the penetrated air is coming from an ambient temperature of 25 °C. It is slightly heated in the fabric nodes and enter to the microclimate air layer leading to significant heat losses from the skin set at a constant temperature of 34 °C.

Table 6: Comparison between the measured and predicted heat losses

Conditions		Arm heat losses (w/m²)		Trunk heat losses (w/m²)	
V_w (m/s)	V_{walk} (m/s)	Experiment	Model	Experiment	Model
0	1.25	61.8 ± 3	61.5	62.96	65.35
1	0	61.22 ± 3	61.01	62.07	64.11
1.2	0	65.28 ± 3	61.37	63.54	65.67
0	0.8	59.77 ± 3	58.12	60.12	60.58
1	0.8	75.02 ± 3	72.87	70.82	73.02
1.2	1.25	79.53 ± 3	78.13	75.91	77.57

CONCLUSION

In this study, an integrated model of heat and ventilation of the upper clothed human body experiencing relative wind due to walking and external wind is developed. The model is based on an electric circuit analogy for the air and heat flow. The model of air flow connects all the human body segments and estimate the inter-segmental ventilation between the segments for the clothed human upper part driven by difference of pressure in the microclimate air of the trunk and the upper arms. The heat circuit model is integrated to a heat resistance network to find the segmental heat losses of the human body to the environment at constant skin temperature. The integrated model is validated by conducting experiments on thermal manikin that gives the segmental heat fluxes associated with the constant skin mode adopted. Good agreement is shown between the experimental and the predicted data with a maximum relative error of 16%.

NOMENCLATURE

DSPM double-steps per minute

e_f thickness of the outer cylinder representing the fabric (m)

walking frequency (Hz)

gravity (m.s^{-2})

h_r radiative heat transfer coefficient (W/m^2·K)

H cylinder length (m)

L electric inductance

R flow resistance

R' thermal resistance

RH relative humidity (%)

r radius of the inner cylinder representing the human body skin (m)

r_f radius of the outer cylinder representing the fabric (m)

t time (s)

T swing period (s)

v_w wind velocity (m/s)

v_{walk} walking velocity (m/s)

axial direction

radial direction

Y microclimate air layer size (m)

$\bar{Y}$ average microclimate air layer size (m)

Greek symbols:

ΔP_m standard pressure difference (Pa)

μ dynamic viscosity (kg/m·s)

ϱ density of air (kg/m^3)

a permeability of fabric (m^3/m^2·s)

θ angular direction

Subscripts:

x axial direction

θ angular direction

r radial direction

d : dry

UF : upper front

LF : lower front

TF : trunk front

SFa : shoulder front connected to the arm

SFt : shoulder front connected to the trunk

UB : upper back

LB : lower back

TB : trunk back

SB : shoulder back

a : arm

t : trunk

skin :skin

env :environment

front :front side

back :back side

(i-f) :inner fabric

(o-f) :inner fabric

REFERENCES

Gómez, F. , Sifre, V. , Montero, L., Vicente, V.D., and L.Gil. 2006. Sustainability in cities: the green areas and climatic comfort as fundamental parameters. *The Sustainable City IV: Urban Regeneration and Sustainability.*

Cotter, J.D., and N.A.S Taylor. 2005. The distribution of cutaneous sudomotor and alliesthesial thermosensitivity in mildly heat-stressed humans: an open-loop approach. *The Journal of Physiology.*

Ghaddar, N., Ghali, K., and B. Jreije. 2008. Ventilation of Wind-Permeable Clothed Cylinder Subject to Periodic Swinging Motion: Modeling and Experimentation. *Journal of Heat Transfer.*

Ghaddar, N., Ghali, K. and M. Othmani. 2010[a]. Effect of Moisture Transport on Mixed Convection in Vertical Annulus of a Heated Clothed Vertical Wet Cylinder in Uniform Cross Wind. *Proceedings of 14th International Heat Transfer Conference (IHTC14).* Washington D.C.

Ghaddar, N., Ghali, K., Al-Othmani, M., Holmer, I. and K. Kuklane. 2010[b]. Experimental and theoretical study of ventilation and heat loss from isothermally heated clothed vertical cylinder in uniform flow field. *Journal of Applied Mechanics.*

Ghaddar, N., Ghali, K., Harathani, J. and E. Jaroudi. 2005. Ventilation rates of micro-climate air annulus of the clothing-skin system under periodic motion. *International Journal of Heat and Mass Transfer.*

Ismail, N., Ghaddar, N., and K. Ghali. 2016[a]. Effect of inter-segmental air exchanges on local and overall clothing ventilation. *Textile Research Journal.*

Ismail, N., Ghaddar, N., and K. Ghali. 2016[b]. Improving Local Ventilation Prediction by Accounting for Inter-segmental Ventilation Rate. *Textile Research Journal.*

Ismail, N., Ghaddar, N., and K. Ghali. 2014. Predicting segmental and overall ventilation of ensembles using an integrated bioheat and clothed cylinder ventilation models. *Textile Research Journal.*

Ismail, N., Ghaddar, N., and K. Ghali. 2016[c]. Determination of segmental and overall ventilation of clothed walking human by means of electric circuit analogy. *Textile Research Journal.*

Ismail, N., Ghaddar, N., and K. Ghali. 2017. A clothing ventilation and heat loss electric circuit model for a walking human. *International Journal of Thermal Sciences.*

Birnbaum, R.R. and G.W. Crockford. 1978. Measurement of the clothing ventilation index. *Applied Ergonomics.*

Lotens, W. 1993. Heat transfer from humans wearing clothing, Doctoral thesis, TNO Institute for Perception, Soesterberg, The Netherlands.

Lotens, W., and G. Havenith. 1991. Calculation of clothing insulation and vapor resistance. *Ergonomics.*

Ke, Y., Li, J., and G. Havenith. 2014. An improved experimental method for local clothing ventilation measurement. *International Journal of Industrial Ergonomics.*

Wan, and J. Fan. 2008. A transient thermal model of the human body–clothing–environment system. *Journal of Thermal Biology*.
Ke, Y., Havenith, G., Zhang, X. Li, X., and J. Li. 2013. Effects of wind and clothing apertures on local clothing ventilation rates and thermal insulation. *Textile Research Journal*.

Numerical Study on PCM-Desiccant Cooling Vest to Improve Cooling and Performance of Workers in Hot Humid Conditions

Mariam Itani

Rana Bachnak

Nesreen Ghaddar, PhD
Member ASHRAE

Kamel Ghali, PhD
Member ASHRAE

ABSTRACT

People working in a hot humid environment are exposed to heat strains due to the high ambient conditions as well as hindered cooling by sweat evaporation from the human body. This causes heat accumulation in the human body and an increase in the core temperature affecting the comfort of the workers and eventually their productivity. Passive cooling vests containing phase change material (PCM) have been designed to provide cooling for the body. However, the presence of PCM packets creates a barrier in the path of sweat evaporation, and in humid conditions, condensation might happen on the surface of the packets. A combination of PCM and solid desiccant packets was used to minimize the risk of condensation and provide a drier microclimate in the cooling vest. The aim of this study is to integrate a validated fabric-PCM-Desiccant model with a validated Bioheat model and assess the human thermal response upon wearing the PCM-Desiccant vest.

The fabric-PCM-Desiccant model is based on heat and mass balances that predict fabric, PCM, desiccant and micro and macro climate air temperatures, as well as micro and macro climate air humidity contents. The bioheat model divides the human body into 25 segments including upper and lower segments of the trunk, that are represented by 8 skin nodes. The bioheat model predicts segmental skin temperatures and heat losses, as well as, mean skin and core temperatures and sweat rate. Thus, an integrated model would allow designing a PCM-Desiccant cooling vest capable of sufficiently cooling the body while maintaining a drier microclimate. The integrated model would also allow assessing the effect of the PCM-Desiccant vest on skin temperature, microclimate temperature, and microclimate humidity content.

In this study, a comparison is done between predicted skin and microclimate temperatures and microclimate humidity content when using a PCM-Desiccant vest and when using a PCM-Only cooling vest in hot humid conditions. Then, the effect of different PCM melting temperatures on the human thermal responses and the performance of the PCM-Desiccant vest is examined to give a recommended vest design. It is expected that the PCM-Desiccant cooling vest would provide a cool and less humid microclimate than that in a PCM-Only cooling vest, thus improving thermal comfort of the worker in hot humid environments.

Key-words: *Bioheat modeling, fabric-PCM-Desiccant modeling; cooling vest; PCM melting temperature.*

Mariam Itani is a graduate student in the Department of Mechanical Engineering, American University of Beirut, Beirut, Lebanon and the corresponding author, email: mmi25@mail.aub.edu. **Rana Bachnak** is a graduate student in the Department of Mechanical Engineering, American University of Beirut, Beirut, Lebanon. **Nesreen Ghaddar** is a professor in the Department of Mechanical Engineering, American University of Beirut, Beirut, Lebanon. **Kamel Ghali** is a professor and chair in the Department of Mechanical Engineering, American University of Beirut, Beirut, Lebanon.

INTRODUCTION

Cooling of the human body can be achieved by sensible heat losses due to temperature gradients between the skin and the ambient and/or by latent heat losses due to sweat evaporation to the surroundings (Zhao et al. 2013). For given clothing and activity level, however, these heat losses are affected by the temperature and humidity at the microclimate of the human body and its surroundings.

In hot dry conditions, sensible heat losses are significantly reduced; therefore, the only means for body cooling is by losing latent heat to the surroundings via sweat evaporation. The reduced heat loss from the body can be overcomed by using personal cooling techniques (Yazdi et al. 2015a). These cooling techniques include phase change material (PCM) cooling vests that were shown to be effective in reducing heat strain and increasing the productivity of subjects in hot environmental conditions (Gao et al. 2011; House et al. 2013 and Yazdi et al. 2016). Studies on human subjects wearing cooling vests while performing activity in hot conditions showed drops in mean and local torso skin temperatures and sweat rates associated with improvement in thermal sensations (Gao et al. 2007; Gao et al. 2012; Hasegawa et al. 2005; Nishihara et al. 2002; Webster et al. 2005; and Yazdi et al. 2015b). However, as reported by Zhao et al. (2013) and Reinertsen et al. (2008), PCM vests hindered sweat evaporation and did not provide additional cooling when tested on a sweating thermal manikin in hot and dry ambient conditions. The effectiveness of PCM vests is constrained due to the reduced moisture transport and the consequent risk of condensation on the PCM packet surface if its melting temperature is below the dew point of micro or macro climate air (Itani et al. 2017). If condensation occurs, heat will be released to the PCM vest and might thus counteract its cooling effect and duration. In addition, sweating and the saturation of the microclimate lead to heat stress and reduction of work efficiency as reported by Houshyar et al. (2015), Dotti et al. (2016) and Wang et al. (2016). Therefore, the use of PCM packets in cooling vests hinders sweat evaporation thus leading to its accumulation on the skin surface and causing feelings of discomfort and skin irritation (Reinertsen et al., 2008; Wang et al., 2016).

Subsequently, hybrid cooling vests have been introduced to justify the use of PCM cooling vests in conditions when the human body is generating more sweat. These vests integrate PCMs and battery-operated fans, used to enhance evaporative cooling, thus improving latent heat loss. Lai et al. (2017) recommended using the hybrid vest when the human is performing heavy work and sweating, otherwise, the use of a vest with only PCM packets is advised. On the other hand, hybrid vests require external power to operate and are ineffective if the ambient conditions are hot and humid (Yi et al. 2017).

A solution was proposed by Itani et al. (2017) to overcome the deterioration of the PCM performance due to condensation in the vest and to overcome the reduction of work efficiency due to sweat accumulation. In order to provide dryer microclimate air conditions and better skin wetness sensation, a thin solid desiccant (Des) packet, such as silica gel, was proposed to be added to the PCM packet resulting with a certain PCM and Des coverage areas on the skin. They developed a mathematical model of the combined fabric-PCM-Des to predict the temperature and moisture content of the microclimate air in the presence of a PCM-Des packet. Their model was validated by experiments conducted on a wet clothed heated cylinder. In the presence of a drier microclimate environment, evaporative heat transfer from the wet skin would be enhanced leading to better skin wetness sensation.

However, during moisture adsorption, the heat released by the desiccants raises the microclimate air temperature and thus the PCM packet melts faster (Itani et al. 2017). To overcome that and maintain similar thermal comfort obtained when using PCM-Only cooling vests, PCM packets of lower melting temperatures can be used in the PCM-Des packet. The preliminary findings of Itani et al. (2017) could be basis for further investigating the effect of adding Des packets to the PCM, and whether it would require using lower PCM melting temperature to produce the same prolonged cooling effect on the human body. In order to predict the effect of using PCM-Des packets with different PCM melting temperatures on human thermal responses, actual skin conditions are needed. Human physiology has not been previously taken into account when assessing the performance of the PCM-Des cooling vest.

The fabric-PCM-Des model of Itani et al. (2017) can be integrated with a bioheat model to predict the cooling performance of the PCM-Des cooling vest. Different bioheat models have been developed in research (Al-Othmani et al. 2008; Avolio et al. 1980; Karaki et al. 2013 and Salloum et al. 2007). The model adopted in this work was that of Karaki et al. (2013) since it has a high level of accuracy in predicting skin blood flow rate, local and overall thermal responses, and segmental and core body temperatures in transient environments. A similar integration has been done in the study of Hamdan et al. (2016), where a model was developed for heat and mass transfer through fabric layers containing PCM packets and then integrated with a bioheat model.

The purpose of this study is to integrate the fabric-PCM-Des model of Itani et al. (2017) with the validated bioheat model of Karaki et al. (2013). Then, the integrated model will be used to compare skin and microclimate temperatures and microclimate humidity content when using a PCM-Desiccant vest and when using a PCM-Only cooling vest in hot humid conditions. The effect of different PCM melting temperatures on human thermal responses and overall thermal comfort, predicted using the model developed by Zhang et al. (2004), will be examined.

INTEGRATED BIOHEAT AND FABRIC-PCM-DES MODEL

In order to predict the effect of wearing a PCM-Des cooling vest on microclimate air conditions near the human skin and human thermal responses, the fabric PCM-Des model of Itani et al. (2017) was integrated with the multi-node segmental bioheat model of Karaki et al. (2013). The fabric-PCM-Des model of Itani et al. (2017) decomposes the cooling vest, as shown in Figure 1(a), into the following parts:

- Outer fabric layer of the vest exposed to the ambient
- Macroclimate air layer consisting of the air between the outer fabric layer and the PCM-Des packets and in between the PCM-Des packets
- PCM-Des packets of specified size and properties
- Microclimate air layer consisting of the air between each PCM-Des packet and the inner fabric layer
- Inner fabric layer near the torso skin

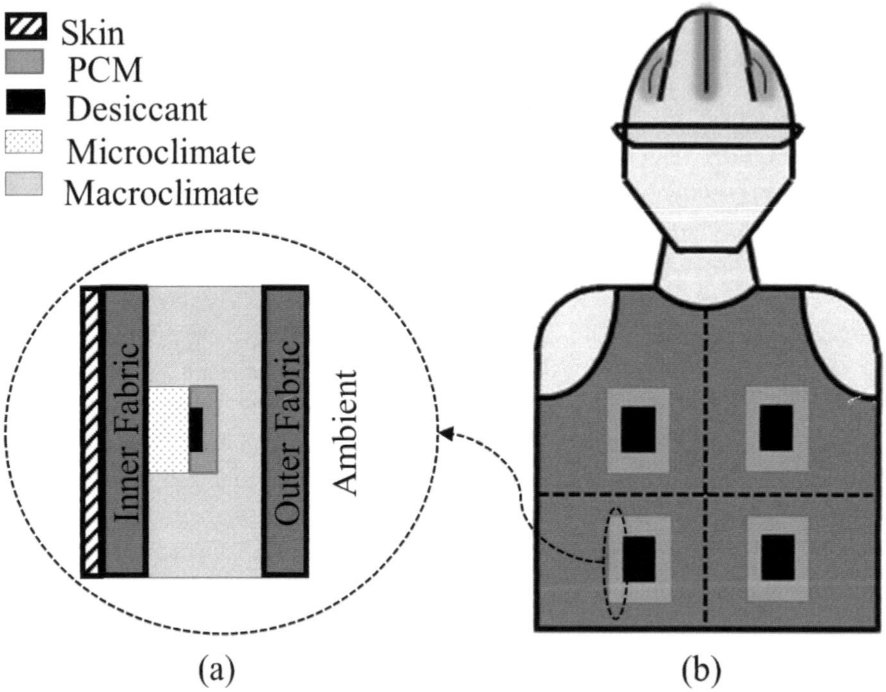

Figure 1: Schematic of the cooling vest showing (a) a side view of the skin layer and the different fabric, PCM, and air layers and (b) a front view of the upper and lower torso segments covered with PCM-Des packets

The bioheat model of Karaki et al. (2013) divides the torso into 4 segments (upper and lower front and upper and lower back), where each segment has two skin nodes; left and right. It can predict segmental and mean skin temperatures, core body temperature, sweat rate and its accumulated mass on the skin under different ambient conditions and activity levels. The mass of accumulated sweat is found based on the model of Jones et al. (1992), as a function of saturated pressure at the skin and sweat rate. The accumulation of sweat on the skin is not desirable since it increases the level of thermal discomfort and causes skin irritation (Reinertsen et al., 2008; Wang et al., 2016). Moreover, overall thermal comfort is predicted using the model of Zhang et al. (2004) which is based on a scale from -4 (very uncomfortable) to +4 (very comfortable). The integration between the models permits covering the 4 torso segments with the cooling vest, in which the number of packets can be defined for every torso segment, as shown in Figure 1(b).

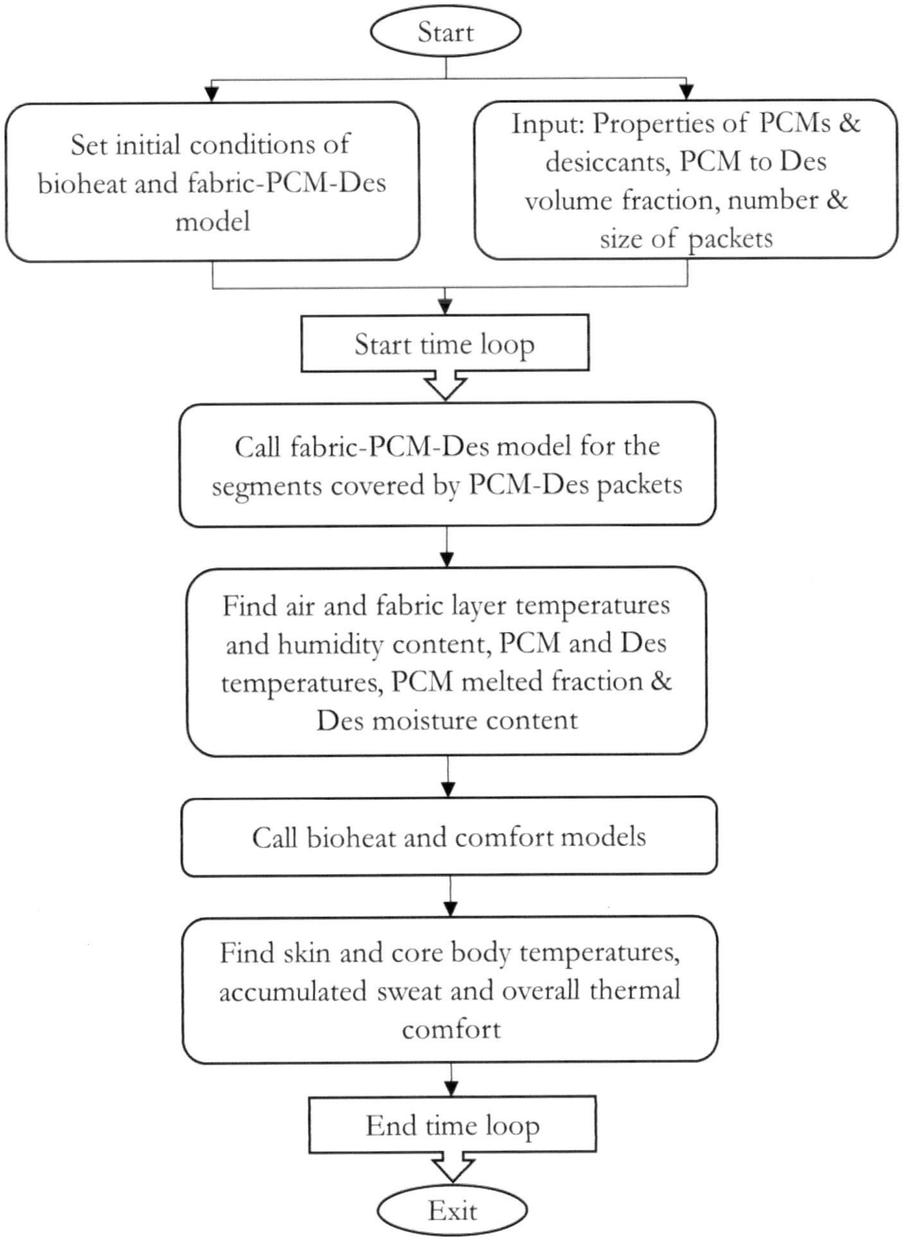

Figure 2: Flowchart showing the coupling methodology between the fabric-PCM-Des and bioheat models

In order to utilize the integrated model, the number, dimensions and PCM-to-Des volume fraction of the PCM-Des packets covering each torso segment need to be defined. In addition, the properties of the fabric layers and the initial conditions of the human body and different vest layers need to be set. The ambient conditions of temperature and relative humidity (RH), the clothing properties, and the metabolic rate are also set in the model. The initial conditions of temperature and pressure of the different vest layers are taken to be at equilibrium with the environment. The fabric-PCM-Des model uses mass and energy balance equations to predict, at a new time-step, the temperatures and humidity ratios of the microclimate and macroclimate air layers, as well as the temperatures, humidity ratios, and vapor pressures of the inner and outer fabric layers. In addition to that, the fabric-PCM-Des model updates the PCM and Des temperatures at every time step, as well as the Des moisture content. Once the PCM reaches its melting temperature, it starts melting at a constant temperature and the model then predicts its melted fraction. The new calculated values of air layer temperatures and humidity ratios from the fabric-PCM-Des model are inputted to the bioheat model to find the new segmental and mean skin temperatures and vapor pressures. Once the segmental and mean skin temperatures, segmental and body core temperatures and their rate of change are found, overall thermal comfort can be predicted. The flowchart, showing the coupling methodology between the fabric-PCM-Des model and the bioheat model is illustrated in Figure 2.

PARAMETRIC STUDY: EFFECT OF PCM-DES PACKETS AT DIFFERENT PCM MELTING TEMPERATURES

The integrated fabric-PCM-Des and bioheat model was used to perform a parametric study and determine the effect of using a PCM-Des cooling vest on human thermal responses and comfort, compared to a PCM-Only cooling vest. In addition, the performance of the PCM-Des vest was studied at different PCM packet melting temperatures, since it affects the microclimate and skin temperatures.

The case study considered for performing the simulations mimics the situation of a human performing activity at 4 Mets for 45 minutes in hot humid ambient conditions of 40°C (104°F) and 45% relative humidity. Two PCM melting temperatures were investigated; 21°C (69.8°F), denoted by PCM21, and 28 °C (82.4°F), denoted by PCM28. For each PCM melting temperature, two cases were simulated: with Des (PCM21-Des and PCM28-Des) and without Des (PCM21-Only amd PCM28-Only). All PCM packets had the same initial temperature of 20 °C (68°F). The four vests contained 20 PCM/PCM-Des packets covering the front and back torso segments. The total PCM/PCM-Des coverage area was 0.16m^2 (1.72ft^2), representing 34% of the whole torso area. The properties of the PCM/PCM-Des packets, such as the PCM latent heat of fusion, the PCM mass, the PCM and Des coverage areas and the volume fraction of the PCM in the PCM-Des packet are tabulated in Table 1.

After performing the simulations, a comparison was done between the microclimate temperature and humidity ratio, mean torso and skin temperatures, mass of accumulated sweat, PCM melted fraction, and overall thermal comfort of the different cases.

Table 1. Characteristics of the PCM/PCM-Des packets

Case	PCM melting temperature in °C (°F)	PCM latent heat of fusion in kJ/kg (BTU/lb)	PCM mass in grams (lb)	PCM packet coverage area facing skin in cm^2 (in^2)	Des packet coverage area facing skin in cm^2 (in^2)	PCM volume fraction in PCM-Des packet in %
PCM28-Only	28 (82.4)	159 (68.4)	87 (0.19)	80 (12.4)	0	100
PCM28-Des	28 (82.4)	159 (68.4)	87 (0.19)	47.9 (7.42)	32.1 (4.98)	76.7
PCM21-Only	21 (69.8)	134 (57.6)	87 (0.19)	80 (12.4)	0	100
PCM21-Des	21 (69.8)	134 (57.6)	87 (0.19)	47.9 (7.42)	32.1 (4.98)	76.7

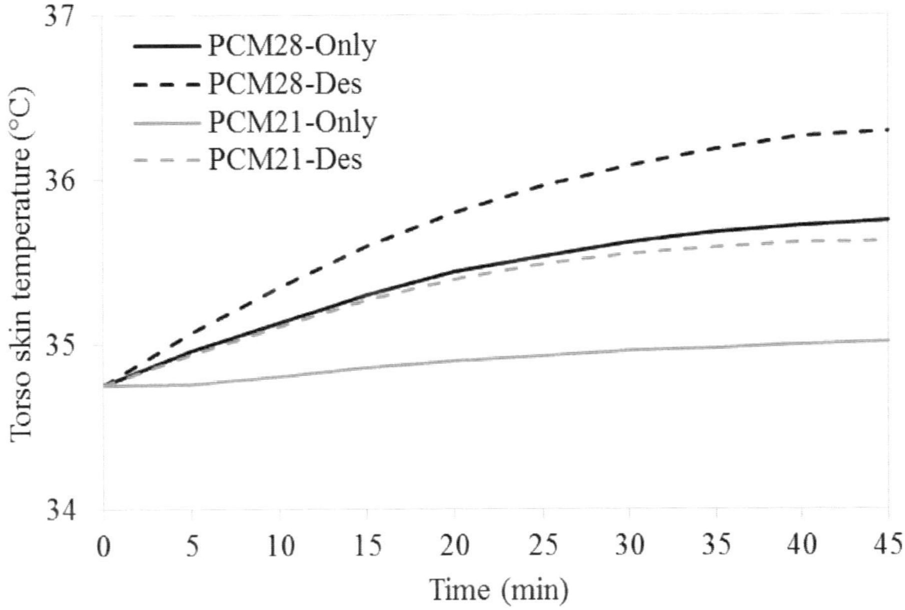

Figure 3: Plots of the variation of the torso mean skin temperature with time for the PCM/PCM-Des cooling vests at different PCM melting temperatures

RESULTS AND DISCUSSION

The torso skin temperature variation with time is plotted in Figure 3 for the four simulated cases. As seen from the figure, the torso skin temperature in all the conditions started from the same initial value of 34.75°C (94.55°F). After that, the torso skin temperature started to increase due to the heat generated under the moderate activity level. It can be noticed that the vests with the higher PCM melting temperature had higher torso skin temperatures. For example, PCM28-Only reached higher torso skin temperature of 35.75°C (96.35°F) by the end of the 45-min working duration, than that reached in the PCM21-Only case of 35.02°C (95.04°F). Moreover, the addition of Des caused an increase in the torso skin temperature due to the release of heat during moisture adsorption. At the end of the simulation time, the torso skin temperature in the PCM28-Des case reached a maximum value of 36.03°C (96.85°F), while that in the PCM21-Des case reached a maximum value of 35.63°C (96.13°F).

A comparison was done between the time-averaged microclimate air temperature, torso skin temperature, and mean skin temperature of the four vests, as shown in Figure 4. It can be noticed that introducing the Des to the PCM caused an increase in the time-averaged microclimate air temperature due to the heat released by the Des as it adsorbs moisture. Thus, an increase in the torso skin temperature and eventually the mean skin temperature was noticed. The negative effect of adding a Des to the PCM packet on the torso skin temperature can be eliminated by using PCM packets of lower melting temperatures as shown in case PCM21-Des, which resulted in similar torso and mean skin temperatures to the PCM28-Only case.

It can be concluded from this comparison that to maintaion a similar cooling effect between the PCM-Only and PCM-Des cooling vest, a PCM with lower melting temperature must be used to compensate for the increase in the microclimate and skin temperatures caused by the Des.

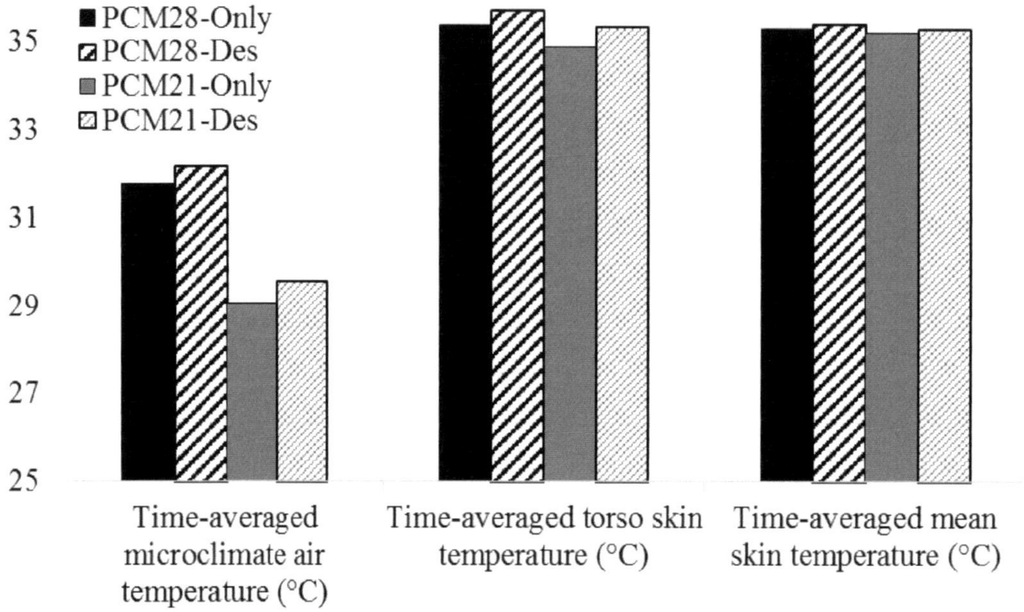

Figure 4: Results of the time-averaged microclimate air temperature, time-averaged torso skin temperature, and time-averaged mean skin temperature for the PCM/PCM-Des cooling vests at different PCM melting temperatures

Table 2. Simulated cases on the integrated model

Case	Time-averaged microclimate air humidity ratio in g/kg (lb/lb)	Final mass of accumulated sweat in g/m² (lb/in²)	PCM melted fraction in %	Time-averaged overall thermal comfort
PCM28-Only	21.36 (21.36e-3)	21.40 (3.04e-5)	14.70	-0.80
PCM28-Des	21.19 (21.19e-3)	15.80 (2.25e-5)	19.70	-0.83
PCM21-Only	21.36 (21.36e-3)	22.07 (3.14e-5)	29.30	-0.76
PCM21-Des	21.18 (21.18e-3)	15.77 (2.24e-5)	41.60	-0.78

Table 2 shows a comparison between the time-averaged microclimate air humidity ratios, the final mass of accumulated sweat on the human skin, the PCM melted fraction, and the time-averaged overall thermal comfort of the four cooling vest cases. It can be noticed that the addition of Des caused a decrease in the microclimate humidity ratio and mass of accumulated sweat. For example, the presence of the Des reduced the microclimate humidity ratio from 21.36 g/kg (21.36e-3 lb/lb) in case PCM28-Only to 21.19 g/kg (21.19e-3 lb/lb) in case PCM28-Des. Moreover, the mass of accumulated sweat was reduced from 21.40 g/m² (3.04e-5 lb/in²) in case PCM28-Only to 15.80 g/m² (2.25e-5 lb/in²) in case PCM28-Des. However, the presence of the Des caused an increase in the PCM melted fraction, where it increased from 14.7 % to 19.7 % for the PCM of 28 °C (82.4°F) melting temperature and from 29.3 % to 41.6 % for the PCM with 21°C (69.8°F) melting temperature. It can be noticed that the packets of lower melting temperature had higher PCM melted fraction, since they started melting at earlier times than the higher melting temperature ones and have lower latent heat of fusion.

In addition, the presence of the Des lead to a less comfortable thermal state due to the higher torso, and mean skin temperatures, as shown in Figure 4. The use of PCM packets of lower melting temperature with the Des, as in case PCM21-Des, resulted with drier microclimate air conditions and less mass of accumulated sweat but with similar overall thermal comfort at about -0.78 (slightly uncomfortable) to case PCM28-Only. Thus, in order to achieve drier microclimate conditions and reduce sweat accumulation on the skin, a Des can be added to the PCM packet. However, in order to maintain the same overall thermal comfort level, PCM packets of lower mleting temperatures need to be used to eliminate the effect of the heat released by the desiccant during moisture adsorption.

CONCLUSION

In this study, a validated fabric-PCM-Des model was integrated with a bioheat model to predict the cooling performance of a PCM-Des cooling vest and compare it with a PCM cooling vest. The model was used to predict the microclimate temperature and humidity ratio, the mean and torso skin temperatures, the PCM melted fraction, and the mass of accumulated sweat on the skin at hot humid ambient conditions and moderate activity level. The results revealed that the vest with Des caused a decrease in the microclimate humidity ratio and mass of accumulated sweat, but it reduced overall thermal comfort dure to the increase in microclimate, toso, and mean skin temperatures. A parametric study was performed to investigate whether the use of PCM packets with lower melting temperature would overcome the negative effect of the Des while decreasing the microclimate moisture content and the mass of accumulated sweat on the skin. The results showed that decreasing the PCM melting temperature from 28 °C (82.4°F) to 21°C (69.8°F) improved the overall thermal comfort and still achieved drier microclimate conditions.

NOMENCLATURE

Des	=	desiccant
Met	=	metabolic equivalent of task
PCM	=	phase change material
PCM28-Only	=	vest contains only PCM with a melting temperature of 28 °C
PCM28-Des	=	vest contains PCM with a melting temperature of 28 °C and desiccants
PCM21-Only	=	vest contains only PCM with a melting temperature of 21 °C
PCM21-Des	=	vest contains PCM with a melting temperature of 21 °C and desiccants

REFERENCES

M. Zhao, C. Gao, F. Wang, K. Kuklane, I. Holmér, J. Li, The torso cooling of vests incorporated with phase change materials: a sweat evaporation perspective, Textile Research Journal 83(4) (2013) 418-425.

M. M. Yazdi, M. Sheikhzadeh, S. E. Chavoshi, Modeling the performance of a PCM cooling vest considering its side effects, International Journal of Clothing Science and Technology 27(4) (2015a) 573-586.

C. Gao, K. Kuklane, I. Holmér, Cooling vests with phase change materials: the effect of melting temperature on heat strain alleviation in an extremely hot environment, European Journal of Applied Physiology 111 (2011) 1207-1216.

J.R. House, H.C. Lunt, R. Taylor, G. Milligan, J.A. Lyons, C.M. House, The impact of a phase-change cooling vest on heat strain and the effect of different cooling pack melting temperature, Eur J Appl Physiol 113 (2013) 1223-1231.

M. M. Yazdi, M. Sheikhzadeh, A. Dabirzadeh, E. Chavoshi, Modeling the efficiency and heat gain of a phase change material cooling vest: The effect of ambient temperature and outer isolation, Journal of Industrial Textiles 46(2) (2016) 436-454.

C. Gao, K. Kuklane, I. Holmér, Cooling effect of a PCM vest on a thermal manikin and on humans exposed to heat, Environmental Ergonomics 12 (4) (2007) 146-149.

C. Gao, K. Kuklane, I. Holmér, Personal cooling with phase change materials to improve thermal comfort from a heat wave perspective, Indoor air 22(6) (2012) 523-530.

H. Hasegawa, T. Takatori, T. Komura, M. Yamasaki, Wearing a cooling jacket during exercise reduces thermal strain and improvesendurance exercise performance in a warm environment, The Journal of Strength and Conditioning Research 19(1) (2005) 122-128.

N. Nishihara, S. I. Tanabe, H. Hayama, M. Komatsu, A cooling vest for working comfortably in a moderately hot environment, Journal of physiological anthropology and applied human science 21(1) (2002) 75-82.

J. Webster, E. J. Holland, G. Sleivert, R. M. Laing, B. E. Niven, A light-weight cooling vest enhances performance of athletes in the heat, Ergonomics 48(7) (2005) 821-837.

M. M. Yazdi, M. Sheikhzadeh, S. Borhani, Modeling the heat transfer in a PCM cooling vest. The Journal of The Textile Institute 106(9) (2015b) 1003-1012.

R.E. Reinertsen, H. Færevik, K. Holbø, R. Nesbakken, J. Reitan, A. Røyset, M. Suong Le Thi, Optimizing the performance of phase-change materials in personal protective clothing systems, International Journal of Occupational Safety and Ergonomics (JOSE) 14(1) (2008) 43–53.

M. Itani, N. Ghaddar, K. Ghali, Innovative PCM-desiccant packet to provide dry microclimate and improve performance of cooling vest in hot environment, Energy Conversion and Management 140 (2017) 218-227.

S. Houshyar, R. Padhye, O. Troynikov, R. Nayak, S. Ranjan, Evaluation and improvement of thermo-physiological comfort properties of firefighters' protective clothing containing super absorbent materials, The Journal of the Textile Institute 106(12) (2015) 1394-1402.

F. Dotti, A. Ferri, M. Moncalero, M. Colonna, Thermo-physiological comfort of soft-shell back protectors under controlled environmental conditions, Applied ergonomics 56 (2016) 144-152.

H. Wang, S. Hu, Experimental study on thermal sensation of people in moderate activities, *Building and Environment 100* (2016) 127-134.

D. Lai, F. Wei, Y. Lu, F. Wang, Evaluation of a hybrid personal cooling system using a manikin operated in constant temperature mode and thermoregulatory model control mode in warm conditions, Textile Research Journal 87(1) (2017) 46-56.

W. Yi, Y. Zhao, A. P. Chan, Evaluating the Effectiveness of Cooling Vest in a Hot and Humid Environment, Annals of Work Exposures and Health 61(4) (2017) 481-494.

M. Al-Othmani, N. Ghaddar, K. Ghali, A multi-segmented human bioheat model for transient and asymmetric radiative environments, International Journal of Heat and Mass Transfer 51(23) (2008) 5522-5533.

A. P Avolio, Multi-branched model of the human arterial system, Medical and Biological Engineering and Computing 18(6) (1980) 709-718.

W. Karaki, N. Ghaddar, K. Ghali, K. Kuklane, I. Holmer, L. Vanggaard, Human thermal response with improved AVA modeling of the digits, International Journal of Thermal Sciences 67 (2013) 41-52.

M. Salloum, N. Ghaddar, K. Ghali, A new transient bioheat model of the human body and its integration to clothing models, International journal of thermal sciences 46(4) (2007) 371-384.

H. Hamdan, N. Ghaddar, D. Ouahrani, K. Ghali, M. Itani, PCM cooling vest for improving thermal comfort in hot environment, International Journal of Thermal Sciences 102 (2016) 154-167.

H. Zhang, C. Huizenga, E. Arens, D. Wang, Thermal sensation and comfort in transient non-uniform thermal environments, European Journal of Applied Physiology 92 (2004) 728–733.

B. W. Jones, Transient interaction between the human and the thermal environment, ASHRAE Trans. 98 (1) (1992) 189-195

Quantifying Losses Due to Thermal Discomfort: An Agent Based Modeling Approach

Mohamad Awada, M.S **Issam Srour, PhD**

ABSTRACT

This paper presents a framework that employs an agent based model (ABM) to examine and understand the impact of occupants' thermal behavior in a commercial building. The proposed simulation framework has two main objectives. The first objective is to model and understand the behavior of occupants in their workplace environment while taking into account the influence of thermal conditions on the behavioral decision-making. The four main behaviors that are studied correspond to the agent's adjusting of clo level, shades use, thermostat use, and personal fan use. The second objective is to quantify the financial and profitability losses due to thermal discomfort in an air-conditioned office by examining the effect of thermal dissatisfaction on productivity, absenteeism and complaint rates. Simulated agents in the proposed ABM framework behave according to the Fogg Behavior Model, where two main elements must converge at the same time for a certain behavior to occur; motivation and ability.

INTRODUCTION

Human resources are a vital asset for any company. Since people spend around 80 to 90 percent of their time indoors, they strive to establish indoor environments in which they feel comfortable and productive (Dounis and Caraiscos, 2009). This is in-line with the overarching objective of any for-profit firm, which is to maximize productivity and, therefore, profitability while maintaining a healthy environment for its employees. The literature provides a wide range of papers that expound how the indoor environmental quality (IEQ) of a certain workplace can affect the comfort and performance of the employees (Mahbob et al., 2011; Wong et al., 2008). For instance, Fisk and Rosenfeld (2000) estimated that improved indoor environment can results in a direct increase in productivity, ranging between 0.5% and 5 percent.

Thermal comfort is considered an essential component of IEQ, thus building designers and operators try to ensure that the thermal environment live up to the expectations of the end users. Thermal satisfaction associated with proper thermal conditions has shown to boost the performance of employees while reducing the frequency and duration of absenteeism (Lan et al., 2011). Unfortunately, satisfaction with thermal conditions is not always guaranteed. Complaints about thermal discomfort are very common among office occupants, perhaps more than other complaints such as burned light bulb or plumbing leaks etc. (Federpiel et al., 1998).

This paper presents a unique simulation method using Agent Based Modeling (ABM) to achieve two objectives. The first objective is to imitate the behavior of occupants in their workplace environment while taking into account the influence of the thermal conditions on the behavioral decision-making. The second objective is to investigate the loss of profitability due to thermal discomfort in an air-conditioned office, by examining the effect of thermal

Mohamad Awada is a Master's student in the Department of Civil Engineering,, American University of Beirut. **Issam Srour** is an associate professor in the Departmenet of Civil Engineering, American University of Beirut.

dissatisfaction on productivity, absenteeism and complaint rates.

LITERATURE REVIEW

Thermal Comfort

According to ASHRAE (2004), the thermal comfort of an occupant in an indoor environment is defined as "*the condition of mind that expresses satisfaction with the thermal environment and is assessed by subjective evaluation*". In general, human bodies tend to produce heat when performing a mechanical effort. A human body's thermal equilibrium with the surroundings is maintained when the body reaches a state of thermal neutrality; i.e. when the heat generated by a human metabolism dissipates from the body.

The Predicted Mean Vote (PMV) is commonly used model describing thermal comfort. It was first proposed by Fanger (Fanger, 1970) and was developed using heat balance theories and a wide range of experimental data that was collected in a controllable climate chamber. This model is primarily used in air-conditioned indoor environments. Thermal comfort surveys typically use a seven-point scale from -3 (too cold) to +3 (too hot) to examine the thermal sensation of the surveyed occupants. Fanger's equations for the PMV are used to compute the thermal sensation of occupants given a certain combination of personal and environmental factors. ASHRAE standard 55 (2004) assigns a value of zero to the ideal PMV score representing the thermal neutrality and the comfort zone is within a PMV range of [-0.5, 0.5].

According to Fanger's Model, there are six primary factors that can directly influence the thermal comfort and determine heat gains and losses; these factors are divided into two major categories:

- Personal factors: characteristics of the occupants such as their metabolic rate and clothing level.
- Environmental factors: conditions of the thermal environment such as the air temperature and speed, the mean radiant temperature, and the relative humidity.

Agent Based Modeling Overview

This paper uses ABM to examine thermal comfort in an office setting. ABM is considered a relatively new method for simulating the actions and interactions among autonomous agents. It has been implemented in a broad range of domains and disciplines and especially in non-computing related scientific domains such as biology, ecology, health and social sciences. Recently, ABM has been applied in the context of simulating building occupants' behavior. Most of these studies focused on the effect of adaptive behaviors of individual agents on energy consumption. For instance, Azar and Menassa (2011) proposed a model to estimate energy consumption in commercial buildings. The model aims at establishing realistic energy consumption predictions during the building design phase, while taking into consideration different and dynamic energy consumption patterns among the occupants. Soo Lee and Malkawi (Soo Lee and Malkawi, 2014) presented an ABM approach that aims to mimic a real-world occupant behavior that interacts with both its environment and other occupants. In this model, the occupant tends to act and make decisions based on its level of thermal comfort. Using simulation coupling, the authors were able to study the influence of behavior on both the thermal conditions and energy consumption in the building. Linkola et al. (2013) introduced an ABM model that simulates the household water consuming behavior in residential buildings. The proposed approach replicates major water-using activities of household members within a rich technological and behavioral context, calibrated with appropriate statistical data.

Despite the contribution of the aforementioned studies, most of the ABM models presented in the literature that investigate human behavior in buildings focused on two main outcomes: energy and water consumption. The paper at hand tries to present a unique simulation that employs an ABM framework to model thermal comfort and to investigate the financial losses imposed on companies due to thermal discomfort of employees in air-conditioned offices.

METHODOLOGY

General Description

According to ASHRAE standard 62.1 (2004), the occupant density in office spaces is typically 5 occupants per $100m^2$. The proposed ABM model complies with this standard by simulating the behavior of 6 occupants in a $120m^2$ workspace office.

A human behavior model is usually based on three main processes that are considered vital in the field of behavior research: Behavior definition, trigger recognition and measurement of the effects associated with the accomplished behavior. The proposed model in this paper focuses on behaviors that are triggered by thermal discomfort and can only affect the thermal satisfaction of the occupants. Given that the focus of the paper is on air conditioned spaces, a list of predefined behaviors, along with the thermal comfort factor affected by each option is shown in Table 1.

Table 1. Behavior Options

Behavior option	Affected Factor
Changing the Clo Level	Clo Level
Pulling up/down shades	Mean Radiant Temperature
Higher/Lower thermostat set point	Air Temperature
Turning On/Off personal Fan	Air Speed

The simulation takes place over a whole year, taking into consideration the effect of the different seasons over the mean radiant temperature and the occupants' choice for their initial clo level at the beginning of each day. Each day starts at 8:00 AM and ends at 6:00 PM. Saturdays and Sundays are considered as holidays and hence are not included in the simulation process.

Human Behavior Modeling

This section discusses two aspects of the human behavior modeling adopted in this paper. The first subsection illustrates the approach implemented for choosing the action initiator among the occupants. The second subsection tackles the decision making process that will be employed for the action initiator to choose a certain behavior over others so that s/he can reach satisfaction. The proposed methodology for describing the human behavior is based on the Fogg Behavior Model (FBM) (Fogg, 2009).

Action Initiator

The proposed office under simulation comprises of 6 employees. To keep the model simple, at each time step one out of the 6 occupants behaves by altering the conditions of the environment based on an assessment of thermal discomfort. To identify the action initiator, each agent calculates a corresponding cost function called "*Initiation Cost Function*" (ICF); the occupant with the highest initiation cost score becomes the action initiator. The ICF is based on the idea of Motivation adopted form the FBM. Motivation is then divided into two main pillars: Sensation and Belonging Motivation.

A. Sensation Motivation

According to Fogg, Sensation Motivation is a dimension that has sides: pleasure and pain. Fogg explains that Sensation Motivation is the most powerful stimulus to most of our actions and behaviors, because the results of such a motivator are usually immediate, with little thinking or anticipating. In this paper, the occupants' Sensation Motivation is based on their pleasure or degree of comfort with their surrounding environment. The comfort zone described by ASHRAE falls within a PMV range of [-0.5, 0.5], the higher the deviation of an occupant's PMV score from that range the more the agent feels the urge to behave and change the conditions of the environment in a way that grants him/her comfortable thermal sensation. Equation 1 represents a mathematical translation of the Sensation

Motivation.

$$\alpha = \frac{|PMV| - 0.5}{2.5} \quad (1)$$

This equation allows for quantifying sensation on both sides of the PMV scale: cold (PMV < -0.5) and hot (PMV > 0.5) edges. For instance, consider two agents occupying a certain office; Agent 1 has a PMV score of -1 and Agent 2 has a PMV score of +2. In this case, $\alpha_1 = 0.2$ whereas $\alpha_2 = 0.6$. Thus, Agent 2 is feeling more uncomfortable which boosts his/her sensation motivation more than Agent 1. The "too hot" feeling of Agent 2 dominated the "too cold" feeling of Agent 1.

B. Belonging Motivation

Fogg describes this motivation part as the social perspective of his proposed model. According to FBM, it is very intuitive that people behave in a way that increases their social acceptance among others and reduces their chances of being socially rejected. In the paper at hand, this is also taken into consideration to identify the agent that will take action. The Belonging Motivation is reflected by the corresponding work position of each agent. A senior employee with the highest work position is not going to think twice before executing a certain behavior that will alter the workspace environmental conditions. On the other hand, a junior employee is going to hesitate before taking an action that will change the conditions of the indoor environment. This is the core of the Belonging Motivation concept; employees with a low work position feel uncomfortable to take an action that would make other occupants of higher work position dissatisfied with their thermal environment. Equation 2 represents a mathematical translation of the Belonging Motivation. In this paper, employees are ranked into 3 different positions: Ranks 1, 2, and 3 with ascending order of seniority.

$$\beta = \left| \frac{Employee's\ current\ position - Highest\ rank}{Highest\ rank} \right| \quad (2)$$

The lower the belonging motivation of a certain employee is, the less guilty s/he feels while changing the indoor environmental conditions of the office.

C. Initiation Cost Function Equation

The governing equation that indicates the action initiator at every time step is given by Equation 3. This equation calculates the overall Sensation and Belonging Motivation for each employee. Agent i with the highest score is considered as the action initiator.

$$\gamma_i = \alpha_i - \beta_i \quad (3)$$

Behavior Utility Function

Agents under simulation in the proposed model have four main options to choose from (Table 1), which can increase their level of satisfaction with the environment and shift those who are dissatisfied to the comfort zone. After indicating the employee responsible for taking the action, the action initiator has to identify the best behavior option that can increase her/his satisfaction with the thermal conditions of the office. The following subsections represent the approach adopted to tackle the following decision making process.

A. Overall Mechanism Description

In their paper, Andrews et al. (2011) presented an agent based modeling approach that describes the behavior of occupants in buildings. The authors suggested a multi-attribute utility function that describes the performance level of a given option. They attributed the use of such a function to three main reasons:

1- Occupant behavior choice which depends on four components: Benefits to self, costs to self, benefits to others and cost to others. A thorough description of these factors is presented in the following subsections.

2- The strong simplifying assumption that occupants perceive the four components of the utility function as having preferential, utility, and additive independence. Therefore, it is reasonable to specify an additive formulation of the four utilities for each behavior.

3- It is reasonable to associate each of the four utilities with a weighting constant. Weights should be based on occupants' responses to a survey questioning the importance of each attribute described above.

Equation 4 shows the additive multi-attribute utility function proposed by Andrews et al.

$$Occupant\ Utility\ for\ a\ behavior = U = \sum k_i U_i \tag{4}$$

B. Benefit to self

The "Cost to self" idea is associated with what Fogg described as Ability to perform a certain action. In this paper, the concept of "Cost to self" is going to be linked with the distance necessary to conduct an action. The longer the distance is, the harder it is to perform the corresponding option thereby reducing the motivation of the agent to complete that behavior and increasing the "Cost to Self". The mathematical equation governing the single attribute utility expression for "Cost to self" notion is given by Equation 5.

$U_1 = 0$ if the agent stays dissatisfied when the option under consideration gets implemented

$U_1 = 1$ if the agent gets satisfied when the option under consideration gets implemented

C. Cost to self

The "Cost to self" idea is associated with what Fogg described as Ability to perform a certain action. In this paper, the concept of "Cost to self" is going to be linked with the distance necessary to conduct an action. The longer the distance is, the harder it is to perform the corresponding option thereby reducing the motivation of the agent to complete that behavior and increasing the "Cost to Self". The mathematical equation governing the single attribute utility expression for "Cost to self" notion is given by Equation 5.

$$U_2 = \frac{Distance\ of\ option\ under\ consideration}{Longest\ distance\ associated\ with\ an\ option} \tag{5}$$

The distance associated with the options of changing the clo level and turning on/off the fans is considered to be equal to zero. Depending on the location of the occupant in the office, the longest distance of an option could be that corresponding to the thermostat or the shades.

D. Cost and Benefits to others

The remaining two concepts "Benefits and Cost to others" are a representation of the Belonging Motivation. When performing a specific behavior that can change the thermal conditions of the environment, agents would anticipate how many dissatisfied occupants are becoming satisfied after performing the behavior under consideration. Similarly, agents will look up if the considered behavior option will shift people outside of the thermal comfort zone. These two metrics reflect the concepts of "Benefits and Cost to others" respectively, represented by Equations 6 and 7.

$$U_3 = \frac{Number\ of\ agents\ that\ became\ satisfied\ after\ implementing\ the\ option}{Total\ number\ of\ agents\ in\ the\ office} \quad (6)$$

$$U_4 = \frac{Number\ of\ agents\ that\ became\ dissatisfied\ after\ implementing\ the\ option}{Total\ number\ of\ agents\ in\ the\ office} \quad (7)$$

Quantifying the losses

Three types of losses are studied in this paper: Productivity or performance losses, losses associated with thermal sensation complaints and absenteeism losses.

Productivity Losses

Numerous studies examined how environmental factors such as the air temperature affect the performance and productivity of office employees. Of particular interest is Roelofsen's work (2002), which takes the analysis a step further by presenting a direct relation between the productivity loss of office workers and the corresponding PMV. Equation 8 gives the mathematical representation of Roelofson's model:

$$P = b_0 + b_1 \times PMV + b_2 \times PMV^2 + b_3 \times PMV^3 + b_4 \times PMV^4 + b_5 \times PMV^5 + b_6 \times PMV^6 \quad (8)$$

Table 2 Regression Coefficients for Equation 8

Regression Coefficient	PMV < 0	PMV > 0
b_0	1.280207	-0.15397397
b_1	15.995451	3.8820297
b_2	31.507402	25.176447
b_3	11.754937	-26.641366
b_4	1.4737526	13.11012
b_5	0	-3.1296854
b_6	0	0.2926092

The cost of the lost productivity per employee can be estimated in terms of staff wages. This can be estimated using Equation 9:

$$Productivity\ Loss = \left[\frac{1}{100} \times \frac{\sum_{i=1}^{i=hours\ occupied} P}{Hours\ occupied}\right] \times Wages \quad (9)$$

Complaints Losses

In his paper, Federpiel et al. (1998) presented the survey results of 23,500 office occupants in 690 commercial buildings, investigating the thermal sensation complaints of the employees. Federpiel presented a probability distribution for complaints about thermal sensation and differentiated between two types of complaints (too cold, too hot), shown in Figure 1. For instance, if an agent has a PMV < -0.5 (cold), the model calculates the chance of filing a cold complaint at that step time, from the cold complaint density function based on the current air temperature.

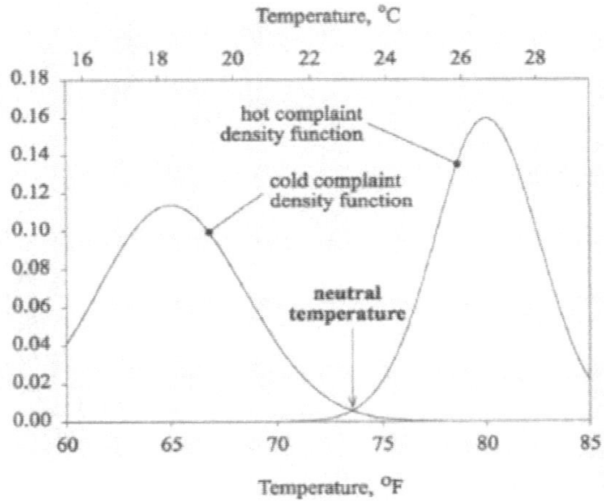

Figure 1 Hot and cold complaints density functions (Federpiel et al., 1998).

The proposed model treats complaints as an alarm for building operators. The significance of this model is that it relates building temperature, which impacts occupants' thermal satisfaction, with operation and maintenance costs because complaints become corrective maintenance work orders in commercial buildings.

In his paper, Martin (2004) presented a methodology to quantify the maintenance cost incurred by thermal complaints in commercial buildings. Equation 10 presents this relation:

$$C_k = H \times L \quad (10)$$

Absenteeism Losses

Research into people's satisfaction with indoor environmental quality (7,000 responses) showed that employees in building are absent on average 2.5 days a year (Roelofsen et al., 2002). In other words, the probability of asking for a sick leave the next day is equal to 0.9 percent. To make the numbers more significant, two scenarios are taken into consideration:

1- If an agent spends three hours (step time) dissatisfied on a certain working day then the likelihood of asking for a sick leave due to bad thermal conditions is equal to the aforementioned probability.

2- If an agent spends less than three hours dissatisfied the probability of asking for a sick leave is 0.

Absenteeism losses are calculated easily: A day off from work because of sickness will cost the company the corresponding wage of the absent employee for that specific day.

CONCLUSION

The paper at hand presents a framework that simulates multiple occupants' behaviors in an operating office. The simulation employs an agent based modeling approach in a unique way to study the financial losses incurred by companies due to thermal discomfort of employees in air conditioned offices.

The proposed framework mimics the human behavior of employees within an office. The main trigger for these behaviors is thermal satisfaction, where agents behave in a way that grants them the maximum number of thermal comfort hours. Also, the model investigates the losses of companies due to occupants' thermal discomfort in air conditioned offices by examining three key factors related to thermal sensation: Productivity losses, losses incurred by thermal complaints and absenteeism. Thermal comfort and behavior of the agents in the proposed model establish emergent results that vary in a very complex way based on the dynamics of the agents' thermal conditions.

NOMENCLATURE

α = Sensation Motivation

$|PMV|$ = Absolute value of the agent's PMV score

0.5 = Absolute value of the edges of the comfort zone

2.5 = Maximum difference between the edges of the comfort zone and the PMV scale

β = Belonging Motivation

Employee's current position = Position of the employee under study [1 or 2 or 3]

Highest rank = Highest work position an employee can have, i.e. [3]

γi= Initiation Cost function for an agent i

U_i = Single attribute utility for attribute i with a range from 0 to 1

k_i = Weighting constant for the utility of attribute i (range is 0 to 1 with $\sum k_i$ =1).

i has 4 values: 1 represents benefits to self, 2 represents cost to self, 3 represents benefits to others and 4 represents cost to others.

U has a range from -1 to 1.

P = Productivity level corresponding to a certain PMV score

$b_i s$ = Regression coefficients presented in Table 2are presented in Table 2.

C_k = Complaint cost

H = Average number of hours in response to a thermal sensation complaint, H = 1.85 hours

L = Total hourly labor rate for technicians who handle complaints

REFERENCES

Andrews, C. J., Yi, D., Krogmann, U., Senick, J. A., & Wener, R. E. (2011). Designing buildings for real occupants: An agent-based approach. *IEEE Transactions on Systems, Man, and Cybernetics-Part A: Systems and Humans, 41*(6), 1077-1091.

ASHRAE (2004) "ANSI/ASHRAE Standard 55R – Thermal Environmental Conditions for Human Occupancy", Atlanta: American Society of Heating, Refrigerating and Air-Conditioning Engineers, Inc.

Azar, E., & Menassa, C. C. (2011). Agent-based modeling of occupants and their impact on energy use in commercial buildings. *Journal of Computing in Civil Engineering, 26*(4), 506-518.

Dounis, A. I., & Caraiscos, C. (2009). Advanced control systems engineering for energy and comfort management in a building environment—A review. *Renewable and Sustainable Energy Reviews, 13*(6-7), 1246-1261.

Fabbri, K. (2015). Indoor thermal comfort perception. A Questionnaire Approach Focusing on Children; Springer: New York City, NY, USA.

Fanger, P. O. (1970). Thermal comfort. Analysis and applications in environmental engineering. Thermal comfort. Analysis and applications in environmental engineering.

Federspiel, C. C., Bridges, B., & Langkilde, G. (1998). Statistical analysis of unsolicited thermal sensation complaints in commercial buildings/Discussion. *ASHRAE Transactions*, *104*, 912.

Fisk, W. J. (2000). Health and productivity gains from better indoor environments and their relationship with building energy efficiency. *Annual Review of Energy and the Environment*, *25*(1), 537-566.

Fogg, B. J. (2009, April). A behavior model for persuasive design. In Proceedings of the 4th international Conference on Persuasive Technology (p. 40). ACM.

Lan, L., Wargocki, P., & Lian, Z. (2011). Quantitative measurement of productivity loss due to thermal discomfort. *Energy and Buildings*, *43*(5), 1057-1062.

Lee, Y. S., & Malkawi, A. M. (2014). Simulating multiple occupant behaviors in buildings: An agent-based modeling approach. *Energy and Buildings*, *69*, 407-416.

Linkola, L., Andrews, C. J., & Schuetze, T. (2013). An agent based model of household water use. *Water*, *5*(3), 1082-1100.

Mahbob, N. S., Kamaruzzaman, S. N., Salleh, N., & Sulaiman, R. (2011). A correlation studies of indoor environmental quality (IEQ) towards productive workplace.

Martin, R. A. (2004). Optimal prediction, alarm, and control in buildings using thermal sensation complaints (Doctoral dissertation, University of Califoria, Berkeley).

Roelofsen, P. (2002). The impact of office environments on employee performance: The design of the workplace as a strategy for productivity enhancement. *Journal of facilities Management*, *1*(3), 247-264.

Standard, A. S. H. R. A. E. (2004). Standard 62.1-2004,". *Ventilation for acceptable indoor air quality.*

Wong, L. T., Mui, K. W., & Hui, P. S. (2008). A multivariate-logistic model for acceptance of indoor environmental quality (IEQ) in offices. *Building and Environment*, *43*(1), 1-6.

A Comparative Assessment of the Performance of Cooling Systems for Large-Scale High-Density Data Centers Using CFD Simulations

Khaled Abou Hweij **Adnan Akhdar**

ABSTRACT HEADING

The increasing demand for storage, networking and high-performance computing has led to a significant growth in the size, energy density and complexity of data centers which drives the need for efficient cooling strategies to cope with the rising energy demands. Thus, thermal performance and efficiency of cooling systems in data centers has become indispensable to ensure that the servers operate within reliable thermal limits for the ever-increasing servers' power density while minimizing as much as possible the energy consumption of the cooling system which constitutes a large portion of the total consumption of data center.

This paper will use computational fluid dynamics (CFD) simulations in order to study and compare the performance of current and emerging cooling strategies in efficient air distribution and thermal management of the heat dissipation from the servers. Specifically, the performance of a traditional and widespread computer room air conditioning (CRAC) system with underfloor plenum will be compared against emerging systems such as Chiller-Door Units (CDU) and In-Row system without raised floors. For this purpose, a detailed investigation of airflow delivery is carried out on full-scale three-dimensional models of a large and high-density data center to comparatively assess the resultant thermal distributions of the considered cooling systems and establish the fundamental benefits and drawbacks of each system.

In summary, the findings indicate that there is significant potential to increase efficiency while still maintaining uniformly delimited hot and cold regions in front of the sever cabinets by moving the cooling energy closer to the heat source to avoid losing cooling capacity by mixing with the hotter air inside the stratified layer above the cabinets.

INTRODUCTION

Significant challenges had emerged due to increased demand on technological advancements, such as, data storage, numerical simulations and telecommunications connections [Capozzoli and Primiceri, 2015]. Such advancements may require a very fast processing and data collection at a localized place [Patankar, 2010]. A suitable solution for the aforesaid requirements is the use of computer servers deployed in a single center [Patankar, 2010], where such a facility is known as Data Center [Capozzoli and Primiceri, 2015; Hassan et al., 2013; Patankar, 2010]. Indeed, it is well known that the servers and racks produce an immense amount of heat while consuming electrical energy during data processing [Patankar, 2010]. The heat generated from processors in servers and racks will affect their performance and disrupt their normal operation [Hassan et al., 2013; Patankar, 2010], and sometimes lead to their critical failure due to excessive heat [Patankar, 2010]. Thus, cooling down these processors to allow them to operate within set acceptable thermal limits is vital to maintain servers' proper functionality [Nada et al., 2016; Patankar, 2010] and to optimize the overall energy efficiency of the data center.

Khaled Abou Hweij, M.E., is a mechanical engineer and member of the CFD Unit in the Mechanical Department at Dar Al-Handasah. **Adnan Akhdar, M.E.,** is the head of CFD Unit in the Mechanical Department at Dar Al-Handasah.

Traditionally, air-cooled systems have been widely adopted as cooling strategies for all sizes of data centers due to their ease of implementation [Capozzoli and Primiceri, 2015]. The most widely used system is based on using Computer Room Air Conditioners (CRACs) where server cabinets are arranged into organized rows vertically arrayed on the top of an underground plenum [Hassan et al., 2013]. Cold air is introduced and distributed horizontally into the whole underground raised floor plenum volume. Perforated tiles located near the bottom edge of rack cabinets draw cold air out into the unconditioned space in the Data Center [Dunlap and Ramussen, 2012; Patankar, 2010]. Cold air's temperature will increase after exchanging heat with servers where hot air will be returned into (CRACs) exhaust ducts [Mulay, 2009; Vangilder, 2005]. One of the uses of the (CRACs) systems is to eliminate the non-uniform air distribution throughout the server cabinets, where the upper parts of the cabinets will be better conditioned than the lower parts. These type of challenges appear significantly in large scale data centers where mixing is prominent. In turn, this will lead to over-design of cooling energy in this system.

Row-based cooling Units (RCUs) is another air based cooling method used in data centers that is gaining more adoption due to its inherent advantages. Specifically, row coolers are inserted in each row of racks and servers where the premise of this technique is to draw hot air from the hot aisle region into cold aisle one minimizing mixing patterns with the hot aisle. Guided vanes are used in coolers to deliver cold air to the opposite or adjacent racks [Lin and Avelar, 2014] enhancing heat removal rates.

A more recent cooling alternative that is gaining significant traction in large scale applications is cabinet liquid cooling which aims to locally extract the heat release from the server cabinets at the source using chilled water pipes and to prevent the heat from spreading into the room. Chilled Door Units (CDUs) represent an industrial implementation of this concept whereas chilled doors are deployed inside the Data Center space within a close proximity to the rear end of the server cabinets by means of interface frames. These Chilled Doors are constituted of thermally effective heat sinks to promote direct contact between the hot expelled air from servers and the chilled water pipes inside the apparatus.

This paper will illustrate the use of CFD simulations to investigate and compare the performance of three different cooling strategies used in large scale data centers using CFD simulations by conducting a detailed investigation of airflow delivery and thermal conditions in a full-scale three-dimensional model of a real data center.

METHODOLOGY AND CFD MODELS

Achieving an effective strategy for large scale and high-density data centers necessitates the accurate prediction of the flow pattern inside the server room for various design scenarios. CFD simulations using ANSYS FLUENT are used to simultaneously predict the airflow pattern and the heat distribution inside the server room; and this detailed information can be used to gain invaluable insight into the underlying physical phenomena and to quantitatively assess the performance of different cooling strategies in a typical large scale data center. 3D full-scale representations of the three cooling modes were simulated and the resulting flow patterns and thermal distributions were established for cooling assessment.

- In **Model 1** a 1.35 m (4.429 ft) raised plenum was used to deliver cold air into the server room using perforated tiles (cold aisles) while hot air was returned via ducts mounted directly overhead the (CRACs) units as shown in **Figure 1**. The computational grid of **Model 1** consists of approximately 6.4 million tetrahedral cells.
- Row-based cooling was implemented in **Model 2** where (RCUs) were introduced between the servers as depicted in **Figure 2.** The computational grid of **Model 2** consists of approximately 11.8 million tetrahedral cells.
- As for **Model 3,** the (CDUs) were employed and mounted in the data center as illustrated in **Figure 3.** The computational grid of **Model 3** consists of approximately 5.2 million tetrahedral cells.

Local mesh refinement was implemented in all models at supply and return outlets, within the server cabinets and for critical flow components such as the guiding vanes in Model 2.

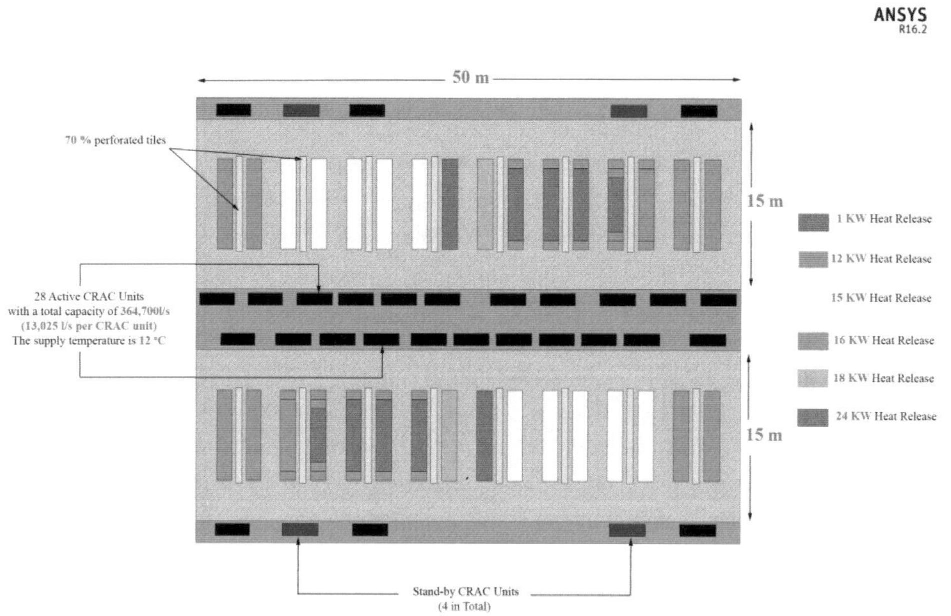

Figure 1: Load components and cooling system components for CRACs based system (Model 1).

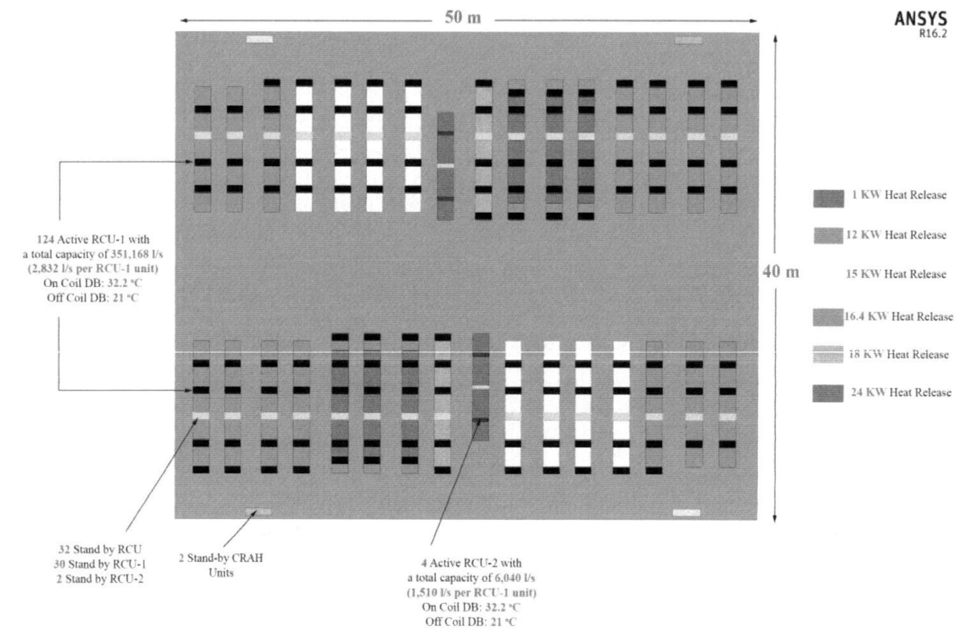

Figure 2: Load components and cooling system components for RCUs based system (Model 2).

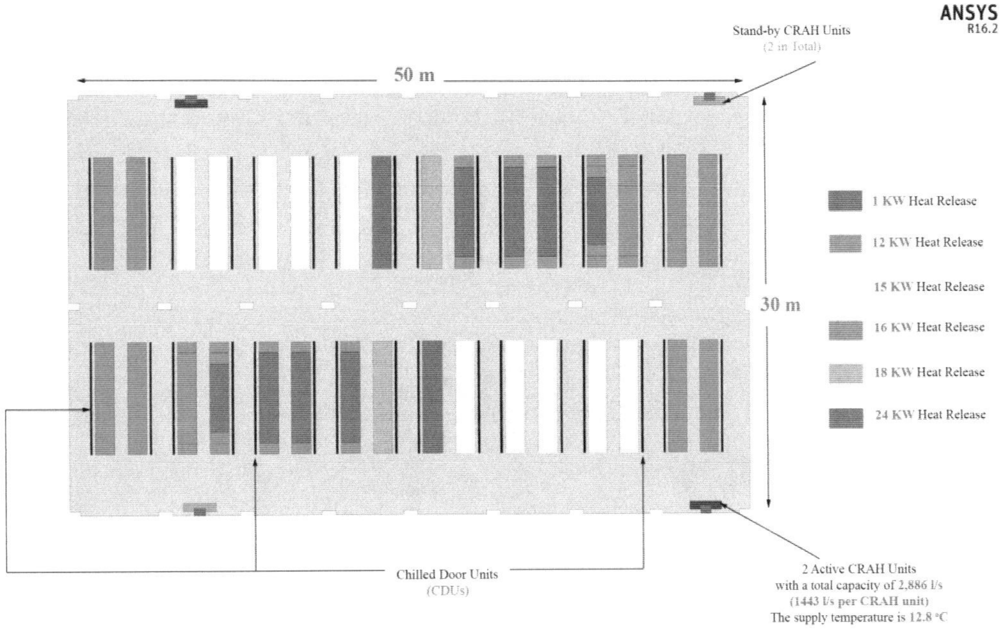

Figure 3: Load components and cooling system components for CDUs based system (Model 3).

DESIGN CRITERIA

It should be noted that identical load components (heat dissipation from cabinets 4,700 kW (16,037,065.67 BTU/h) and lightings 15 W/m^2 (1.41 BTU/hr.ft^2)) have been considered in all models as detailed in **Figures (1-3),** given that the design indoor temperature is 23 °C (73.4°F) and 100 % return is maintained in all of the three models. The main design parameters for Models 1, 2 and 3 are listed below for reference:

- In **Model 1,** cold air at 12°C (53.6°F) is pumped into a 1.35 m underground plenum with a total supply flow rate of 364,700 l/s (5,780,630 gpm) from 28 CRAC units (13,025 l/s (206,451.1 gpm) per CRAC Unit, which is delivered to the space by passing through 70% perforated tiles mounted in the cold aisle region between servers' rows as shown in **Figure 1**.
- In **Model 2,** A total of 357,328 l/s (5,663,781.89 gpm) was supplied via 130 (RCUs) introduced between the servers as shown in **Figure 2.** Guiding vanes (45°) are provided to better direct and distribute the discharge of the (RCUs) into cold aisles. Two Computer Room Air Handlers (CRAH) units supply cold air at 12.8°C (55.04 °F) with a total supply flow rate of 2 886 l/s (45 664 gpm) for room ventilation purposes.
- In lieu of using (RCUs) positioned flush with the servers, active (CDUs) are arrayed in **Model 3** near the rear end of the servers as shown in **Figure 3**. It is worth noting that (CDUs) have been considered operating at full load, and the heat sink from any (CDU) matches exactly the heat dissipation from the corresponding adjacent cabinet. Similarly to Model 2, two Computer Room Air Handlers (CRAH) units supply cold air at 12.8°C (55.04 °F) with a total supply flow rate of 2 886 l/s (45 664 gpm) for room ventilation purposes.

RESULTS

Figure 4a and 4b, depicting contours of temperature and air speed at 1.5m (4.92 ft)AFFL for **Model 1,** indicate that while overall adequate average conditions are achieved in the data hall (average room temperature is 23.3°C (73.94 °F)) , the thermal distribution is uneven with visible cold and hot spots. Indeed, several hot areas with temperatures reaching up to 30°C (86 °F) are observed near the cabinets with the higher heat release 24KW (81,891 BTU/hr). In fact, this is to be expected since the cold air is uniformly introduced into the hall (the perforated tiles are uniformly distributed between cabinets) irrelevant of the heat dissipation of the cabinets. As such, higher temperatures are observed near the higher power density cabinets and lower temperatures are reported near the lower heat dissipation ones. As no containment is considered, some of the cooling capacity is lost by mixing with the hotter air inside the stratified layer above the cabinets. In addition, Figure 4b clearly shows a very active flow pattern with relatively high air speeds inside the room (ranging between 1.4 and 2m/s (4.59 and 6.56 ft/s), denoting significant detrimental turbulence and mixing patterns.

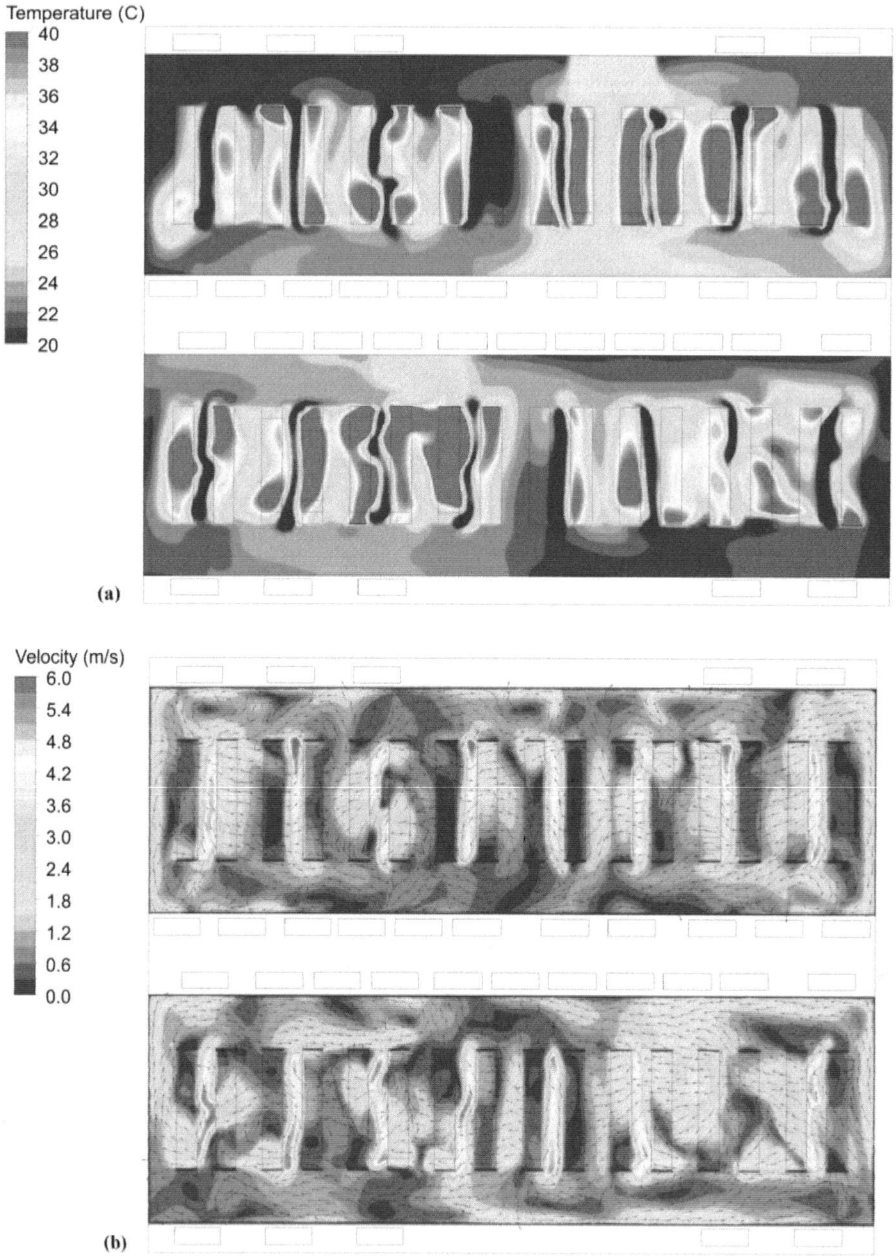

Figure 4 (a) Temperature Contours at 1.5 m (4.92 ft) Above Finished Floor Level **(b)** Velocity Contours at 1.5 m (4.92 ft) Above Finished Floor Level in Rack-Based Cooling for Data Hall

Figures 5a and 5b showing similar contours for Model 2 clearly indicate that the proposed cooling system using (RCUs) is able to effectively establish uniformly delimited cold aisles in front of the server cabinets. Indeed, cold air flow discharged from the (RCUs) is observed to be adequately distributed by the vanes ensuring that uniform temperatures are reached in all cold aisles and thus efficiently extracting the heat released from the server cabinets. Figure 5(a) shows that thermal uniformity is achieved in data hall with similar temperatures observed in all areas of the hall irrespective to the cabinet loads. This further shows the potential of the (RCUs) in locally extracting the heat release of the servers without any impact on the room thermal conditions where the average reported room temperature is 24.2°C (75.56 °F). Moreover, Figure 5b shows a much less active flow pattern inside the room as compared to Model 1 indicating that mixing patterns are effectively mitigated with the use of RCUs.

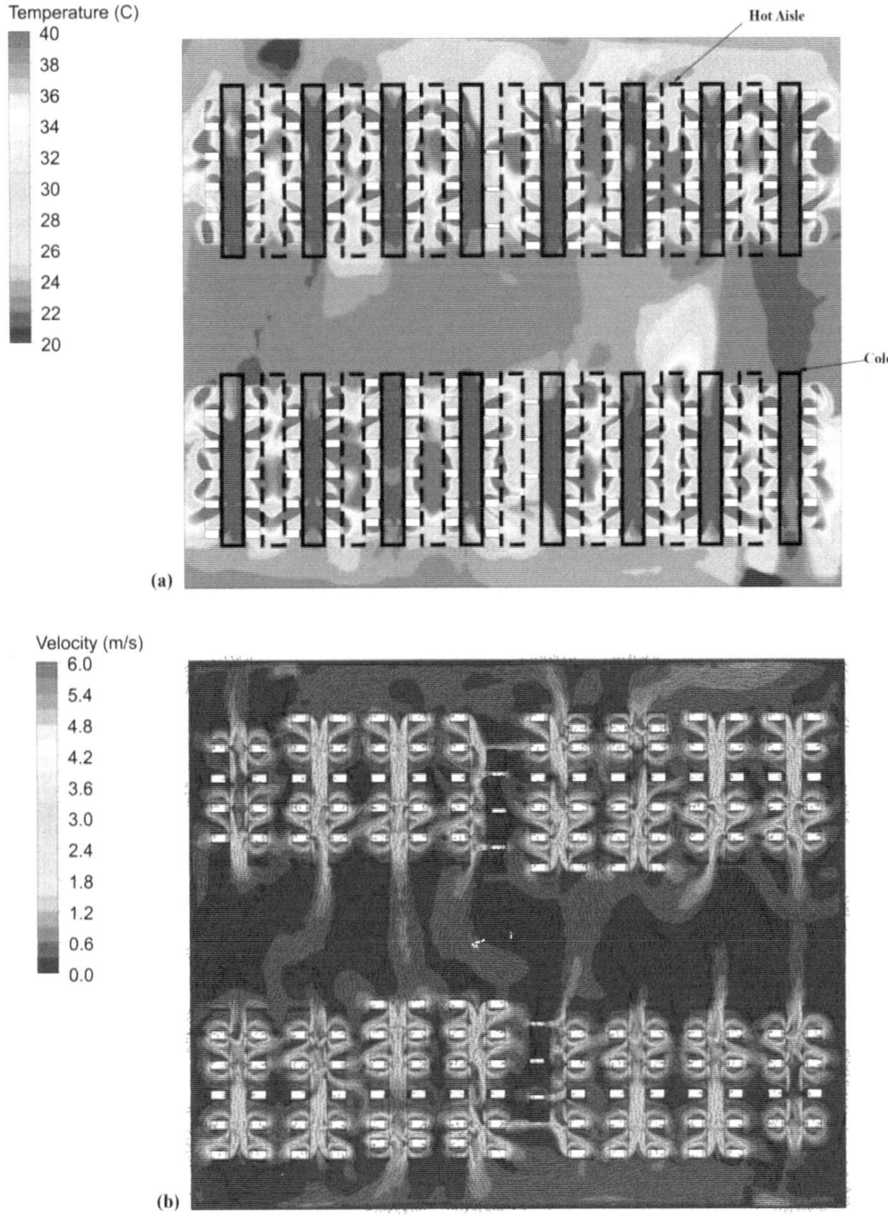

Figure 5: (a) Temperature Contours at 1.5 m (4.92 ft) Above Finished Floor Level **(b)** Velocity Contours at 1.5 m (4.92 ft) Above Finished Floor Level in Row-Based Cooling for Data Hall

Finally, **Figures 6a and 6b** show that, similarly to the previous models, (CDUs) are also effectively extracting the heat released by the cabinets at the source with minimal spread into the room and that a quasi-uniform thermal distribution is achieved in the data hall with similar temperatures observed in all areas irrespective of cabinets' loads. In fact, the average reported room temperature is 24.5°C (76.1 °F).

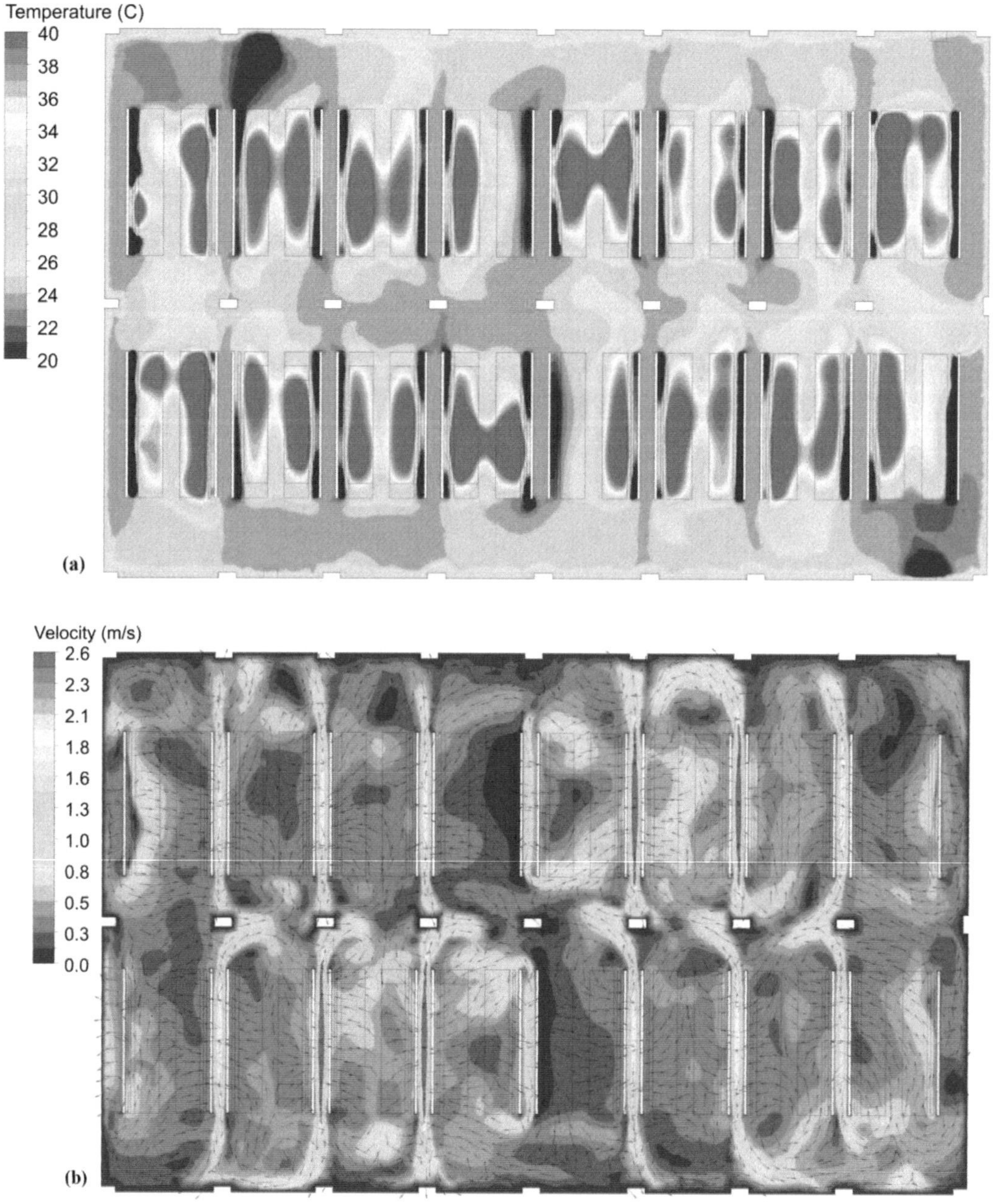

Figure 6: (a) Temperature Contours at 1.5 m (4.92 ft) Above Finished Floor Level **(b)** Velocity Contours at 1.5 m (4.92 ft) Above Finished Floor Level in Chilled Doors Cooling for Data Hall

CONCLUSION

This study investigates the thermal performance of different cooling strategies in a large scale and high density data center using CFD simulations. These simulations have provided invaluable insight in predicting air flow and temperature distributions to comparatively assess the performance of different cooling strategies. The simulation results indicate that while adequate average thermal conditions are achieved with the 3 cooling strategies, the temperature distributions achieved in Models 2 and 3 are quasi-uniform with smoother gradients and no hot spots or mixing patterns. These strategies should help achieve satisfactory thermal conditions to ensure reliable operation of server's racks at reduced energy costs as compared to traditional HACA systems without containment.

ACKNOWLEDGMENTS

We would like to dedicate a special thank you to the design teams at Dar Al-Handasah for their collaboration assistance, and valuable contribution.

NOMENCLATURE

CFD	=	Computational Fluid Dynamics
CDUs	=	Chilled Door Units
CRACs	=	Computer Room Air Conditions
CRAH	=	Computer Room Air Handlers
RCUs	=	Row-based Cooling Units

REFERENCES

Capozzoli, A., & Primiceri, G. (2015). Cooling systems in data centers: state of art and emerging technologies. *Energy Procedia, 83*, 484-493.

Hassan, N. M. S., Khan, M. M. K., & Rasul, M. G. (2013). Temperature monitoring and CFD analysis of data centre. *Procedia Engineering, 56*, 551-559.

Lin, P., & Avelar, V., (2012). How Row-based Data Center Cooling Works. *APC cooling*.

Mulay, V. (2009). Analysis of data center cooling strategies and the impact of the dynamic thermal management on the data center energy efficiency, PhD thesis, The University of Texas

Nada, S. A., Said, M. A., & Rady, M. A. (2016). CFD investigations of data centers' thermal performance for different configurations of CRACs units and aisles separation. *Alexandria Engineering Journal, 55*(2), 959-971.

Patankar, S. V. (2010). Airflow and cooling in a data center. *Journal of Heat transfer, 132*(7), 073001.

VanGilder, J. W., & Schmidt, R. R. (2005, January). Airflow uniformity through perforated tiles in a raised-floor data center. In *ASME 2005 Pacific Rim Technical Conference and Exhibition on Integration and Packaging of MEMS, NEMS, and Electronic Systems collocated with the ASME 2005 Heat Transfer Summer Conference* (pp. 493-501). American Society of Mechanical Engineers.

A Full Three-dimensional Simulation of an Industrial Baking Oven

Mohamad Al Nasser **Iyad Fayssal, PhD** **Fadl Moukalled, PhD**

ABSTRACT

Forced convection baking ovens are widely used in the baking industry for their high efficiency and effective heat transfer over radiation-based ovens. The aim of this study is to model and simulate an industrial baking oven as a prior step to optimize the operation of the oven (airflow management) and baking process, and reduce the required fuel consumption. This may help in compensating for the expected rising energy costs and meeting the mandatory industrial standards. A full scale three-dimensional model of an industrial oven is developed and analyzed using the ANSYS FLUENT computational fluid dynamics package that adopts a finite volume approach for the discretization of the governing conservation equations. The oven, which is composed of a heat exchanger, a fan, a baking chamber, and rotary racks, was first constructed using the SOLIDWORKS software and then imported to the design modeler in ANSYS. The computational domain was covered with a mesh of size 14 million elements. Initially, single phase simulations to predict the hydrodynamic and thermal fields were conducted for the sake of establishing an idea about the velocity and temperature distribution of the air circulating inside the oven. The results were validated with available experimental data. Temporal and spatial variations of the velocity field, recirculation zones, and temperature distribution will be presented and analyzed in terms of contour plots. Preliminary results indicate that some improvements in performance may be achieved by slightly modifying the design of the oven.

INTRODUCTION

Developments in the bread baking process progressed through time. The industrial revolution of the seventeenth century shifted bread baking from being conducted in the open air through burning of wood to the use baking ovens. Nowadays, ovens are used heavily for a variety of industrial food applications.

Computational Fluid Dynamics (CFD) is a very powerful numerical tool for predicting fluid flow, heat transfer and other transport phenomena. Even though the drive behind its development was the aeronautic and aerospace industries, it has grown to become an essential tool in a range of industries [1]. CFD found its way to food processing in the past decade due to the increase in computational power and the existence of reliable softwares [2]. Nowadays, CFD has become an essential tool for predicting combustion processes and optimizing the performance of ovens.

The many developed types of ovens can be broadly classified as either continuous or discontinuous. Continuous ovens are commonly used for baking and are of interest in this study. These ovens can be further sub-

Mohamad Al Nasser, graduate research assistant, Department of Mechanical Enegineering, American University of Beiruit, Beirut, Lebanon. **Iyad Fayssal** (PhD), Department of Mechanical and Mechatronics Engineering, Rafik Hariri University, Mechref, Lebanon. **Fadl Moukalled** (PhD), professor in the Department of Mechanical Enegineering, American University of Beiruit, Beirut, Lebanon.

categorized, based on their heat transfer mode, as forced convection ovens, radiative infrared ovens, or combined ovens (i.e. involving a combination of both convection and radiation heat transfer modes). The aim of the different designs is to achieve the optimum operating conditions by controlling the heat transfer rate in order to produce the best product. Purlis [3] conducted a theoretical study to optimize the baking process inside ovens. In his work, he combined forced convection with infrared radiation to obtain the best efficiency.

Electrical ovens have been extensively studied in the literature [4-6] and are mainly used for the advantage of requiring less time for baking. Those relying solely on radiation suffer the drawback of having limited baking space. This shortcoming arises because baking requires a direct exposure between the source of radiation and the bread. This drawback is not found in forced convection ovens permitting a larger number of breads to be baked at the same time. The most commonly used convection-based type is the continuous rotary rack oven. As reported in [7], energy optimization of convective ovens is critically dependent on the position of heat source and blower. Therdthai et al. [10] further stated that variations in bread quality can be minimized by good design of the oven, and insuring uniform distribution and constant heat transfer with exact baking time. Despite their wide use, optimizing the operation of convective ovens is usually determined experimentally [8], which renders the optimization process difficult and costly. Therefore there is a need to conduct numerical simulations in order to improve oven designs (by producing uniform heat distribution inside the oven) and optimize their performance.

This study aims at building a full three-dimensional model of an industrial baking oven. A detailed simulation is performed to analyze the flow and temperature distribution inside the oven. The study models the bread as a solid object allowing to analyze the effect of bread geometry and thermal properties on the flow and heat transfer fields in the oven.

PHYSICAL MODEL

The oven model consists of the following two main compartments: (i) the clean air compartment, and (ii) the flue gases compartment. Communication between the two sections occurs by heat transfer only through a heat exchanger. No transfer of mass or mixing between the clean and flue gases takes place. The industrial bread baking oven configuration considered in this study is depicted in Figure 1.

The clean air section includes the blower, air-side heat exchanger, baking chamber, and inlet and return airflows. The blower drives the air with a speed of 6 m/s. The air passes over the heat exchanger and exchanges heat with the flue gases flowing inside. Then, air enters the baking chamber after being heated. In the baking chamber, the hot air transfers heat by convection to the bread causing it to bake. During baking, water vapor and carbon dioxide (CO_2) are released to the hot clean air. This process is continuous and is stopped once the bread is cooked and has become ready to eat.

In the combustion gases compartment, which is shown in Figure 2, flue gases enter the heat exchanger with a temperature of 1073 K at a rate of 0.002 kg/s. While moving inside the heat exchanger, flue gases lose heat to the air flowing on the outer side of the heat exchanger and exit with a lower temperature.

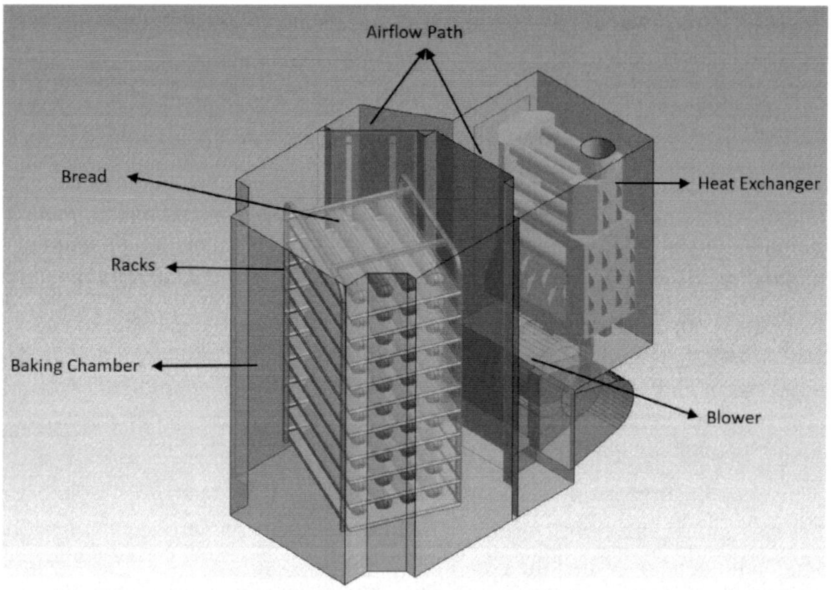

Figure 1: Oven Assembly

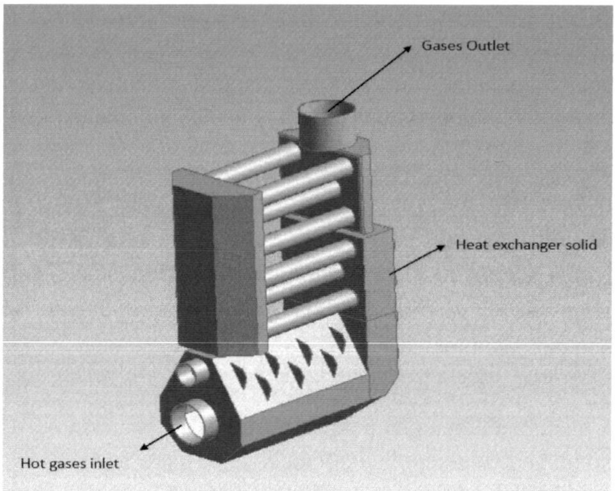

Figure 2: Combustion gases flow path

CONSERVATION EQUATIONS

The water content of the bread is not accounted for in this study. Thus, the flow in the oven is single phase composed of air . The combustion gases inside the heat exchanger is a mixture of CO_2, NO_x, SOOT, etc. The equations governing the turbulent flow and heat transfer fields inside the oven are those expressing the conservation of mass, momentum, energy, and the turbulence model equations and are given by:

$$\frac{\partial \rho}{\partial t}+\nabla.(\rho \mathbf{u})=0 \tag{1}$$

$$\frac{\partial(\rho \mathbf{u})}{\partial t}+\nabla.(\rho \mathbf{u u})=-\nabla \mathrm{p}+\nabla.(\tau)+\rho \mathbf{g}+\mathbf{F} \tag{2}$$

$$\frac{\partial(\rho E)}{\partial t}+\nabla \cdot[\mathbf{u}(\rho E+p)]=\nabla \cdot\left(k_{eff} \nabla T\right)+S_h \tag{3}$$

$$\frac{\partial(\rho k)}{\partial t}+\nabla.(\rho \mathbf{u} k)=-\nabla.\left[\left(\mu+\frac{\mu_t}{\sigma_k}\right) \nabla k\right]+G_k+G_b-\rho \grave{o}+S_k \tag{4}$$

$$\frac{\partial \rho \varepsilon}{\partial t}+\nabla \cdot(\rho \mathbf{u} \varepsilon)=\nabla \cdot\left[\left(\mu+\frac{\mu_t}{\sigma_\varepsilon}\right) \nabla k\right]+\rho C_1 S \varepsilon-\rho C_2 \frac{\varepsilon^2}{k+\sqrt{v \varepsilon}}+C_{1 \varepsilon} \frac{\varepsilon}{k}+C_{3 \varepsilon} G_b+S_\varepsilon \tag{5}$$

The expressions for the coefficients of the turbulence model can be found in [12].

The percentages of the different components composing the flue gase, as supplied by the manufacturer, are shown in Table 1. In addition, the thermo-physical properties of the flue gase, heat exchanger, and bread are displayed in Table 2.

Table 1: Composition of flue gases

Element	Percentage Composition
Oxygen (O2)	6
Carbon Dioxide (CO2)	7
Water (H2O)	5
Nitrogen (N2)	78
Nitride Oxide (NOx)	1
Sulfur Oxide (SOx)	2
SOOT	1

Table 2. The thermo-physical properties of materials

property	flue gas	oven-heat exchanger	bread
Density (Kg/m3)	0.3379896	7750	161.4
Cp (J/Kg-K)	1190	500	2730
Thermal Conductivity (W/m-K)	0.000467226	16	0.0989
Viscosity(Kg/m-s)	0.0000427	-	-

At inlet to the flue gas compartment, the mass flow rate and temperature are assigned the values of 0.002 kg/sec and 1073 K, respectively. At outlet, a fixed pressure boundary condition is applied. The envelop of the industrial oven is well insulated, and hence, the conditions at the outer walls are assumed to be adiabatic. The provided properties of the bread correspond to a bun of French baguette. The blower rotates at a speed of 1400 rpm. The racks where the bread is placed, rotate at 5 rpm to insure even distribution of air around the bread.

RESULTS AND DISCUSSION

The hydrodynamics and thermal fields in the oven (Figure 1) are predicted by solving numerically equations (1)-(5) using the ANSYS FLUENT commercial CFD package. The exact geometry of the blower rotor is used and its rotational effects are modeled using a rotating reference frame. In order to avoid divergence and due to the high angular speed of the blower, the time step size is limited to 0.02 s.

The first step in the solution process is to mesh the domain by generating a suitable grid system. To accomplish this task, each part of the oven is meshed individually and then the full mesh (Figure 3) is acquired by assembling the individual meshes. A final mesh with a size of $14x10^6$ elements is obtained.

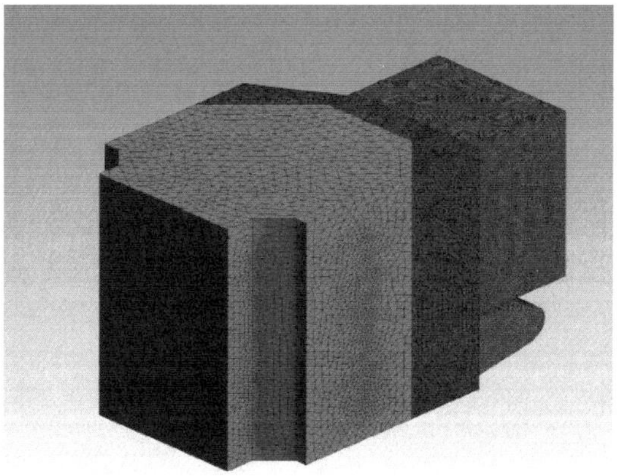

Figure 3: Meshed oven.

Velocity vectors and streamlines showing the flow in the blower are depicted in Figure 4. The magnitude of the air velocity around the fan is in the range of 16-17 m/sec, which is in excellent agreement with the experimental value of 16.8 m/s reported by the manufacturer.

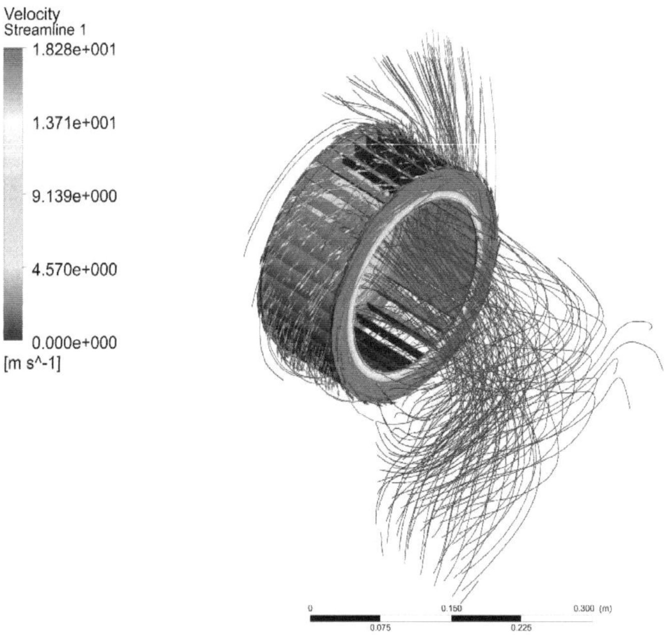

Figure 4: Velocity vectors and streamlines for the air flow in the blower.

Predicted values of the flow inside the Heat exchanger are in good agreement with the average experimental value of 0.7 m/sec at inlet reported by the manufacturer. The baking of the bread starts by preheating the oven to a temperature of 463 K. Then the bread, which is at the room temperature of 300 K, is placed in the oven. The progressive change in temperature and flow contours of the oven for the first two minutes recorded at 30 seconds intervals are presented in Figure 5.

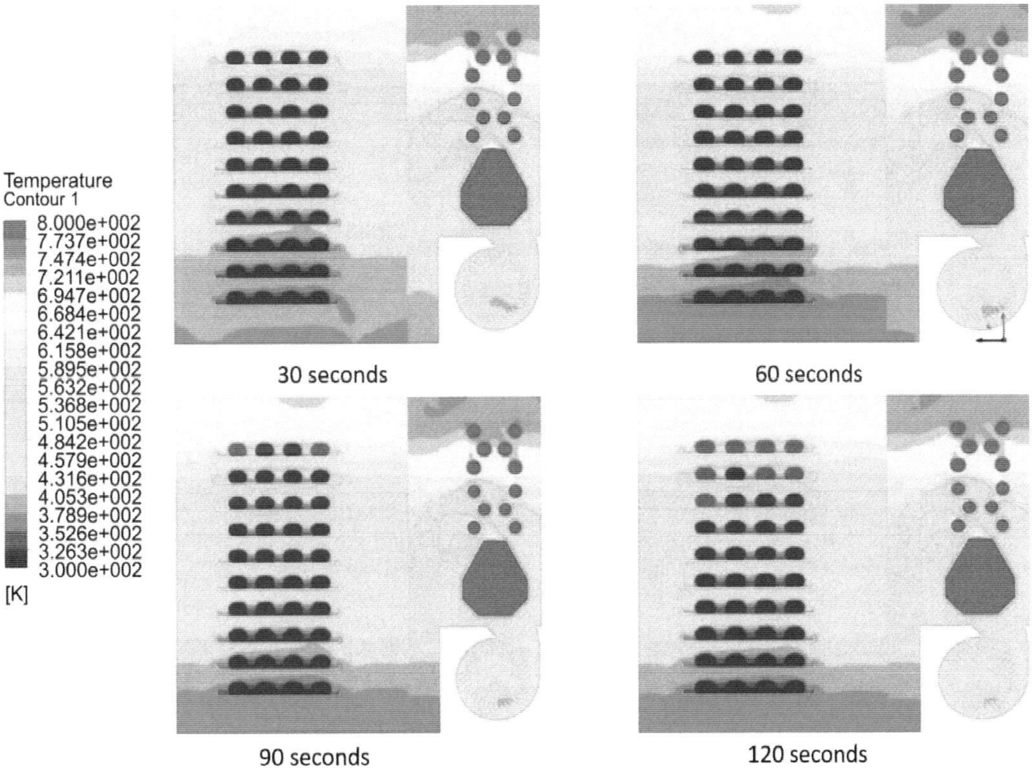

Figure 5: Flue gases Streamlines

The contours show a non uniform temperature distribution in the baking chamber which causes the bread in the upper racks to be baked much faster than the lower ones. The reason causing this temperature difference can be observed in the velocity distribution in the baking chamber (Figure 6). This can be resolved through increasing the flow of the blower, and redirecting the flow in the baking chamber to ensure a uniform heat distribution all over the chamber.

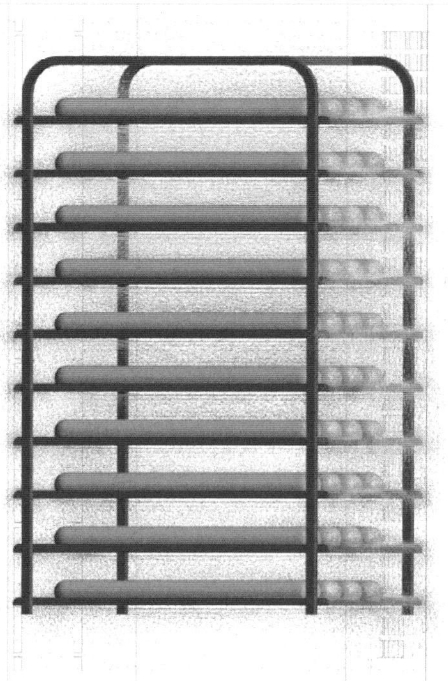

Figure 6: Velocity Vectors in Chamber.

Figure 7(a) presents the temporal variation of baking chamber temperature upon introducing the bread into the oven. Upon the placement of the breads at initial temperature of 300 K in the baking chamber of temperature 463 K, the average temperature of the chamber decreases drastically till it reaches 445 K after which it rises again as a result of the hot air flow driven from the heat exchanger. Figure 7(b) shows the increase of the temperature of the bread upon putting them in the baking chamber. This increase is shown initially (for the first 2 minutes) to be linear and equivalent to 9.5 K increase for each minute. Further simulations are still in progress to generate more details about the baking process which is expected to endure for 15 minutes.

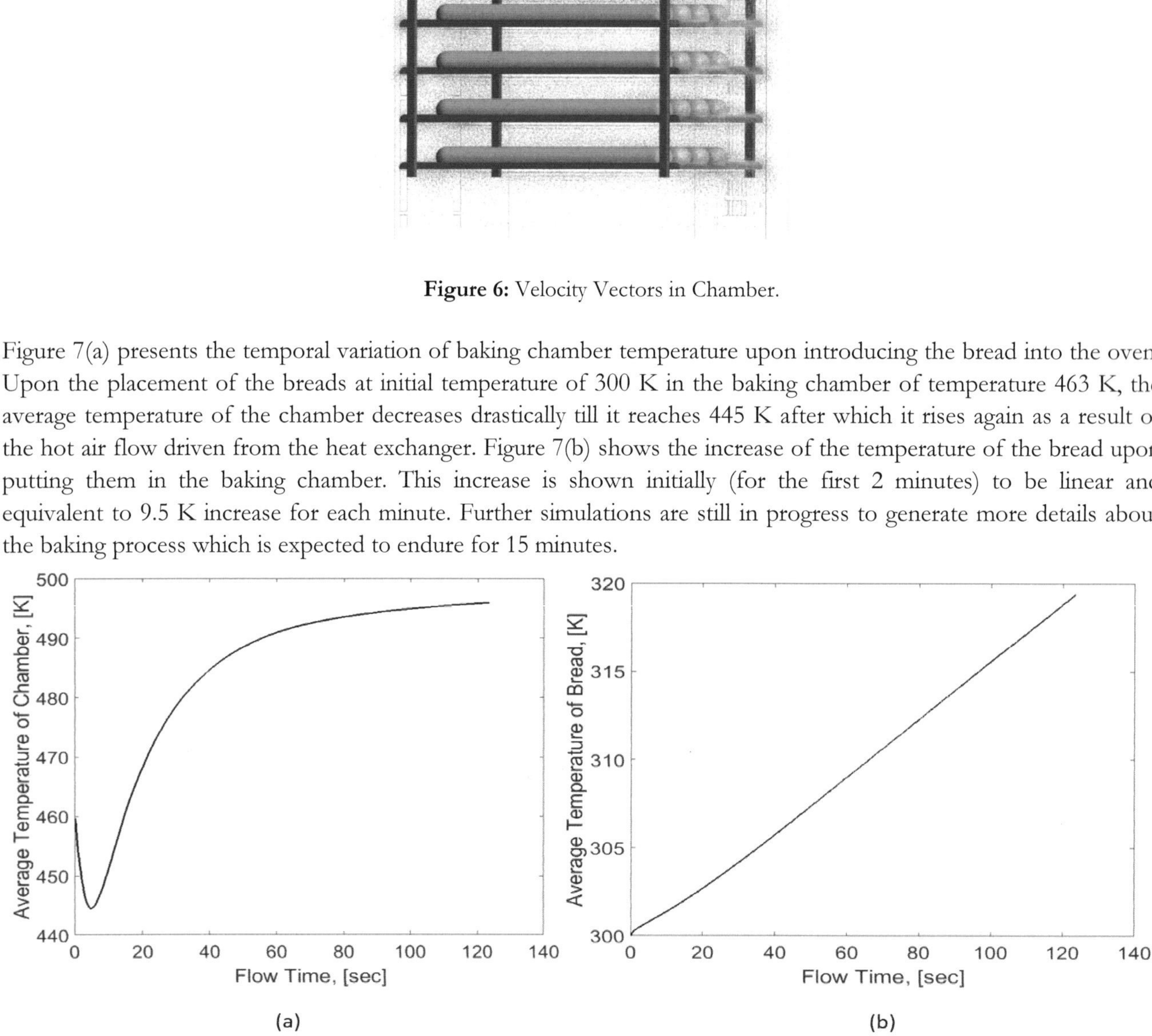

Figure 7: Varation of (a) Bread and (b) Chamber Temperature.

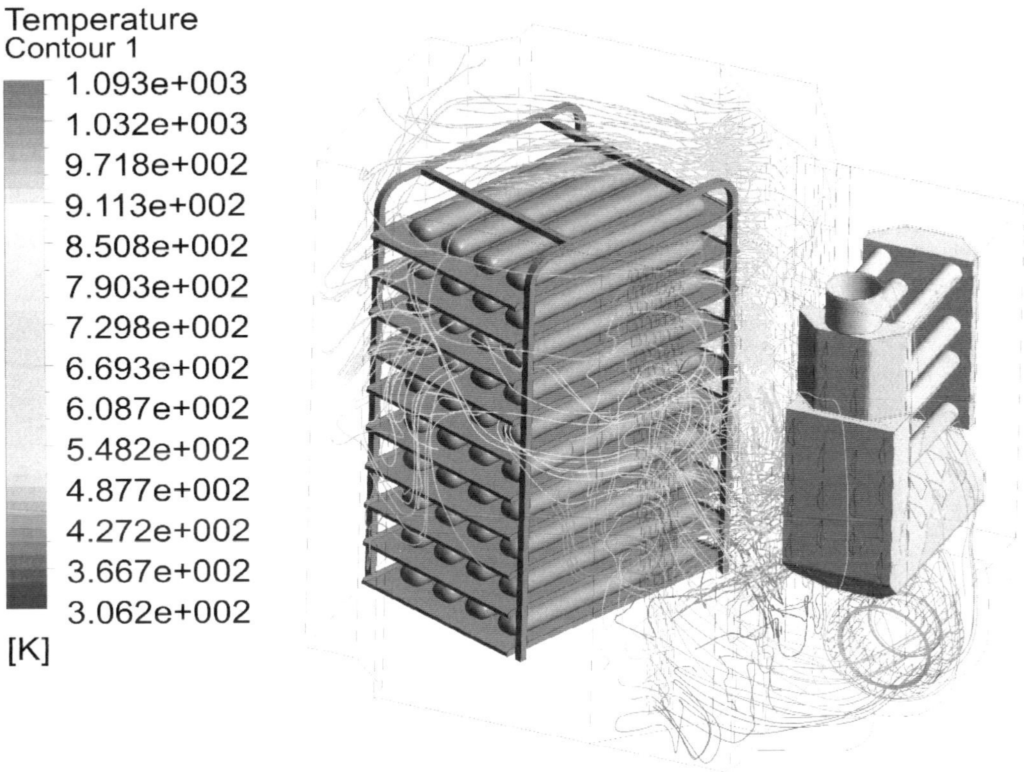

Figure 8: Velocity Streamlines with temprature variation

Closing Remarks

A full 3D numerical study for predicting and optimizing the performance of an industrial baking oven is being conducted. The study considers the variations of the hydrodynamic and thermal fields in the oven. The exact three-dimenstional geometry of the oven is modeled. The only approximation in the flow is the MRF but it can still can efficiently predict the flow in the oven. The generated flow field and temperature distribution maps reveal that some major improvements should be made for optimizing the design of the oven. Therefore, additional simulations are still warranted.

NOMENCLATURE

ρ = density

$\mathbf{u}$ = velocity vector velocity

τ = Stress tensor

g = gravitational acceralation

$\mathbf{F}$ = external body force

μ = viscosity

δ = Chronrcker's delta

T = temperature

Γ = thermal diffusivity

β = thermal expansion coefficient

k = turbulence kinetic energy

σ_k = turbulent Parndtl number for k

σ_ε = turbulent Prandtl number for ε

C_p = specific heat

G_b = generation of turbulence kinetic energy due to buoyancy

G_k = generation of turbulence kinetic energy due to mean velocity gradient

S = source term per unit volume

Ω = mean rate-of-rotation tensor

REFERENCES

1. Moukalled, F., L. Mangani, and M. Darwish, *The finite volume method in computational fluid dynamics.* 2016.
2. Williamson, M. and D. Wilson, *Development of an improved heating system for industrial tunnel baking ovens.* Journal of Food Engineering, 2009. **91**(1): p. 64-71.
3. Purlis, E., *Optimal design of bread baking: numerical investigation on combined convective and infrared heating.* Journal of Food Engineering, 2014. **137**: p. 39-50.
4. Chhanwal, N., et al., *Computational fluid dynamics modeling of bread baking process.* Food Research International, 2011. **44**(4): p. 978-983.
5. Abraham, J.P. and E.M. Sparrow, *A simple model and validating experiments for predicting the heat transfer to a load situated in an electrically heated oven.* Journal of food engineering, 2004. **62**(4): p. 409-415.

6. Chhanwal, N. and C. Anandharamakrishnan, *Temperature-and Moisture-Based Modeling for Prediction of Starch Gelatinization and Crumb Softness during Bread-Baking Process.* Journal of Texture Studies, 2014. **45**(6): p. 462-476.
7. Chhanwal, N., et al., *Computational fluid dynamics (CFD) modeling for bread baking process—a review.* Food and Bioprocess Technology, 2012. **5**(4): p. 1157-1172.
8. Pinelli, M. and A. Suman, *Thermal and fluid dynamic analysis of an air-forced convection rotary bread-baking oven by means of an experimental and numerical approach.* Applied Thermal Engineering, 2017. **117**: p. 330-342.

CFD-Optimized Radiant Cooling with Dedicated Outdoor Air System (DOAS) for High Ceilinged Spaces in Hot and Dry Climates

Youssef Ghoussoub **Adnan Akhdar**

ABSTRACT

The increasing demand for sustainable and energy efficient buildings, and fast-track delivery of projects with limited design time, has driven the need for advanced design tools to accurately assess and optimize ever more complex and diversified design strategies to find ways to reduce loads, boost efficiency, and utilize renewable resources. Computational Fluid Dynamics (CFD) proved to be a solid and flexible platform for design development and validation to meet such challenges within tight schedules, and deliver state-of-the art designs for high performance buildings.

This paper will present the use of computational fluid dynamics (CFD) simulations at Dar Al-Handasah to develop an innovative air-conditioning system for high-ceilinged spaces in hot and dry climates based on radiant cooling coupled with a dedicated outdoor air system (DOAS). Specifically, the performance effectiveness of the proposed cooling strategy in establishing a comfortable thermal environment is compared against a conventional overhead supply ceiling diffusers system by conducting a detailed investigation of airflow delivery and temperature distribution in a typical transient high-ceilinged assembly hall during full load and part load conditions. Moreover, to analyze accurately the performance of the compared VAC designs, several parameters have been considered such as the supply air flow rate and temperature, the air distribution layout, radiant floor temperature and heat loads.

The findings indicate that the proposed cooling strategy based on a combination of radiant cooling with DOAS helped achieve a thermally uniform environment with markedly improved thermal comfort conditions and reduced energy use as compared to the conventional design and the study concludes with a set of recommendations to optimize the air conditioning system energy use and thermal comfort levels in such spaces.

INTRODUCTION

In recent years, radiant systems using large surface floors, walls, or ceilings for heating or cooling indoor spaces are getting an increasing interest in search for improved, sustainable, and more energy-efficient HVAC systems. As defined by ASHRAE, a radiant surface is 'a temperature controlled surface if 50% or more of the design heat transfer on the temperature-controlled surface takes place by thermal radiation' (ASHRAE Handbook, HVAC Systems and Equipment, 2016) [Karimi & Ghias, 2017].

Natural resources and environment are being severly depleted by the building sector. In United States, 33% of energy and 66% of electricity usage are being consumed by residential buildings [Zhai, 2006]. In Iran, buildings' energy consumption constitutes around 40% of the total energy demand [Moslehi et al., 2016]. Based on building type, energy

Youssef Ghoussoub, B.E., is a mechanical design engineer in the CFD Unit of the Mechanical Department at Dar Al-Handasah. **Adnan Akhdar, M.E.,** is the leader of the CFD Unit in the Mechanical Department at Dar Al-Handasah.

usage of conventional air systems varies between 32-55% in a building in India [Pérez-Lombard et al., 2011; Khan et al., 2015] and Indonesia [Gunawan, 2012; Satrio et al., 2017]. Conventional heating and cooling systems use fans to transport conditioned air into the space. These traditional systems account for a significant amount of the total buildings' energy consumption. [Moslehi et al., 2016; Khan et al., 2015; Satrio et al., 2017]. In radiant cooling systems, pipes and radiant panels are used in the building structure to allow for low temperature chilled water to flow through, thus cooling the surface structure and enhancing the radiant heat transfer from the human body and the building as heat is absorbed and carried out by the cooler chilled water [Khan et al., 2015; Satrio et al., 2017]. Compared to conventional HVAC systems, research on the energy consumption of radiant systems indicated about 40% less electrical energy usage. [Watson & Chapman, 2002; Feustel & Stetiu, 1995; Khan et al., 2015]. Studies on radiant cooling systems for buildings conducted by [Henze et al., 2008] not only pointed to a 20% lower consumption, but also presented a better thermal comfort. [Stetiu, 1999] discovered that radiant systems can establish best results in dry climates and proved through numerical simulation that such systems in United States can achieve around 30% savings on energy consumption and 27% reduction in power requirements. [Tian & Love, 2009; Khan et al., 2015]. The advantage of a radiant cooling system over a conventional HVAC system in energy reduction resides in its ability to separate the thermal conditioning load from the ventilation task. In fact, the majority of sensible load is taken away by the radiant system as the ventilation system handles the latent load and the rest of the sensible load [Feustel & Stetiu, 1995; Hui & Leung, 2012; Khan et al., 2015]. Therefore, most of the time, the radiant system must be coupled with a displacement ventilation system [Rhee & Kim, 2015]. In addition, better comfortable thermal conditions are obtained in occupied indoor spaces when coupling radiant and displacement systems as a uniform temperature distribution and a very low velocity field are established [Saelens et al., 2011]. Such comfortable conditions are mainly due to the lower temperature difference needed to achieve the cooling capacity due to the presence of large radiant surface areas. [Karimi & Ghias, 2017]

Appropriate tools are a must when assessing the performance of a ventilation system in a building and for optimizing indoor air parameters [Chen, 2009]. Prior to the introduction of CFD into HVAC systems design, lab diagrams of diffuser behavior as well as analytical and empirical models of the jet were used to calculate the thermal comfort parameters such as temperature, velocity field, density and relative humidity. These methods caused considerable variations between tests and reality and weren't accurate in predicting stratification [Banjac, 2014; Chen, 2009], as well as being not suitable for irregular and complex geometries. Beyond conventional methodologies, CFD is a numerical simulation process in which the governing conservation equations of mass, momentum, energy and others are used to describe the flow in a computational domain. The process involves splitting the fluid region into a large number of smaller volumes, called cells, and applying the appropriate conservation equations to each of these cells. The equations are then simultaneously solved using numerical techniques. The rapid acceleration of computer technology in recent decades has resulted in computer hardware and software that is now capable of numerically solving these equations and problems to time scales that fit within project schedules. The versatility of the tool and the unprecedented ability to visualize and quantify the flows has made CFD an indispensable tool whenever practical analysis and engineering design work involving fluids is required. Knowledgeable application of CFD modeling allows engineers to create accurate models of project conditions. These models are a cost-effective way to experiment with design approaches and determine the best design to meet the project's team needs and goals. A comparative assessment of cold air drafts, temperature stratification and velocity fields for two different HVAC systems was conducted by [Myhren & Holmberg, 2008] using CFD simulations. CFD was also used to investigate indoor thermal conditions in industrial premises by [Rohdin & Moshfegh, 2007] and in an Olympic Arena by [Stamou et al., 2008] [Catalina et al., 2009].

The paper presents the results from the Computational Fluid Dynamics (CFD) simulations for the analysis and assessment of the air conditioning system of a typical Transient High-Ceilinged Assembly Hall Station in a hot and dry climate. The simulations investigate the performance of the designed air conditioning system where cooling is provided in two ways: from conventional overhead supply ceiling diffusers and a radiant floor cooling system. The simulations verify the adequacy of the current VAC design and the performance effectiveness of the combination of a radiant cooling system and a conventional overhead cooling air supply to maintain optimum thermal comfort conditions throughout the station. Three models have been chosen to analyze the VAC system performance during full load and part load i.e. no solar loads conditions. The analysis concludes with a set of recommendations to optimize the air conditioning system energy use and thermal comfort levels in the station.

What follows is a description of the methodology, design criteria, the geometry and boundary conditions of the CFD model, the simulations conducted and the results from the assessments for the three models.

METHODOLOGY

The CFD model is built according to the station drawings received from the architectural team and consists of a high-ceilinged assembly hall in a transient bus station. It is cooled via a radiant floor cooling system coupled with overhead ceiling diffusers. To analyze accurately the performance of the proposed cooling system, several parameters have been considered such as the supply air flow rate and temperature, the air distribution layout, radiant floor temperature and heat loads.

Model 1 simulates air conditioning system at full load with the radiant floor exposed to sun. The radiant floor cooling system is ON and the supply air flow rate is at full capacity with 18,432 L/s (39 054 cfm).

With the results indicating that the occupant zone is slightly overcooled, the same system is re-assessed in Model 2 with the radiant floor in the shade (solar load is zero and floor temperature is 20.3°C (68.54°F)). This model aims to confirm that the VAC system must be optimized during partial load conditions while still achieving comfortable thermal conditions at full load. Accordingly, and similar to Model 1 with the radiant system kept ON and supply air flow rate at full capacity with 18,432 L/s (39 054 cfm), the results proved that significant overcooling is achieved in the occupant zone.

Seeking to establish higher space temperatures, Model 3 re-investigates the performance of the hybrid cooling system at partial load with a reduced air supply. Radiant floor cooling system is ON and supply air flow rate is 10,772 L/s (22 824 cfm), which coincides with the total fresh air supply.

DESIGN CRITERIA

The design criteria that the CFD simulations have been based on are listed in Tables 1 and 2 below:

Table 1. Design Criteria

	Model 1	Model 2	Model 3
Cooling System	**Overhead Supply + Radiant Floor Cooling**	**Overhead Supply + Radiant Floor Cooling**	**Overhead Supply + Radiant Floor Cooling**
Radint Floor Exposure	Exposed to Sun	In the Shade	In the Shade
Radiant Floor Cooling		ON	
Radiant Floor Average Cooling Capacity		75 W/m² (23.78 BTU/hr.ft²)	
Radiant Floor Temperature		In Shade: 20.3°C (68.54°F) Under Skylights: 30.8°C (87.44°F) In the area 4 to 5 m away from the Glazed Façades: 25.6°C (78.08°F)	
Overhead Supply Diffusers		128 Supply Linear Bar Diffusers (SLBD) L = 1 m (3.28 ft) with 2 Slots W = 25 mm (0.082 ft) each	
Supply Air Temperature		11.7°C (53.06°F) DB / 8°C (46.4°F) WB	
Ambient Temperature		46°C (114.8°F) DB / 18.7°C (65.66°F) WB	
Design Temperature		24°C (75.2°F) DB / 15.7°C (60.26°F) WB	
Supply Air Flow Rate	18,432 L/s (39 054 cfm)	18,432 L/s (39 054 cfm)	10,772 L/s (22 824 cfm)
Return Air Flow Rate	7,660 L/s (16 230 cfm)	7,660 L/s (16 230 cfm)	0 L/s (0 cfm)

Toilet Exhaust Flow Rate	1,310 L/s (2 776 cfm)		
Exfiltration from Doors	9,462 L/s (20 048 cfm)		
Number of Occupants	1673 persons		
People Density	0.63 m^2/person (6.78 ft^2/person)		
Sensible Load	82 W/person (280 BTU/hr.person)		
Latent Load	79 W/person (269.6 BTU/hr.person)		
Lighting Load	18.69 kW (63 773 BTU/hr) (70% to Space and 30% to Plenum)		
Equipment Load	8.01 kW (27 331 BTU/hr)		
Total Solar Load (Skylights & Glazed Façades)	35.88 kW (122 428 BTU/hr)	0 kW (0 BTU/hr)	0 kW (0 BTU/hr)
Total Heat Transmission Load (Building Envelope)	176.68 kW (602 857.2 BTU/hr)		

Table 2. Building Envelope Loads at Peak Conditions

Exposure	Element	Cooling Transmission Load W [BTU/hr]	Cooling Solar Load W [BTU/hr]
SE	Walls	1,970 [6 721.92]	-
	Fenestration	69,452 [236 980.1]	13,249 [45 207.5]
SW	Walls	1,330 [4 538.15]	-
	Fenestration	18,878 [64 414.42]	3,984 [13 594]
NW	Walls	1,961 [6 691.21]	-
	Fenestration	69,452 [236 980.1]	14,477 [49 397.6]
NE	Walls	1,290 [4 401.66]	-
	Fenestration	0 [0]	0 [0]
HOR	Roof	2,623 [8 950]	-
	Skylight	9,728 [33 193.32]	4,170 [14 228.63]

CFD MODEL

A full-scale three-dimensional model of the Transient High-Ceilinged Assembly Hall Station is analyzed. Only the open space is included in the model. The adjacent enclosed spaces are excluded as well as the return air plenum. Along with Tables 1 & 2 above, Figure 1 displays the detailed geometry and boundary conditions used for all three models where several parameters have been considered such as the supply air flow rate and temperature, the air distribution layout, radiant floor temperature and heat loads, in order to accurately analyse the performance of the proposed hybrid cooling system.

The computational grid consists of approximately 3.3 million tetrahedral cells. Special attention was given to the precise modeling and high grid resolution of the supply, return and exhaust air diffusers, as well as the occupant region up to 2 m from radiant floor level. The conformal mesh is shown in Figure 2 and has a minimum cell size of 0.025 x 0.025 x 0.025 m (0.082 x 0.082 x 0.082 ft) at the diffusers.

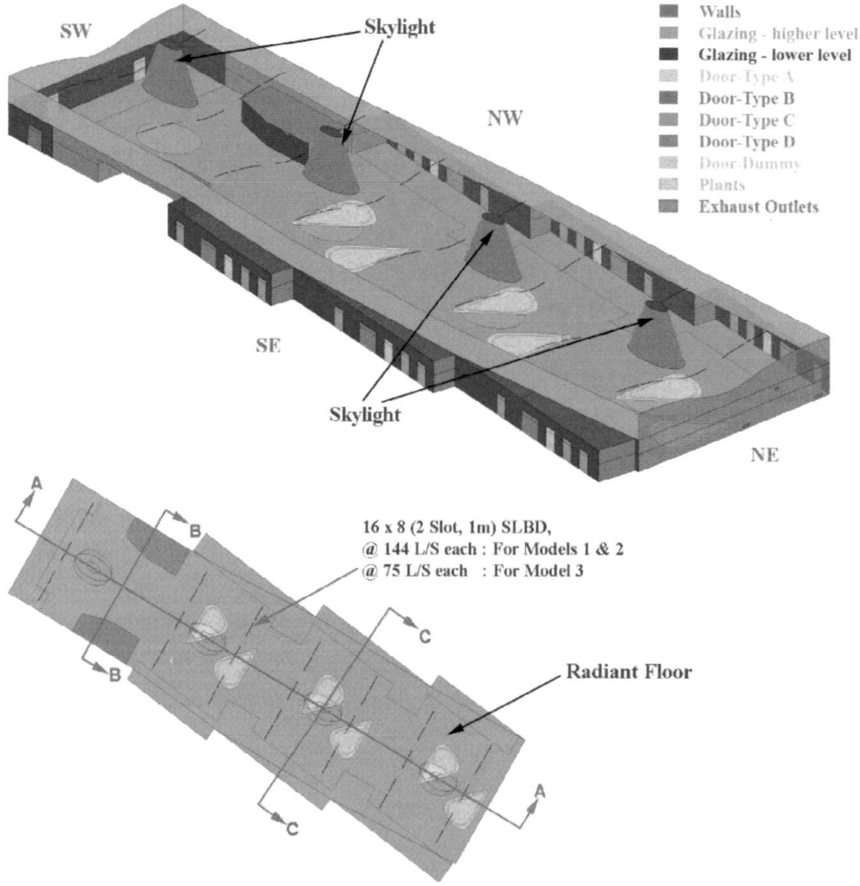

Figure 1 Full-scale 3-D Geometry and Boundary Conditions for Models 1, 2, and 3.

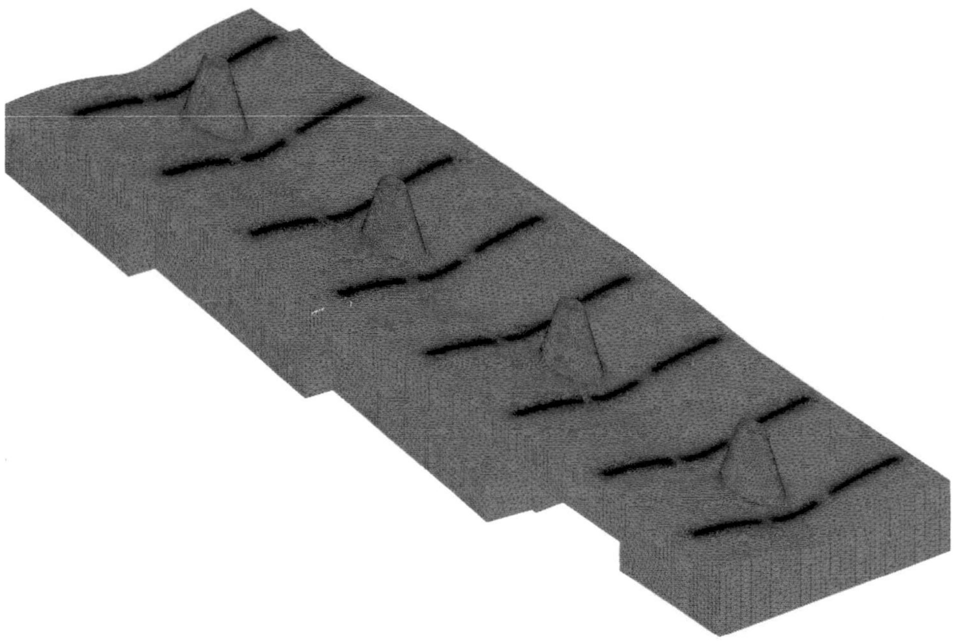

Figure 2 Simulated mesh for the CFD Models of the High-Ceilinged Assembly Hall.

RESULTS

The temperature contours at ankle, waist and head levels are shown in Figures 3, 4, and 5, respectively. Letters a, b, and c in the figures refer to Models 1, 2, and 3, respectively.

For Model 1, the temperature at ankle level is close to that of the radiant floor temperature as seen in Figure 3a. Figures 3a, 4a, and 5a show that temperature is lower in the shaded regions of the floor in Model 1 and varies from a minimum of 20°C (68°F) to a maximum of 23°C (73.4°F). Since the variation in temperature from ankle to head level does not exceed 3°C (37.4°F), the design is in compliance with ASHRAE Standard 55. The average temperature in the occupant zone is 23°C (73.4°F) and increases to around 26°C (78.8°F) close to ceiling level. Since the target design temperature is 24°C (75.2°F), the station is slightly overcooled. Average air speeds are around 0.2 m/s (0.656 ft/s) as shown in Table 3 except under the supply diffusers where air speeds are as high as 0.5 m/s (1.64 ft/s). Average relative humidity is around 50% and decreases under the skylights and close to the glazed façades due to the relatively higher temperatures at these locations.

A similar flow pattern is observed in Model 2 with temperatures in the occupied zone lower than that of Model 1. Indeed, tempratures vary from 20°C (68°F) at ankle level to 22°C (71.6°F) at head level as displayed in Figures 3b, 4b, and 5b. Relative humidity is higher than that of Model 1 because of the lower temperatures in the occupant zone in the absence of solar loads and at full system capacity with an average relative humidity value in the occupant zone of around 55%.

To achieve higher space temperatures, Model 3 re-simulates the air conditioning system with a reduced supply flow rate equal to the fresh air supply. Since the radiant floor cooling capacity is still at full capacity, the average temperature in the station is about 21°C (69.8°F) as shown in Figures 3c, 4c, and 5c. Due to the reduced supply flow rate, air speeds are lower than those of Models 1 and 2 with an average value of 0.15 m/s (0.492 ft/s) as tabulated in Table 3. However, relative humidity increases and is higher than 50% throughout the occupant zone while it is lower towards the west side due to a lower occupant density and a higher exfiltration rate from the doors.

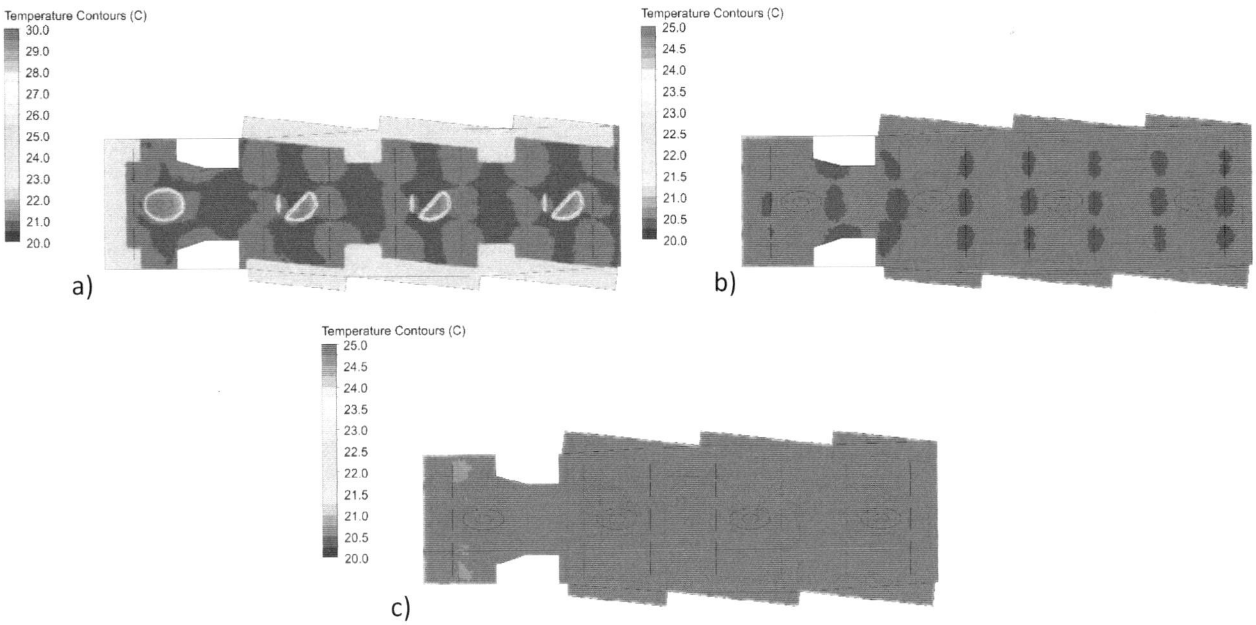

Figure 3 Temperature Contours at ankle level i.e. at 0.1 m (0.328 ft) above FFL (a) Model 1 and (b) Model 2 and (c) Model 3.

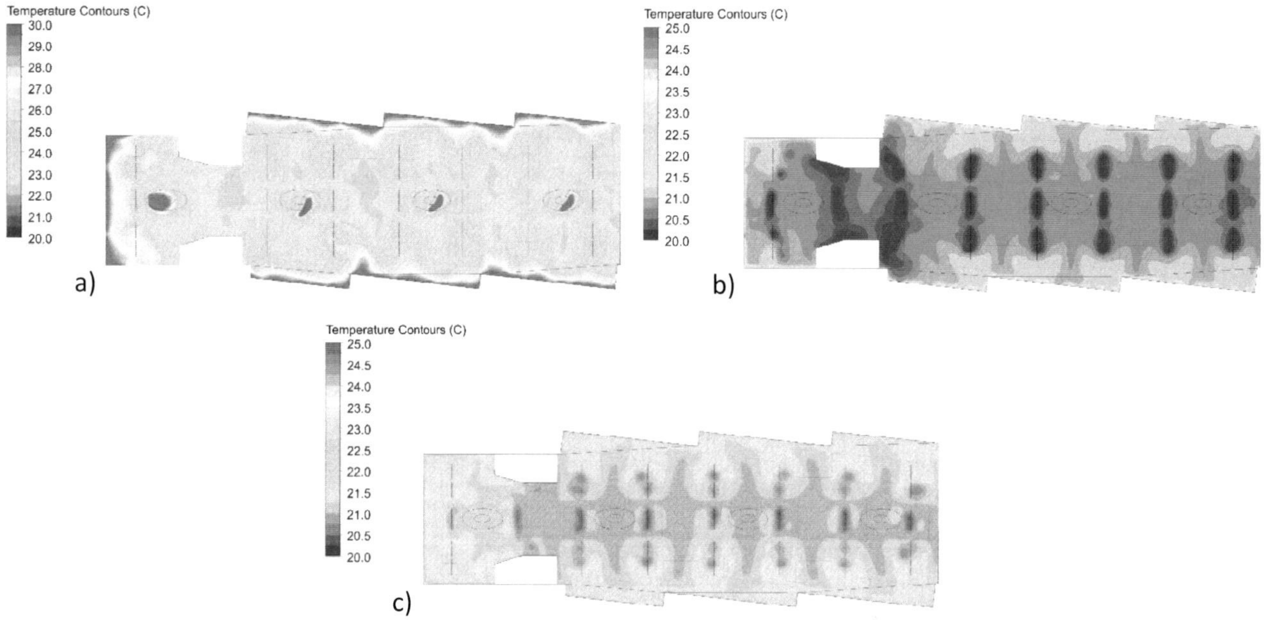

Figure 4 Temperature Contours at waist level i.e. at 1.1 m (3.609 ft) above FFL (a) Model 1 and (b) Model 2 and (c) Model 3.

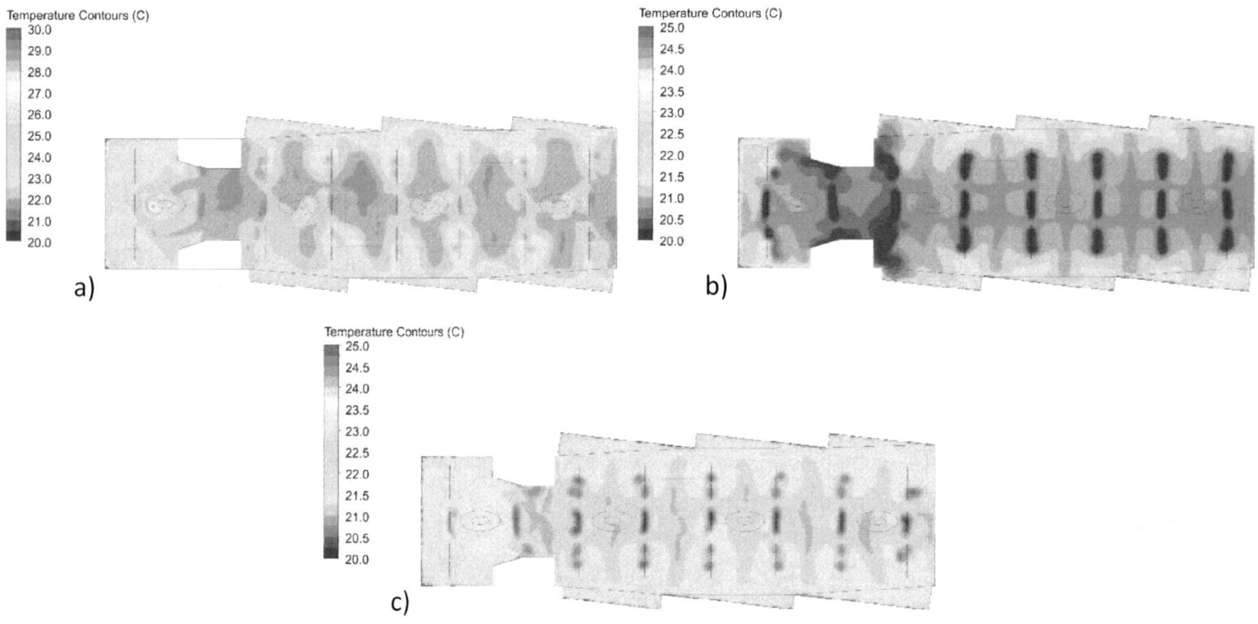

Figure 5 Temperature Contours at head level i.e. at 1.7 m (5.578 ft) above FFL (a) Model 1 and (b) Model 2 and (c) Model 3.

Finally, **Table 3** below summarizes the average temperature, air speed and relative humidity values in the occupant zone. Additionally, the temperature histogram **in Figure 6** below compares between the temperature distributions of Models 1 to 3. In all models, a significant percentage of the occupant zone is below 24°C (75.2°F) thus, the cooling design capacity can be reduced.

Table 3. Average Temperature, Velocity, & RH in Occupant Zone

Model	Average Temperature °C [°F]	Air Speed m/s [ft/s]	Relative Humidity %
Model 1 – Full Supply of 18,432 L/s [39 054 cfm] Full Load	22.3 [72.14]	0.21 [0.689]	49
Model 2 – Full Supply of 18,432 L/s [39 054 cfm] Partial Load	20.6 [69.08]	0.22 [0.722]	54
Model 1 – Fresh Air Flow of 10,772 L/s [22 824 cfm] Partial Load	21 [69.8]	0.15 [0.492]	64

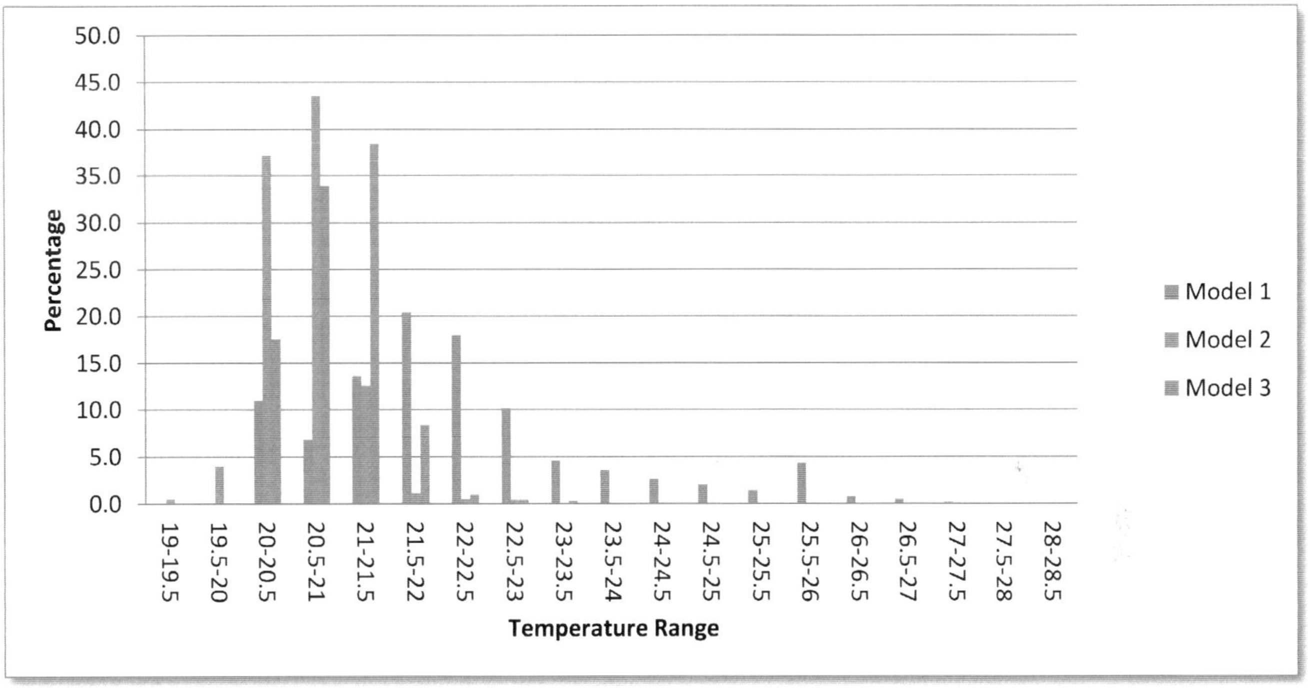

Figure 6 Temperature Distribution in Occupant Zone.

CONCLUSION

The paper summarizes the CFD simulation results for the air conditioning system of a typical Transient High-Ceilinged Assembly Hall Station in a hot and dry climate taking into account the coupling of the radiant cooling system with the overhead air supply. The performance of the system has been analyzed for both peak load conditions and at part load, i.e. no solar loads.

In fact, the simulations confirm that comfort conditions are maintained throughout the station, although space is slightly overcooled even at peak load conditions and a substantial volume of the occupant zone is below 24°C (75.2°F). Reducing the air supply flow rate at part load conditions to match that of the fresh air supply is not enough as the average occupant zone temperature is around 21°C (69.8°F) and relative humidity exceeds 50%.

It is recommended that the air supply flow rate be reduced at peak design conditions while maintaining the space relative humidity within 50%. This will also allow a reduction in fan energy. Finally, the radiant cooling capacity should be modulated at part load conditions to match the space load.

ACKNOWLEDGMENTS

We would like to thank all the members of the CFD team at Dar Al-Handasah who established a solid foundation in numerical simulations with their commitment and continuous learning curve leading to a wide technical expertise able to brilliantly extract design innovations and systems optimization. Special thanks to all members of the design teams for their cooperation and design integration, boosting efficiency and assisting in fast-track delivery of successful projects.

NOMENCLATURE

CFD	=	Computational Fluid Dynamics
DOAS	=	Dedicated Outdoor Air System
VAC	=	Ventilation and Air Conditioning
FFL	=	Finished Floor Level
DB	=	Dry bulb temprature
WB	=	Wet bulb temprature
RH	=	Relative Humidity

REFERENCES

ASHRAE. 2016. *ASHRAE Handbook—HVAC Systems and Equipment.* Atlanta: ASHRAE.

Banjac, M. J. (2014). Application of Computational Fluid Dynamics in cooling systems design for special purpose objects. FME Transactions, 42(1), 26-33.

Catalina, T., Virgone, J., & Kuznik, F. (2009). Evaluation of thermal comfort using combined CFD and experimentation study in a test room equipped with a cooling ceiling. Building and Environment, 44(8), 1740-1750.

Chen, Q. (2009). Ventilation performance prediction for buildings: A method overview and recent applications. Building and Environment, Vol. 44, pp. 848-858.

Feustel, H. E., & Stetiu, C. (1995). Hydronic radiant cooling—preliminary assessment. Energy and buildings, 22(3), 193-205.

Gunawan, B. (2012). Buku pedoman energi efisiensi untuk desain bangunan gedung di Indonesia. Energy Efficiency and Conservation Clearing House Indonesia.

Henze et al. [8] Henze, G. P., Felsmann, C., Kalz, D. E., & Herkel, S. (2008). Primary energy and comfort performance of ventilation assisted thermo-active building systems in continental climates. Energy and Buildings, 40(2), 99-111.

Hui, S. C., & Leung, J. Y. (2012). Thermal comfort and energy performance of chilled ceiling systems. In Proceedings of the Fujian-Hong Kong Joint Symposium 2012. The Conference.

Karimi, A., & Ghias, R. Cooling Performance Analysis and Optimization of a Room with Radiant Panel using CFD.

Khan, Y., Khare, V. R., Mathur, J., & Bhandari, M. (2015). Performance evaluation of radiant cooling system integrated with air system under different operational strategies. Energy and Buildings, 97, 118-128.

Moslehi, S., Maerefat, M., & Arababadi, R. (2016). Applicability of radiant heating-cooling ceiling panels in residential buildings in different climates of Iran. Procedia Engineering, 145, 18-25.

Myhren, J. A., & Holmberg, S. (2008). Flow patterns and thermal comfort in a room with panel, floor and wall heating. Energy and Buildings, 40(4), 524-536.

Pérez-Lombard, L., Ortiz, J., Coronel, J. F., & Maestre, I. R. (2011). A review of HVAC systems requirements in building energy regulations. Energy and Buildings, 43(2-3), 255-268.

Rhee, K. N., & Kim, K. W. (2015). A 50 year review of basic and applied research in radiant heating and cooling systems for the built environment. Building and Environment, 91, 166-190.

Rohdin, P., & Moshfegh, B. (2007). Numerical predictions of indoor climate in large industrial premises. A comparison between different k–ε models supported by field measurements. Building and Environment, 42(11), 3872-3882.

Saelens, D., Parys, W., & Baetens, R. (2011). Energy and comfort performance of thermally activated building systems including occupant behavior. Building and Environment, 46(4), 835-848.

Satrio, P., Sholahudin, S., & Nasruddin. (2017, March). Performance evaluation of radiant cooling system application on a university building in Indonesia. In AIP Conference Proceedings (Vol. 1826, No. 1, p. 020018). AIP Publishing.

Stamou, A. I., Katsiris, I., & Schaelin, A. (2008). Evaluation of thermal comfort in Galatsi Arena of the Olympics "Athens 2004" using a CFD model. Applied Thermal Engineering, 28(10), 1206-1215.

Stetiu [9] Stetiu, C. (1999). Energy and peak power savings potential of radiant cooling systems in US commercial buildings. Energy and buildings, 30(2), 127-138.
Tian, Z., & Love, J. A. (2009, July). Application of radiant cooling in different climates: assessment of office buildings through simulation. In Elev. Int. IBPSA Conf.
Watson, R. D., & Chapman, K. S. (2002). Radiant heating and cooling handbook (pp. 5-251). New York: McGraw-Hill.
Zhai, Z. (2006). Application of Computational Fluid Dynamics in Building Design: Aspects and Trends. Indoor and Built Environment, Vol. 15, pp. 305.

Leveraging Digital Transformation in Regional Economies with Concomitant Skills Development of People for More Efficient Operations & Resilience of Buildings over Their Life-Cycle

OM TANEJA, PHD PE, *Fellow ASHRAE*

ABSTRACT

Digital technologies are ubiquitous, but how digital transformation is applied to derive benefits for the built environment requires a critical assessment and execution. It is the need of the times to leverage technologies for buildings to improve IAQ, life-cycle efficiency and to become a flagship amongst integrated smart organized cities. The Challenge in regional economies lies in selecting the appropriate mix of technologies that can be reliably sustained and embraced by the community and workers without creating frustration and digital divide. If regional economies limit to technologies that are too simple, they can miss the opportunities, while overly complex technologies can be unnerving, chaotic and detrimental. In current environment of sharing and social networking, collaboration and information exchange with other buildings, cities and other sectors can strengthen development of a good strategy and execution program.

Digital Transformation & IoT facilitate cost-effective data gathering and analytics to deliver a comprehensive insight and management strategies for any sector. As digitization continues its advance to different building systems and dimensions of the built environment, the analysis of data gathered from equipment operations and occupant behaviors coupled with weather and utilities grid conditions helps save money, minimize costly interruptions , and maximize the occupant experience. A high performance building is founded on harmony of environment with human nature and integrates sensors, controls, measurements and automation to target optimization of how space is used, how equipment is operated, and how behavior modification is directed based on insights garnered through software analytics. Use of alternate energy, water, sewerage, lighting and HVAC systems along with testing, balancing, commissioning and benchmarking complements the operations.

This paper first addresses the elements required for transformation to best in class operations to sustain high life-cycle performance, with varying occupancy and space configurations. How well trained and motivated facilities leaders and operating and maintenance personnel can, if provided with adequately hardened infrastructure and a better toolkit can deliver measureable improvements. Energy use tracking and analytics can guide the operations and maintenance staff in performing their responsibilities and facilitate continuous commissioning. Regional skills development program specific to safe, and efficient operations of buildings needs to be developed and offered to the management and staff. For deeper energy cuts, facilities managers and IT system administrators need to collaborate to be able to dynamically manage the building operating parameters based on occupancy, time of day, and other human, environmental, and business considerations.

INTRODUCTION

There is a real digital transformation occurring all over the world. Technologies are emerging and affecting our lives in ways that indicate we are at the beginning of a fourth industrial revolution, Figure 1, a new era that builds and extends the impact of digitization in new and unanticipated ways (Global Alliance 2016). Leveraging technologies in

Om Taneja is currently MD of Associated Consulting Engineers & past Associate Director of GSA Region 2

an efficient and wholesome manner can overcome digital divide and generate benefits for many, rather than the few. Digital revolution is challenged by rising energy use which has nexus with sustainability and climate change scenarios where debilitating damage to environment is projected by continuing expansion of spaces and services by the buildings. The real estate sector is accountable for more than 20% of carbon emissions, which must be significantly reduced to avoid further destabilizing our climate and to meet the challenges of future demographic change. While this is worrying in the context of health, it is a trend likely to continue. We will see more online retail in the future, an increasing number of people working from their home office and individuals completely immersed in virtual realities. If there is not an external shock or massive behavior change in society, we will spend more than 90% of our time in properties on average (Global Alliance 2017). Since these are increasingly equipped with sensors and connected, granular datasets about our daily life will soon be readily available.

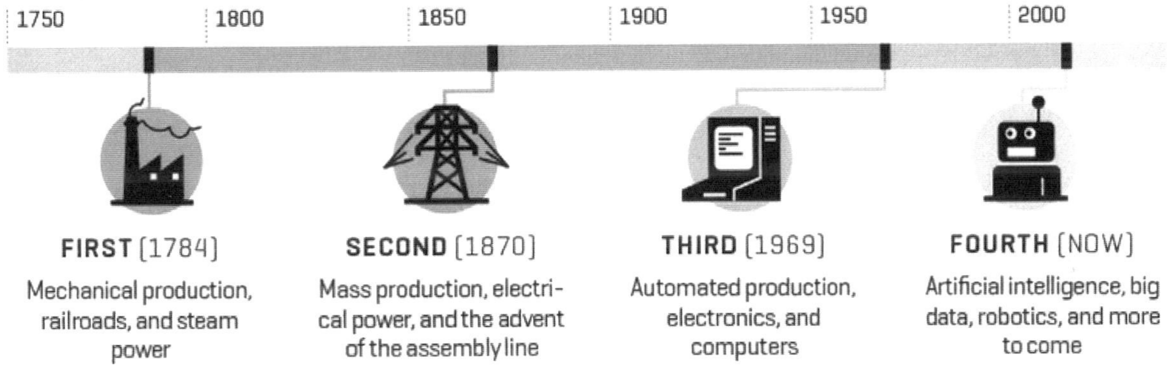

FIGURE 1 - The Four Industrial Revolutions - Leveraging Technologies is the Fourth Revolution

DIGITAL TRANSFORMATION NEEDS LEVERAGING TECHNOLOGIES

Developing countries and other regional economies are experiencing high economic growth, greater social and environmental awareness, and rapid urbanization. Along with global trade, regional economies face specific challenges that need to be focused at a local level. This is against the backdrop that around 1.5 billion people having no access to electricity (IEA 2012, IMF 2013), there is urgent need to decarbonize (Global Alliance 2016) and shift to cleaner fuels, renewable energy sources and "Net Zero Energy" buildings. Migration from rural to urban areas and other projections of changing world stipulate need for transitioning of regional economies towards a more service oriented framework, with much greater demand for skill-development, meaningful employment and advancement opportunities, construction of more residential and commercial, health and education focused buildings and associated infrastructure. Application of technology can be one of the pathway for addressing these challenges. To sustain this, one of the prime resource is to embrace a Future of Facilities with a greater use of Information & Communication technologies that can monitor, control and provide technical, administrative and environmental assessments for required actions, as shown in Figure 1. This requires greater collaboration of the public and private sectors. However, in the industry, there is such a rapid proliferation of

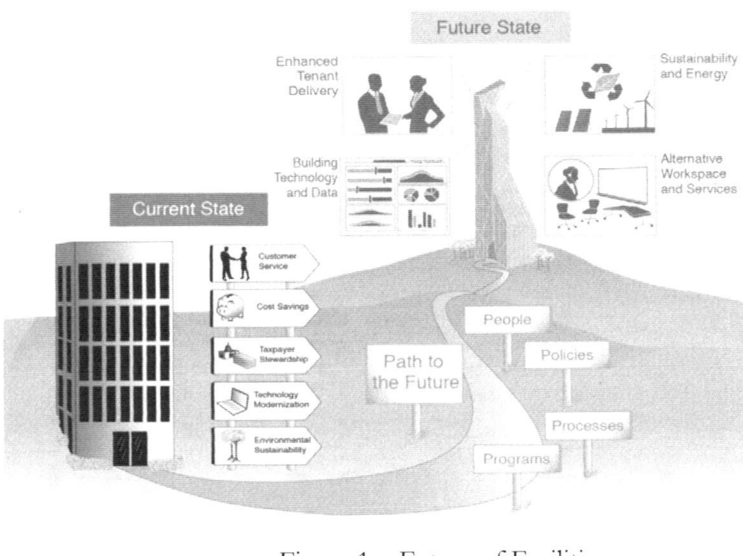

Figure 1 – Future of Facilities

technologies that astuteness is required to select and migrate to technology applications that are contextual, and mission oriented Technology that is too simple may pass up the opportunities, but introduction of too much and too fast technology can be unnerving, chaotic and detrimental. Sharing of data across the facility and enterprise can be done through insertion of a variety of context specific attributed tags that can connect to form a "system of systems " (CBRE 2017).

BUILDING INDUSTRY, ENERGY, WATER & SUSTAINABILITY NEXUS TECHNOLOGY PATHWAY

Due to the need for substantial new construction and renovations of commercial and residential spaces, all countries have the opportunity to progress towards more energy efficient construction, while creating jobs and healthy, productive environment. Technology for buildings have traditionally focused on improving efficiency of equipment, appliances, office equipment, integration of systems to lower parasitic loads of base building systems, such as, chillers, boilers, pumps, fans and distribution systems. Deep energy and water use reduction of 60% or more can only be enabled by fresh thinking and deploying innovative technology and building automation systems. Adopting additional interactive technology tools for a holistic, harmonized approach can uplift towards higher performance. This requires metering, sensors, and analytic tools for data monitoring and compilation, managing energy and water use, benchmarking, retro-commissioning and reporting. Innovations in HVAC systems that can help reduce the energy and water use are to be explored. (IEA World Energy Outlook 2017, 2012, 2013).

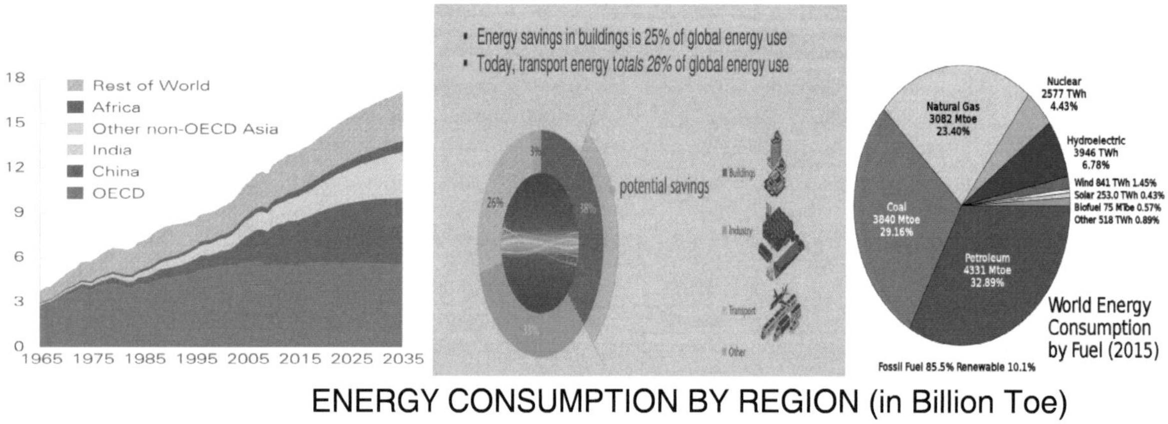

FIGURE 2 - World Energy Consumption 1990-2035 & Potential for Energy Savings in Buildings (Source IEA 2012 - Global Energy Trends)

Need To Retrofit Existing Buildings

Building efficiency is key to cutting energy consumption. The concentration of CO_2 at 403.3 ppm in 2016were the highest in at least 800,000 years and were 45% higher than pre-industrial levels. Last 3 years were the warmest on record and 2 to 4 times as many coastal glaciers are at the risk of accelerated melting. (World Resources Institute 2017). While roughly two percent of commercial floor space is newly constructed each year, and a comparable amount renovated, the majority of opportunities to improve efficiency over the next several decades will be in existing building stock, most of which are constrained by old equipment, aging infrastructure, and inadequate operations resources. Improved efficiency of existing buildings, through building retrofitting and other measures, represents a high-volume, low-cost approach to reducing energy use and greenhouse gas emissions. The challenge for existing building efficiency is to “unlock” that vast potential and realize the benefits of a built environment that is comfortable, efficient, and cost-effective (ASHRAE Standard 100 -2015)

Need to Upgrade the envelope and automation for Residential Buildings

Residential buildings consume more than 60% of total energy used in buildings. Significantly reducing the energy consumption of existing homes is a critical priority if we are to stem greenhouse gas emissions and global warming. It is also very challenging. Achieving 50% energy savings in an existing home is not as easy as it might seem—but it can be done. The costs of committing nations to a cutting-edge technology-enabled smart-homes program to retrofit a significant percent of housing stock would be huge, hundreds of billions of dollars, if not trillions of dollars, over the coming decades (Architecture 2030).

Energy & Non-Energy Socio-Economic Benefits

In terms of energy use, improving building efficiency offers benefits beyond those realized onsite. For every unit of energy actually used by the building, many more units of energy are lost along the way due to inefficiencies inherent in the delivery system. In fact, there may be as many as ***ten units*** of fuel energy wasted for every one unit actually used by the building. Reducing energy demands at the point of use, therefore, produces magnified effects. There are other benefits in improved indoor air quality, greater comfort and tenant satisfaction along with greater occupancy rates and higher rental rates (NRDC 2013, WRI 2013).

HIGH PERFORMANCE BUILDINGS NEED GUIDING PRINCIPLES & LEVERAGING TECHNOLOGY

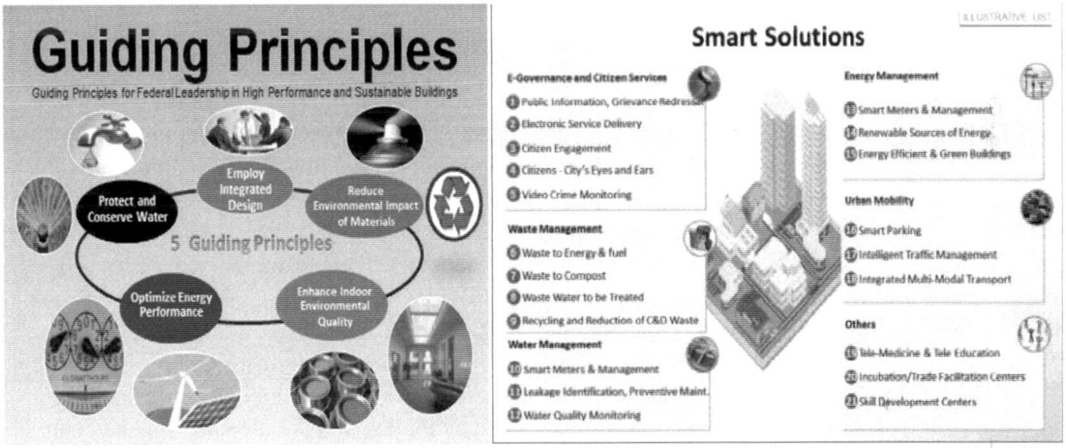

Figure 3 - Leverage Technology To Make Office spaces & Building Systems Better Integrated, And "Intelligent"

High Performance Buildings are deceptively complex. Understanding Whole Building Design concepts enables us to think and practice in an integrated fashion to meet the demands of today's as well as tomorrow's high-

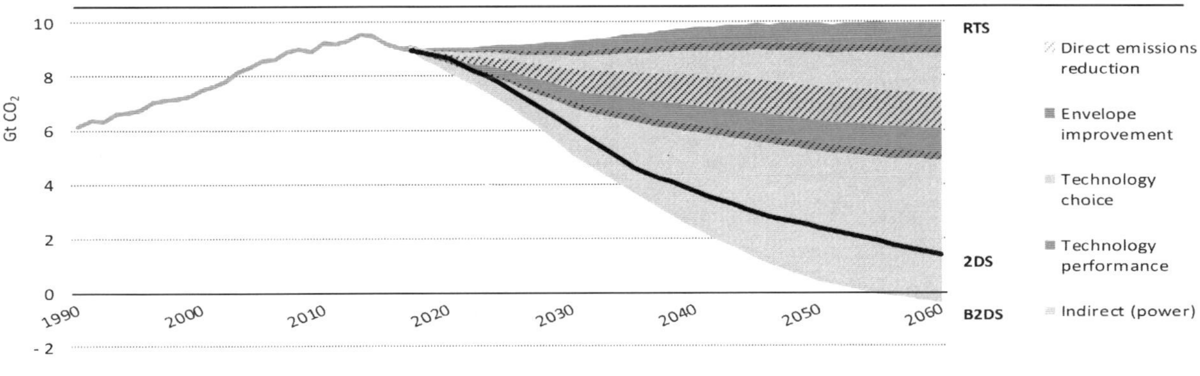

Notes: "Direct emissions reduction" represents a decrease in emissions from reductions in direct fossil fuel consumption in the buildings sector. "Envelope improvements" include measures (including deep energy renovations) that improve the energy intensity of the building envelope. "Technology choice" represents shifts from one type of technology and/or fuel to another (e.g. incandescent lamps to LEDs or gas boilers to electric heat pumps). "Technology performance" represents energy technology efficiency improvements (e.g. higher operational performance for heat pumps). "Indirect (power)" emissions reduction is from improved carbon intensities of power generation, where negative emissions are from carbon capture and storage (CCS) technologies.

Source: IEA (2017), *Energy Technology Perspectives 2017*, IEA/OECD, Paris, www.iea.org/etp.

Figure 4 – Key Contributions to CO_2 emissions reduction in the Global Building Sector to 2060

performance building projects. Integrated design provides evaluation of the cross-component impact of multiple systems as they are changed or upgraded. A high-performance building cannot yield results unless its HVAC, heating and cooling central plants are designed in harmony with pumping, whole building distribution and building management system (NIBS 2017). For all buildings and energy efficiency projects, US General Services Administration & Federal Leadership across agencies employs Government's Guiding Principles that establish performance goals for energy, water, materials and indoor air quality along with other goals that need to be pursued throughout the life-cycle of the buildings (Federal Leadership 2006). Guiding principles, as shown in Figure 3 set a policy for integrating and optimizing building systems, by incorporating operating decisions in a manner to promote sustainable building operations and maintenance practices. Key technology initiatives that contribute to emissions reduction is shown in Figure 4. This requires requisite training and skills development, innovation, customer intimacy, behavioral changes, timely communications, and transformation of middle management and Buildings Managers (Adam Elkington 2012; ACEEE 2017) as an active participatory team.

LEVERAGING TECHNOLOGIES WITH BIG DATA CAN REVITALIZE THE ENERGY AGE

Energy age can be revitalized by leveraging advances in technology by integrating mobile communications, Internet of Things, cloud computing, social media, web tools, and the human dimension into Building Automation & share on enterprise or cloud-based ICT. Most of the modern equipment and devices are embedded with sensors with ability to communicate on different platforms. The resulting information networks promise to, improve business processes, and reduce costs and risks. The predictable pathways of information are changing: the physical world itself is becoming a type of information system. Kevin Ashton coined the word, "Internet of Things" in 1999 to describe the network connecting objects in the physical world to the Internet. It has taken more than 15 years to take hold. In what's called the Internet of Things (IoT), sensors and actuators embedded in physical objects—from roadways to pacemakers—are linked through wired and wireless networks, often using the same Internet Protocol (IP) that connects the Internet. Examples are Home Security Systems, Fire Safety Systems, Smart Cell Phones, smart watches, smart meters, fitness bands, production, shipping and distribution networks. These networks churn out huge volumes of data that flow to computers for analysis. When objects can both sense the environment and communicate, they become tools for understanding detailed operations. Challenge lies in – transition from traditional enterprise IT architecture to one optimized for the IoT is complex, as shown in Figure 5. Elements of current technology stacks may need to be redesigned so that they can support multitude of independent processing events from millions of devices and applications. Massive increases in storage and computing power, some of it available via cloud computing make number crunching possible at very large scale and at declining cost.

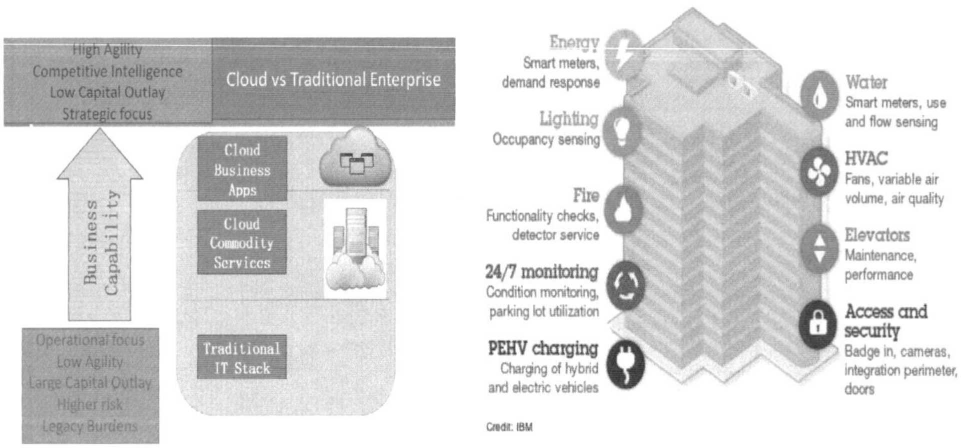

Figure 5 - Traditional IT Stack vs Cloud IT Infrastructure

MOVEMENT TOWARDS INTELLIGENT BUILDINGS AND ENERGY ANALYTICS FOR SMART CITIES

An intelligent building has smart sensors, meters, adaptive BMS, energy analytics and historical data. It integrates building subsystems as shown in Figure 5. The amount of data available to the facility manager continues to increase, and use of emerging intelligent buildings' analytics tools can help organize and effectively convey the data in a useful form[2, 3, 6, 7] . The scope and use of analytic tools can incorporate multi-variant measures to integrate weather, occupancy, HVAC and lighting systems diagnostics and operation (Watson-2012). This tool can become a great companion for building managers and management. To make buildings smart, use of networked technologies to connect Building Management systems to a central, cloud based technology (Ethan et al-2013, GSA-2013).

SUMMARY OF ESSENTIAL FEATURES OF TECHNOLOGY FOR INTELLIGENT BUILDINGS

Deploy Advanced meters and Sub-meters Technologies

Advanced metering systems are designed to monitor and/or store energy consumption data for specific building systems or the entire building. It allows building operators to take corrective action immediately. These systems also comply with programs sponsored by local power companies, enabling utilities to reduce power to participating buildings when demand on the power grid is high. Deploying "smart meters" and related infrastructure greatly reduces the need for manual or walk/drive by meter readings.(GSA 2013, Pike Research & Navigant Consulting 2013).

Provide Building Systems Performance Assessment, Analysis & Diagnostics Technologies

Apply modern technology for sensors, measurement and verification (M&V) devices and controls
Building systems that use a combination of high-efficiency components, combined with adaptive control, advanced logic and algorithms along with skilled O&M team can lower energy use from 20 to 30%..

Proper Measurements and Performance Verification (M&V) and diagnostic tools can find changes in baseline conditions, drifts in equipment performance, changes in after-hours use of utilities, changes in conditions outside the control of Facilities Manager (such as weather, occupancy changes). These systems can provide baseline and on-demand and annual energy use intensity and costs. Remote Analytic capabilities using Cloud Computing can be helpful. This becomes an avenue for early finding of problems and perform diagnostics guiding to recommended actions.

Continual or periodic Benchmarking and commissioning - Energy Benchmarking, is the comparison of a building's energy performance or energy consumption against similar building types while accounting for a variety of factors such as weather, fuel type, building construction, occupancy, etc. Energy benchmarking provides the owner with a priority list of which buildings are performing well and which are under-performing. It can perform periodic commissioning to calibrate sensors and tune building systems so that performance does not degrade over time. USDOE Energy Star Portfolio Manager and **ASHRAE's** Building Energy Quotient (ASHRAE Building EQ Portal) are helpful tools to assess gaps between actual energy use in operation versus design basis energy use.

Include innovative measures for central heating and cooling plant and auxiliary systems to achieve highest possible efficiency not only during full load, but also for part-loads. Most buildings operate at full load less than 5% of time and so part load efficiency is important. Review and adopt innovative HVAC and lighting technologies, such as, LED lighting, decoupling heating and cooling from ventilation, dedicated outdoor air ventilation, demand control ventilation, displacement air conditioning, heat recovery and humidification systems, evaporative and desiccant cooling, geothermal heat pumps, cogeneration systems, thermal storage, ductless cooling, chilled beams , radiant heating and cooling, variable refrigerant flow systems (ASHRAE Standard 90.1-2016 and Standard 189.1-2015 & ASHRAE design Guides)

Manage equipment sequencing and hours of operation – operating equipment when required can save more energy than most other measures. This can also facilitate timely peak load shaving and demand response activation with possible interaction with a smart grid.

Systems should Generate recommended action items on discovery of variance from optimal conditions, use computerized applications to *track and improve service delivery,* collaboration and sharing of best practices.

TECHNOLOGY TO FACILITATE DEVELOPMENT OF A TOOLKIT FOR CLIMATIC RESILIENCE

IT systems, and building structures and systems need to adapt to and quickly recover from inclement weather conditions. Climate change represents one of the greatest challenges of our time. As we act to curb the greenhouse gas pollution that is driving climate change, we must also prepare for the impacts that are too late to avoid. For example, 2017 was the third hottest year in the USA and more than US $ 306 billions was spent on managing natural disasters, while world spent in trillions with major interruptions to services. It involves not only coping with immediate problems more efficiently, but also establishing and maintaining a reserve sufficient to cope with multiple stressors in anticipation of future changes. In essence, climate change adaptation requires building a robust capacity to manage risk. as explained by the united nations development program (WRI 2017,Global Alliance 2016; World Bank 100 Resilient Cities 2016). The process of adaptation is not new; the idea of incorporating future climate risk into policy making is.

LIFE CYCLE ENVIRONMENTAL AND LIFE CYCLE ECONOMIC COST ANALYSIS OF ENERGY EFFICIENCY PROJECTS, CENTRAL PLANT & BUILDINGS

Owners are focusing not only on initial cost, but on the total life-cycle operating cost. The detailed analysis for each of the phases requires substantive analytical focus and is very much a function of local weather, buildings use and operational considerations. However, considering the average span of buildings of 50 to 100 years, and comparing to total life-span energy use and greenhouse gases release, the contribution of demolition is negligible, and that of initial construction is relatively small; transport, operations and maintenance and land development impacts are the major factors (Seppo Junita et al 2004; Om Taneja, et al 2008). Therefore, the buildings of tomorrow should be required to develop measures for total cost burden, total environmental burden, total energy use burden. An organization struggles in justifying investments in intelligent buildings where short term gains seem imperceptible. The Natural Resource Defense Council examined the net costs associated with several carbon reduction strategies and found that, after considering the benefit of reduced energy bills, Carbon abatement through building efficiency actually results in a negative cost to society (NRDC 2013 C40 Green Cities 2016). Post-occupancy evaluation study of 22 "green" federal buildings from across the country; found that,: compared to national average,: Green buildings cost less to maintain, by 19 percent, use less energy and water, by 25 percent, emit fewer carbon dioxide emissions, by 36 percent and have more satisfied occupants, by 27 percent, as shown in Figure 6. (Kim Fowler et al 2011)

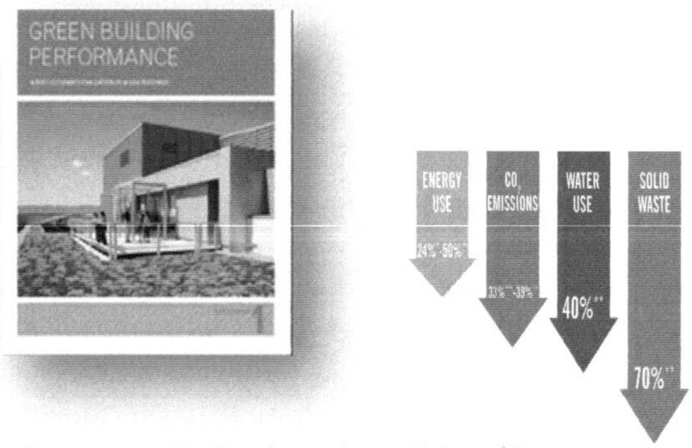

Figure 6 – Post Occupancy Evaluation of 22 High Performance Federal Buildings

MAKING THE CASE FOR TRAINING & DEVELOPMENT OF O&M SUPERVISORS AND STAFF

Operational efficiency represents the life-cycle cost-effective mix of preventive, predictive, and reliability-centered maintenance technologies, coupled with equipment calibration, tracking, and computerized maintenance. Over the life cycle of a building, O&M services cost at least three times more than the initial design and construction costs (Federal Facilities Council-2008, UNEP 2011; Global Alliance 2017) For proper use of technology tools, training, development and continued refresher courses for staff managing the buildings is necessary. In this regard steps taken by the US Federal Agencies can serve as a guide. US Federal Energy Management Program (FEMP) has estimated that O&M programs used with smart technology can save 5% to 20% on energy bills without a significant capital investment. U.S. Federal Facilities Council reviewed the skills a building manager should possess to operate and maintain high performance buildings. Along the recommendations of the Federal Facilities Council, the U.S mandates development of core competencies for facilities managers and O&M staff assigned to Federal Buildings. The required skills base should include Technical, business, inter-personal and enterprise knowledge.

CONCLUSION

Buildings are ecosystems, with far reaching social, economic and environmental impacts. They need to be built and operated so as to harmonize with the natural environment. They can evolve from a drain on our resources, to a vehicle for economic growth and improved productivity. Ultimately, the power of social technologies hinges on the full and enthusiastic participation of employees. By leveraging use of technologies, buildings can become more intelligent and optimize performance under changing occupancy, weather or other situations. In addition, by building capacity of facilities managers with requisite skills for operating high performance buildings, buildings can further sustain high performance over their life cycle. Innovative ideas and belief an delivering excellent services to the tenants, coupled with "tried and tested" industry best practices offer hope for climate change mitigation and resilience. Greater Resilience for buildings to adapt to increasing frequency of debilitating climate needs to be included in buildings design, equipment locations and local and remote operations

REFERENCES

Adam Elkington ,"New Smart Building Technology to increase Federal Buildings Energy Efficiency", US GSA Press Release, May 14, 2012 (http://www.gsa.gov/portal/category/100731).

The American Council for an Energy Efficient Economy (ACEEE) 2017, "smart Buildings save Energy", A deeper dive into energy efficiency market

Architecture 2030 - Alex Wilson and Allyson Wendt, Architecture 2030, www.architecture2030.org

ASHRAE/IES Standard 90.1-2016 - Energy Code

ANSI/ASHRAE/USGBC/IES Standard 189.1-2014 - Sustainability Standard

ASHRAE Standard 100-2015 - Energy Efficiency in Existing Buildings

ASHRAE Ventilation Standards 62.1 2016 (new and existing buildings) and 62.2 - 2016 (Residential)

CBRE 101389 2017 - Top Trends in Facilities management, www.CBRE.eu.

C40 Cities Green Investment Report 2016

Ethan A. Rogers, R. Neal Elliott, Sameer Kwatra, Dan Trombley, and Vasanth Nadadaur, "intelligent Efficiency Opportunities, Barriers and solutions", American Council for Energy Efficient Economy, October 2013

Federal Facilities Council of the National Research Council, "Core Competencies for Federal Facilities Asset Management Through 2020: TransformationalStrategies", 2008, http://www.nap.edu/catalog/12049.html

Federal Leadership in high Performance Buildings – inter-Agency Memorandum of Understanding, February 2006

Global Alliance for Buildings & Construction, "what is the 4th Industial Revolution", Nicholas Davis, Head of Society and Innovation, Member of the Executive Committee, World Economic Forum Geneva, 2016 & December 2017

IEA "World Energy outlook" - 2017, 2013 & 2012

IMF-World Bank Report, "Developing Countries Need to Harness Urbanization to Achieve the Millennium Development Goals(MDGs)" April 17, 2013.

Kim Fowler, et al, Pacific Northwest National Laboratory, "Reassessing Green Buildings Performance: A Post Of occupancy Evaluation of 22 GSA Buildings", September 2011.

National Institute of Building Sciences (NIBS), "Whole Building Design Guide" 2017

Natural Resources Defense Council (NRDC) "Building Green from Principle to Practice- Build your Business Case", 2013.

Om Taneja and Adam Hinge, "Life Cycle Assessment Measures of Energy Efficient Projects" International Conference on Improving Energy Efficiency in Commercial Buildings, IEECB08, Frankfurt, Germany.

Pike Research, Navigant Consulting, "Smart meters for commercial and industrial markets, May 2013.

Seppo Junita, "The Environmental Impact of an Office Building Throughout its Life Cycle, Dissertation for D. SC degree, Helsinki University of Technology, Espoo, Finland, 2004

Todd Watson, "IBM's 2012 Tech Trends Report: Skills, Skills, and More Skills!" Internet Evolution, Dec 2012

US General Services Administration, "GSA's Smart Buildings & advanced Metering Technologies", 2013

World Bank, 100 Resilient Cities 2017

World Resources Institute 2017, "Climate - a Billion Dollar Disaster to Disappearing Ice"

High Solar Combi-Plus System using PCM Storage: KSA Case Study

Mohamed Hmadi **Adel Mourtada, PhD, PE**

ABSTRACT

Based on the increasing worldwide tendency towards sustainability and the 'go green' movement, the idea was targeted towards developing a standalone "High Solar Combi Plus System" (HSC+) that can provide all thermal energy loads of cooling, heating, and domestic hot water for any building throughout the year without the use of an auxiliary source of energy. The high solar combi-plus system consists of evacuated tube heat pipe solar collector panels, thermal energy storage in the form of a phase change material (PCM) to maintain the stability of the system, a hot water driven lithium bromide absorption chiller, the heating, and the domestic hot water systems. The sequence of operation and control for such a system was described in four modes: solar regeneration mode, solar source only mode, solar and storage mixing mode, and the storage source only mode. In order to study its validity and reliability, a mathematical model was developed for its different components. These models were then coupled with a fully programmed calculator developed using "Visual Basic ©". This calculator was capable of finding the optimized combination between the solar collectors and PCM thermal energy storage, together with all annual hourly simulation results from the consumption of the building and the collected solar radiation, to the PCM thermal energy storage level. Furthermore, a villa in Riyadh KSA designed according to Saudi Arabia Thermal Standard was taken as a case study. The thermal and electrical consumption baselines were simulated hourly using the "Design Builder" Program, and then input in the calculator. The results showed that only 80m² (861 ft²) solar collectors and 3.5 m³ (925 gal) PCM storage were enough for running this system continuously throughout the year without the need for any other auxiliary source of energy. Finally, the feasibility of the novel system shows that although the very low electrical tariff in KSA deters its usage, adopting this system in the Country's Energy Policy will help the government decrease 12.7% of the overall electrical consumption in the country.

INTRODUCTION

Buildings have the highest potential to reduce carbon emissions, since they are responsible for the highest percentage of worldwide energy consumption. This makes them the main contiburing factor in warming the planet, polluting the environment, raising sea levels, and endangering our survival.

Based on the increasing worldwide tendency towards sustainability and the 'go green' movement, a lot of research has been conducted with the goal of reducing the electrical consumption of buildings. Shane Colclough et al. [1] investigated how nearly zero energy heating could be achieved for a low-energy building through the addition of a solar domestic hot water (DHW) and space heating system with a Seasonal Thermal Energy Store. Based on the DHW and space heating demand, they presented a cumulative life cycle energy and carbon analysis for the installation. Florides et al. developed a solar cooling system simulation model for a Cypriot building using TRNSYS [2-3]. Furthermore, an analyzed optimization for a solar cooling prototype located in Beirut was performed by Gaddhar et al. [4]. Francis Agyenim et al. [5] investigated by an experimental setup the performance of a 4.5 kW (1.3 Tons) absorption system (COP of 0.66) and a PCM system. They studied the heat transfer enhancement techniques by conducting four PCM experiments. Mateus et al. [6] studied a solar heating and cooling system in order to provide the

Mohamed Hmadi is a PHD student research fellow at Lebanese University, Hadath, Lebanon. **Dr. Adel Mourtada** is the head professor of the research unit "Energy and Sustainable Development", Scientific Research Center in Engineering (CRSI) in the Lebanese University, Hadath, Lebanon.

cooling, heating, and domestic hot water user demand for the United Europe (UE) climates; they developed a simulated model in TRNSYS and investigated the energetic and economic performance of such a system in different UE cities.

Most of the studies mentioned above analyze the performance of introducing individual solar systems in buildings, for example solar cooling only system, solar heating only system, etc. while papers studying combined systems with thermal storage show that another conventional auxiliary energy source is still needed to operate the system continuously throughout the year. These papers also studied the performance in a specific location in the world with specific demand baseline. They didn't provide a tool to generalize the study and determine the new system's size, if the location or demand changes.

It was from here the that idea towards developing a HSC+ System that could provide the cooling, heating and domestic hot water for any building all over the world was born. This will be achieved by designing a fully programmed calculator which can find the optimized combination between the solar collector area and the PCM volume in a way to provide the thermal need of the building throughout the year without the use of any other conventional auxiliary source of energy.

SYSTEM DESCRIPTION: THE STATE OF ART

Figure 1 shows the different components of the HSC+ system. The different elements are outlined below:

1. Evacuated tube heat pipe solar collectors with a dry connection that lowers the maintenance and installation cost and provide high supply temperatures. These collectors are equipped with built-in overheat protection.
2. Phase Change Material storage that uses its high latent solid-liquid phase change energy to store a large amount of energy in a relatively small volume. The energy stored will be used during periods where no solar energy is available.
3. Hot water Lithium Bromide (LiBr-H2O) absorption chiller, where the condenser and absorber are cooled using a closed loop geothermal pit.
4. Heating Hot water heat exchanger tank.
5. Domestic hot water tank.

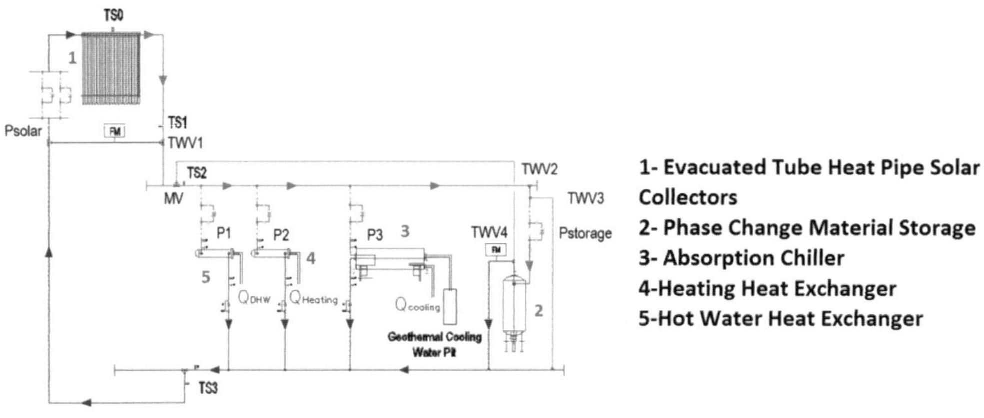

Figure 1 HSC+ System components

Sequence of operation:

Circulating Mode: As the sun appears after a long non-solar period the temperature sensor TS0 in the header of the panels will detect an increase in temperature and will thus run the pumps in circulating mode until the supply temperature TS1 is reached.

Solar Only Mode($Q_{solar.useful} \geq Q_{load} > 0$): When TS1 is reached and the energy collected by the solar collectors is greater than that needed by the loads of the building, only the solar panels will run the system by directly supplying the loads. The excess energy will be stored in the PCM.

Mixing Mode ($0 < Q_{solar.useful} < Q_{load}$,and $(Q_{solar.useful} + Q_{storage}) \geq Q_{load}$): If the Energy needed by the building is greater than that that could be collected by the solar panels, the PCM storage will interfere to maintain the stability of the system and thus provide the remaining necessary energy.

Storage Only Mode ($Q_{solar.useful} = 0$, and $Q_{storage} \geq Q_{load} > 0$): When there is no solar radiation, the PCM storage only will run the system. If the storage is empty an auxiliary source of heat is needed.

MATHEMATICAL MODELING:

Evacuated Tube Solar Panels:

After calculating the total global solar radiation on the tilted solar collector G, the solar radiation absorbed is calculated on an hourly basis based on the following equation [7]:

$$Q_{absorbed} = F.A_a.G.(\tau * \alpha) \tag{1}$$

Where $Q_{absorbed}$ is absorbed power by the collector, F is the collector efficiency factor, A_a is the absorber area in m^2, τ is the transmittance coefficient of cover tube, and α is the solar absorptance coefficient of the solar collector. The heat losses from the evacuated tube are by convection and radiation. The theoretical equation for calculating these losses is summarized in the following equation:

$$Q_{thermal\_loss} = F.U_L.A_a.(T_{m_i} - T_a) \tag{2}$$

Therefore, the useful power (P_u) is calculated using the following equation:

$$P_u = Q_{absorbed} - Q_{Thermal\_loss} \tag{3}$$

Where $Q_{thermal\_losses}$ is the thermal power lost by the evacuated tube collector, U_L is the overall heat transfer coefficient of the collector, T_{m_i} is the mean temperature of the fluid inside the solar collector in degrees Celsuis. Evaluated as shown in equations (4) and (5):

During Normal operation periods with solar radiations:

$$T_{m_i} = T_m = \frac{T_{in} + T_{out}}{2} \tag{4}$$

During Non-operation periods where there is no supply from solar collectors:

$$T_{m_i} = T_{m_{i-1}} + \frac{P_u}{1.16 * V_w} \tag{5}$$

Where T_{in} and T_{out} are respectively the hot water temperatures entering and leaving the collector, T_a is the ambient exterior dry bulb air temperature, T_m is the mean average operating temperature of the solar panels, T_{m_i} and $T_{m_{i-1}}$ are respectively the mean temperature inside solar collector at "i" and "i-1" hour, and V_w is the volume of water inside the solar collector system.

When the useful power is negative, T_{m_i} will decrease during the night where there is no solar radiation with a lower limit $T_{m_i} \geq T_a$. When the useful power is positive, the T_{m_i} will increase during the day where there is solar radiation until it reaches the upper limit $T_{m_i} = T_m$, the average operating temperature of the solar panels.

Based on Europe (EN 12975 norm) the three parameter model for efficiency η_{sc} could be summarized as follows:

$$F.U_L = k_1 + k_2(T_{m_i} - T_a) \quad (6)$$

$$\eta_{SC} = k_0 - k_1.x - k_2.G.x^2 \quad (7)$$

Where:

$$k_0 = F.(\tau.\alpha) \quad (8)$$

$$x = \frac{(T_{m_i} - T_a)}{G} \quad (9)$$

Where k_0 is the efficiency coefficient and k_1 and k_2 are the coefficients for the chosen solar collectors. Based on the experimental studies on evacuated tube solar collectors, conducted by S. Brunold et al. [14], the characteristics of the solar collectors and the constant coefficients (k_0, k_1, k_2) are collected.

Then the hours and useful amount of energy $Q_{solar.useful}$ that the panel can provide the system or storage with after regenerating the losses during non-solar hours and maintaining the operation temperature of hot water T_m, can be found as follows:

$$Q_{solar.useful} = P_u - Q_{solar.regenerating} \quad (10)$$

The absorbed power, overall heat transfer coefficient, thermal losses power, useful power, useful solar energy after regenerating, and efficiencies are all calculated on an hourly basis.

Phase Change Material Thermal Storage:

After calculating Q_1, Q_2, and Q_3 the portion of domestic hot water, heating, and cooling load respectively from the hot water source, the Q_{load} needed at each hour is calculated as follows:

$$Q_{load} = Q_1 + Q_2 + Q_3 \quad (11)$$

The system will work in a way that whenever $Q_{solar.useful}$ from the solar is available and T_m is reached, this load energy Q_{load} will be provided directly from the solar collector system, otherwise the storage will provide the additional load required. The energy needed from storage (negative value) and the energy available (positive value) that could be stored in storage are calculated:

$$Q_{difference} = Q_{solar.useful} - Q_{load} \quad (12)$$

It was assumed that initially the storage is full, after which it enters the discharging and charging phases as follows: Initially: the liquid fraction of melt $f = 1$, then the maximum enthalpy of PCM:

$$H_{max} = \rho_l * \lambda \quad (13)$$

$$V = \frac{Q_{max.stored}}{H_{max}} \quad (14)$$

Where $H_{\max}$ is the maximum enthalpy at the end of the phase change $f=1$, ρ_l is the liquid density PCM, λ is the latent heat of fusion, $Q_{\max.stored}$ is the maximum storage needed, and V is the Storage volume.

Discharging Mode: If $Q_{difference} < 0$, the storage must provide this energy as shown in Figure 2a:

$$Q_{discharged} = \frac{Q_{difference}}{\eta_{storage}} \tag{15}$$

$$V * H_{f_k} = V * H_{f_{k-1}} - \left|Q_{discharged}\right| \tag{16}$$

Charging Mode: If $Q_{difference} > 0$, the storage can store this available excess energy as shown in Figure 2b:

$$Q_{charged} = Q_{difference} * \eta_{storage} \tag{17}$$

$$V * H_{f_k} = V * H_{f_{k-1}} + \left|Q_{charged}\right| \tag{18}$$

Where $Q_{discharged}$ is the energy discharged from the storage to provide the needed load $Q_{difference}$ and $Q_{charged}$ is the energy that could possibly be stored in the storage from the available energy $Q_{difference}$, $\eta_{storage}$ is the efficiency of PCM storage, H_{f_k} and H_{f_k-1} are respectively the PCM's enthalpy at melt fraction $f=f_k$ and $f=f_{k-1}$; and k and k-1 are the number of hours from the beginning of discharging or charging.

A limitation of $H_{\max} \geq H_{f_k} > 0$ is considered, so that the melt fraction will stay between $1 \geq f > 0$, and the PCM will stay within the solid-liquid phase change [8]. If $H_{f_k} = 0$ the storage is empty and an auxiliary heat source is needed:

$$Q_{aux} = Q_{load} - (Q_{solar.useful} + Q_{storage}) \tag{19}$$

The aim of our project is to find the optimal combination between the solar collector area and energy PCM storage such that $Q_{aux} = 0$. Then an iterative function (loop) is programmed in a way to search for the amount of storage (KWh) necessary for each solar collector area (m2), so that no auxiliary energy is needed.

Temperatures:

To choose the operating temperatures at different parts of the system, the logic outline below is followed: The absorption chillers' generating hot water temperature range is between 88-83 °C (190.4-181.4 °F), so it is considered the supply and return hot water working temperatures of the other load systems to be in this range. In order for the storage to supply this hot water temperature range, the melting temperature of the PCM must be greater than 88 °C (190.4 °F) as shown in the Figure 2a.

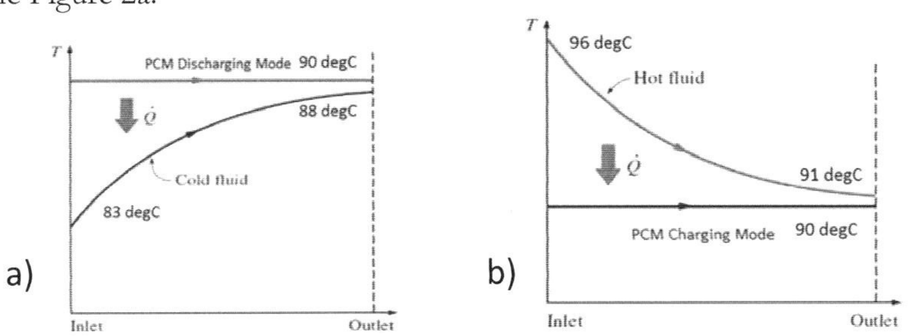

Figure 2 Discharging from PCM to System and Charging of PCM from Solar Hot Water

Mg(NO3)2.6H2O salt has a melting temperature of 90 °C (194 °F) which is suitable for our system [9]. Hence, to charge this PCM, the outlet charging hot water temperature 91 °C (195.8 °F) from the storage tank must be greater than the melting point of this PCM. Therefore, the outlet temperature from solar panels will be 96 °C (204.8 °F).

PROGRAM DEVELOPING CALCULATOR:

An HSC + system calculator with a user friendly interface is developed through the implementation of the above stated modeling equations using "Visual Basic" as shown in Figure 3. This calculator could find the optimized solution for any building all over the world, and could also provide the hourly simulation results all over the year together with the feasibility study. In addition, it has a default page where you can change any of the system component characteristics.

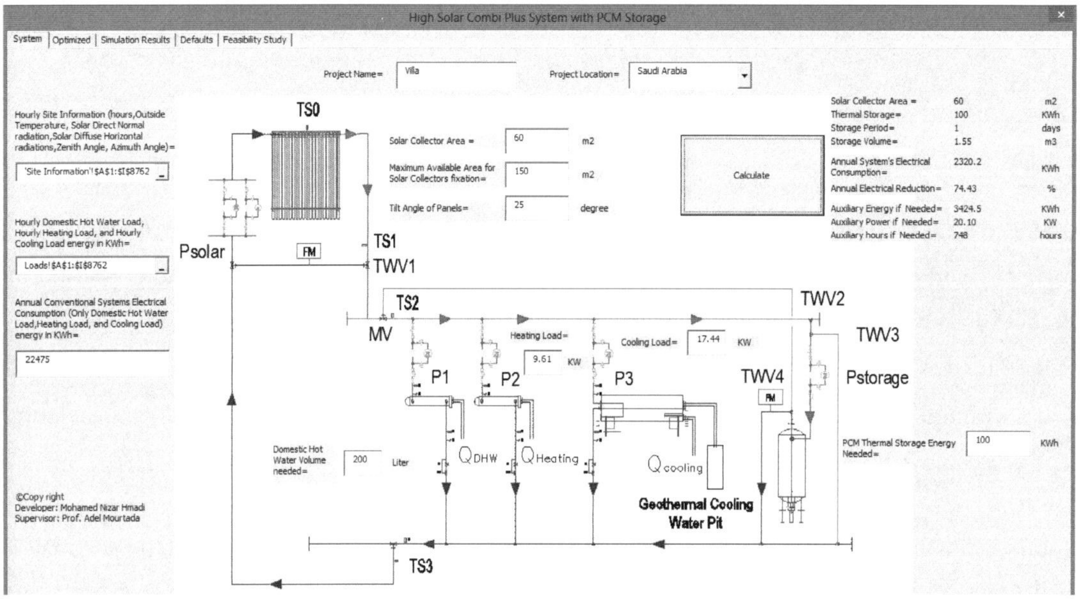

Figure 3 Calculator Interface

CASE STUDY: A VILLA IN KSA

Taking a case study, a villa in Riyadh KSA was chosen since this location has an extreme desert climate which may need cooling and heating in the same day. Furthermore, KSA is witnessing a huge increase in electrical consumption, while it is characterized by high solar radiations. These factors make this system an ideal solution for such a country. The villa is a standard duplex residence of 300 m^2 (3230 ft^2) total occupied area. It is a large enough for a 10-member family. It is designed according to Saudi Arabia Thermal Standards. Using "Design Builder Simulation Program" the hourly site information data was found and the hourly cooling, heating and domestic hot water consumption baselines were calculated all over the year together with the CO_2 emissions. The solar collectors will be placed over a tilted roof directed towards the south at an angle equal to the latitude of Riyadh (25°), while the different components of the system will be located on the roof. The location of the villa; the hourly site information data from a simplified excel sheet, the villa's hourly thermal loads, and the annual conventional electrical consumption are the user inputs to the developed calculator. The calculator provides two options for sizing. Firstly, the sizes of the available solar collector area and the PCM storage volume can be entered manually in order to find the storage period, electrical consumption of this HSC+ system, and the amount of auxiliary energy if needed as shown in Figure 4. Secondly, the optimized solution is found by running a loop. "Loop Run" is an iterative function that finds for each solar collector area available its corresponding PCM storage volume such that the auxiliary energy needed is zero. This loop terminates when the maximum available solar collector area is reached and the resulting combinations are summerized in Fig. 4a. To choose the optimized size, the financial factor is taken into consideration. The minimum overall cost, as illustrated in Fig.4b, represents the desired combination between the solar collectors' area and the PCM storage requirement.

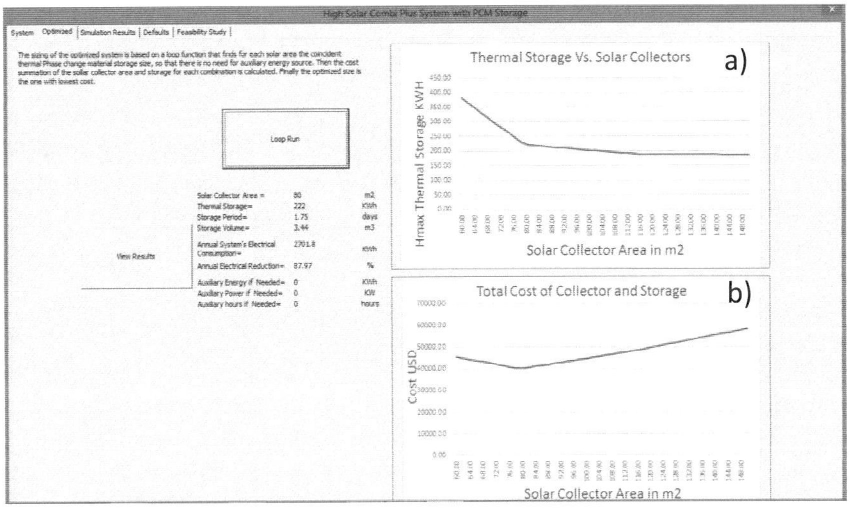

Figure 4 Optimized Page in the Calculator

After finding the optimized solution, the simulation results give the hourly behavior of the system components throughout the year starting from the solar energy collected by the solar panels, thermal energy level stored in the PCM storage, the consumption loads, the efficiencies, flows, and temperatures as shown in Figure 5. These results show that the HSC+ system in a villa in KSA could work normally year-round without the need for an auxiliary source of energy saving 19773 KWH (67.5 Mbtu) of annual electrical consumption.

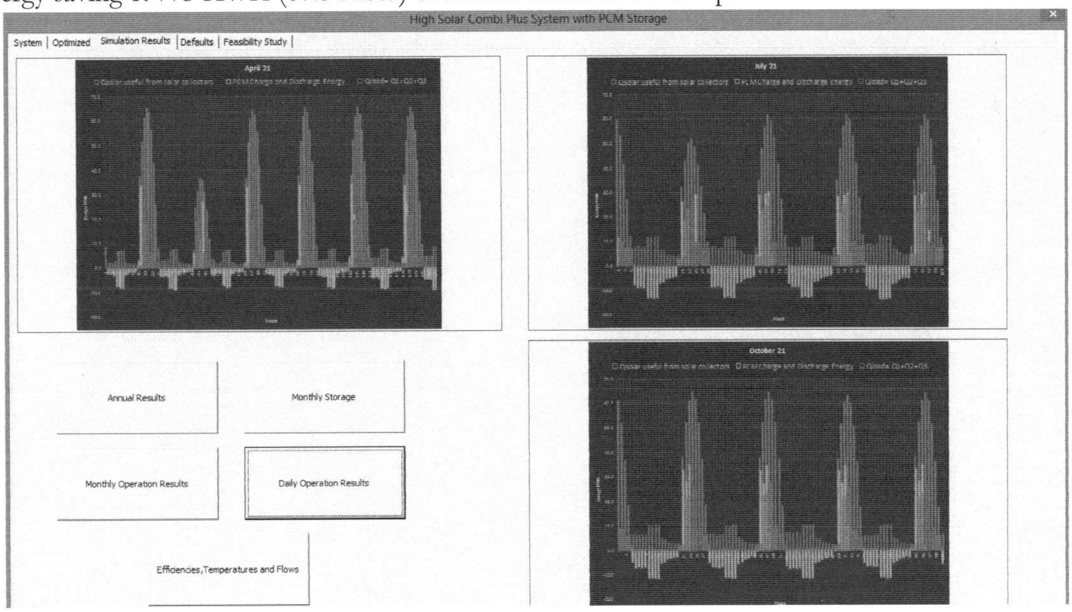

Figure 5 Hourly simulation results

FEASIBILITY STUDY

The initial cost of such a system is estimated to be 61590 USD as reported by name of author [10], together with the conventional electrical bill (1185 USD) and the optimized electrical bill (534 USD) after introducing the HSC+ system. Due to the high subsidies paid by the KSA government on electricity production which lowers the consumer's electrical bill, the payback period of such a system is 32 years, greater than the expected life time of such a system, assumed to be 30 years. This makes the HSC+ system unfeasible.

This result directs towards studying the benefits of this system from the governmental point of view. 50 % of Saudi Arabia's electrical consumption is from residential buildings [11]. 38 % of these residential buildings are villas [12]. 76% of the electrical consumption of villas is from HVAC and DHW systems [13]. The HSC+ system shows that it could reduce up to 88% of the electrical loads of these systems. Hence, introducing the HSC+ system to villas could reduce up to 12.7 % of the KSA's overall electrical consumption. If the government subsidizes this system in the same way it subsidizes electricity production, it could gain, over the lifetime of one system installed, a net present value up to 180 thousand USD, 590 MWh (2013 Mbtu) reduction in electricity, and 440 tons reduction in CO_2 emissions, with a payback period of 11 years. This will reduce the payback period from the owner's point of view to 14 years encouraging more people to invest in such a system.

CONCLUSION

This paper develops a fully programed HSC+ calculator that can find the optimized combination for any residential building all over the world. The case study showed that the system can operate continuously throughout the year without the use of any other auxiliary energy source by using only an 80 m^2 (861 ft^2) solar collector area and 3.5 m^3 (925 gal) PCM storage. The feasibility study shows that although its high initial cost and high KSA governmental subsidies is deterring from using it, adopting the HSC+ system in the Country's Energy Efficiency Policy by subsidizing it with the same percentage as the electricity production will save up to 590 MWh (2013 Mbtu) electrical consumption, 440 tons CO_2 emissions, and 180,000 USD as a net present value over the lifetime of each system installed with a payback period of 11 years and 7 months. Applying this system only on villas will reduce up to 12.7 % of the country's overall electrical consumption.

ACKNOWLEDGMENTS

I would also like to thank Prof. Adel Mourtada for his recommendations, thoughtful ideas and help in completing this work.

NOMENCLATURE

HSC+	=	High Solar Combi-Plus System
DHW	=	Domestic Hot Water
PCM	=	Phase Change Material
COP	=	Coefficient of Performance

Subscripts

λ	=	Latent heat of fusion (KWh/Kg)
ρ	=	Density of water (kg/m3)
g	=	Gravitational acceleration (N/kg)
η_p	=	Efficiency of pumps (%)

REFERENCES

[1] Shane Colclough, Teresa McGrath, Net energy analysis of a solar combi system with Seasonal Thermal Energy Store. Applied Energy 147 (2015) 611–616

[2] Folrides, G.A., Kalogirou, S.A., Tassou, S.A., Wrobel, L.C., Modelling and simulation of an absorption solar cooling system for Cyprus. Solar Energy, 2001. 72(1): p. 43-51.

[3] Folrides, G.A., Kalogirou, S.A., Tassou, S.A., Wrobel, L.C., Modelling, simulation and warming impact assessment of a domestic-size absorption solar cooling system. Applied Thermal Engineering, 2002. 22: p. 1313-1325.

[4] Ghaddar, N.K., Shihab, M., Bdeir, F., Modeling and simulation of solar absorption system performance in Beirut. Renewable Energy, 1996. 10(4): p. 539-558.

[5] Francis Agyenim, The use of enhanced heat transfer phase change materials (PCM) to improve the coefficient of performance (COP) of solar powered LiBr/H2O absorption cooling systems. Renewable Energy 87 (2016) 229-239

[6] Mateus, T., Oliveira, A.C., Energy and economic analysis of an integrated solar absorption cooling and heating system in different building types and climates. Applied Energy, 2009. 86: p. 949-957.

[7] S. Brunold, R. Frey, U. Frei, A comparison of three different collectors for process heat applications. Solarenergie Prüf- und Forschungsstelle Ingenieurschule ITR Oberseestr. 10; CH-8640 Rapperswil.

[8] Atul Sharma, V.V. Tyagi, C.R. Chen, D. Buddhi, Review on thermal energy storage with phase change materials and applications. Renewable and Sustainable Energy Reviews 13 (2009) 318–345.

[9] Belen Zalba, Jose M. Marin, Luisa F. Cabeza, Harald Mehling, Review on thermal energy storage with phase change: materials, heat transfer analysis and applications. Applied Thermal Engineering 23 (2003) 251–283.

[10] WEBCO s.a.r.l is among the global firms in the market developing sustainable energy and energy efficiency products Beirut – Lebanon

[11] Hatim Asim Azzouz, Saving the Black Gold (Saudi Arabia). AUTODESK Sustainability Workshop, May-2013 [Online]. Available: http://sustainabilityworkshop.autodesk.com/project-gallery/saving-black-gold-saudi-arabia

[12] Diana Aljassem, 52% of Saudis in Riyadh rent houses. The Population Statistic Center at King Saud University (KSU) in Riyadh. February 2013. [Online]. Available: http://www.arabnews.com/saudiarabia/52saudisriyadhrenthouses

[13] Heidy Rahman, analyst at Citigroup, Saudi: Oil importer by 2030. www.zawya.com. [Online]. Available: http://www.solarify.eu/2012/09/06/saudiarabienbaldolimporteur/

Sustainable Design in Metro Stations

Steven K.H. LAI
Member ASHRAE

Anne H.E. BEH
Non-Member ASHRAE

William Y.M. XIE, CEng
Member ASHRAE

Hadi WIJAYA, PEng, CEng
Non-Member ASHRAE

K.W. ANG, PEng, CEng
Non-Member ASHRAE

ABSTRACT

In addressing the Global Energy Crisis, engineers have the social responsibility to find and implement practical solutions. This paper discusses the sustainable heating, ventilation and air-conditioning (HVAC) design concepts that can be applied in metro systems.

The use of Platform Screen Door (PSD) along the underground station platform to isolate the station from the tunnel has a positive impact on the overall energy consumption of the metro systems in tropical regions (e.g. Singapore, Thailand, Malaysia, etc.) and subtropical regions (e.g. Hong Kong, Guangzhou, Shanghai, etc.). During cooler months, the concept of free cooling by introducing more outdoor air in seasoned countries can further reduce energy consumption as well as improving indoor air quality.

To contribute on sustainable development, engineers should participate in the overall planning of the metro stations with other stakeholders. Effective use of space is a key strategy in mitigating the challenge of land scarcity in developed urban cities. This planning may include (1) utilizing the space created incidentally due to metro construction by converting it to commercial areas instead of backfilling it, (2) integrate development above metro stations or (3) connecting the metro stations to nearby developments to mitigate impact to the aboveground traffic.

For sustainable design, Central Chiller Plant can be considered for adjacent stations and their connecting developments. Apart from saving initial cost and running cost, this approach also saves the overall land space. To further release space above ground for other usage, open top vent shafts can be designed to reduce the size of the aboveground structure.

There are other sustainable designs in metro systems that are related to HVAC design. Taking the lighting system for example, the use of "Light without Fitting" (LwF) concept with Light-emitting Diode (LED) lighting is an ongoing trend to bring a new aesthetic experience to the public. It not only reduces the lighting energy consumption, but also reduces the cooling load requirement.

Examples from various projects in Singapore and Hong Kong will be referenced in this paper. Other sustainable measures e.g. Variable Speed Drive (VSD) for pumps and cooling towers, outdoor air control, regenerative energy, etc. which have been used in metro systems are also covered in this paper.

1. INTRODUCTION

Large amount of on-road vehicles in densely populated major cities has made cities facing severe traffic congestion problems every day, along with the associated noise and air pollution problems. The rapid depletion of fuel and natural resources has also sent an alert across the world on the importance of environmental sustainability. Governments of many major cities have gone into the construction of underground metro systems to alleviate such problems. This paper will consider the various measures to attain a sustainable design in metro stations.

Mr. Steven K.H. LAI is a Director of Infrastructure for the China Region at WSP (Asia) Limited based in Hong Kong. **Ms. Anne H.E. BEH** is an Engineer at WSP (Asia) Limited based in Hong Kong. **Mr. William Y.M. XIE** is a Senior Associate at WSP Consultancy Pte Limited based in Singapore. **Er. Hadi WIJAYA,** is a Deputy Director at the Land Transport Authority of Singapore. **Er. Kok Wee ANG** is a Manager at the Land Transport Authority of Singapore.

Though the metro system is one of the most energy-efficient means of transporting large number of people at a time, it still consumes huge amount of energy due to the wide expansion of the metro network (especially in major cities) and high frequency services. For example, Hong Kong MTR has consumed over 1,948 GWh in 2016 for railway and property operation, for which property operation is accounted for 21% of that [2]. Several surveys were conducted in various countries to show that amongst all the subsystems in the stations, majority of the energy consumption goes to the Heating, Ventilation and Air-Conditioning (HVAC) subsystems. For example, in Beijing Subway, HVAC system consumed over 40% of the total energy consumption of the station operation [7], while the figure may hike up to about 55% in a tropical city like Singapore [5]. This paper will present some sustainable HVAC design concepts that can be applied in metro systems.

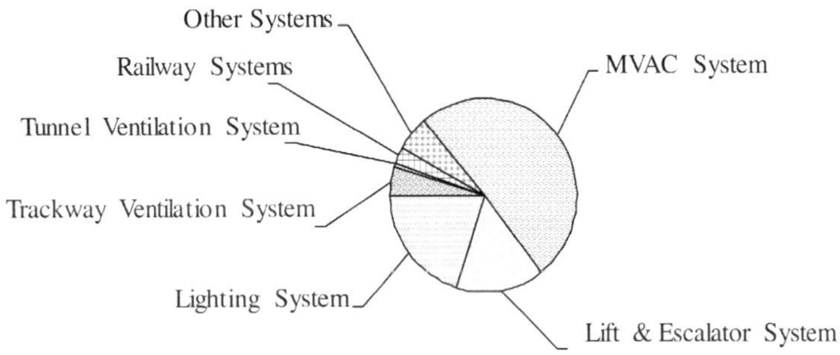

Figure 1: Annual Energy Usage of an Underground Station [5]

Lighting system is the second highest source of energy consumption in the metro station, taking up to 15 to 20% of the station overall energy consumption. Using light emitting diode (LED) lighting for the whole station, the lighting energy can be reduced by about 50%, while also reducing the cooling load for the HVAC system due to the lower heat emission from the LED lights [8].

While scarcity of land is a worsening issue in major cities, the use of Central Chiller Plant for several adjacent stations and their connecting developments not only serves as a solution to this problem but also saves initial and running cost. For example, for the Tsuen Wan Line in Hong Kong, a centralized chilled water plant has been used to serve 5 stations in order to improve the overall energy efficiency of the air-conditioning system and to reduce the station footprint.

It is an on-going trend to integrate development above or around metro stations for the following reasons: (1) utilizing the space created incidentally due to metro construction by converting it to commercial areas instead of backfilling it, (2) integrate development above metro stations or (3) connecting the metro stations to nearby developments to mitigate impact to the aboveground traffic. This paper will discuss on how engineers could participate in the overall planning of the metro with other stakeholders.

2. ENVIRONMENTAL CONTROL SYSTEM (ECS) FOR METRO STATIONS

As early as in the 19th century, Open System has been used for the tunnels and underground railway stations, especially in places with lower outdoor temperature, such as the New York subway. However, with the introduction of air-conditioner and its increasing use over time, Closed System began to be considered in the design of metro stations, mostly in East Asia, where during summer, station temperature of between 3 to 5 degrees Celsius lower than that of the outdoors can be achieved. With the improvement in the signaling system's reliability and accuracy, Platform Screen Doors (PSD) can be installed at platforms to separate the trackway and the station platform area.

This could lower the station temperature by 4 to 8 degrees Celsius relative to the outdoors temperature, and provide a cleaner station environment for the passengers. Following are brief descriptions of the three aforementioned systems:

2.1 Open System

Open System describes a system where the ventilation shafts in the tunnels are in an open mode, such that when the trains move through the tunnels, the associated piston effect will pull outdoor air from the station entrances/exits and ventilation shafts behind the moving train into the station and through the tunnel, while the hot air inside the tunnel will be pushed out through the station entrances/exits and ventilation shafts ahead of the moving train. If air-conditioning system is not provided, the in-station temperature will be higher than outdoors. Thus, this system is more widely used in the early railway systems and in cooler regions (such as the New York subway system).

2.2 Closed System

For Closed System, the ventilation shaft will be closed with the provision of shut-off motorized dampers such that the tunnel environment is isolated from the outside except at tunnel portals. When outdoor temperature is high, the closed ventilation shaft arrangement will reduce the station cooling load compared with the Open System. Ventilation of the tunnels will then rely solely on the train piston effect bringing cool station air into the tunnels, while the hot tunnel air will be cooled when passing through the platform area. During the hot summer seasons (for example, when outdoor temperature is 32 degrees Celsius), the station temperature can be maintained at around 28 to 30 degrees Celsius with the operation of air-conditioning system in the station. This system is widely used in the East Asia metro systems without platform screen doors installed along the station platform (e.g. Tianjin, Beijing, Shenyang, etc.).

2.3 Platform Screen Door (PSD)

The PSD system was implemented in all underground metro stations in Singapore. A full height screen is provided along the platform edge to separate the station platform area and the trackway. Automated doors are provided at the screen which align and synchronize with the train doors to allow for passenger to board and alight. The installation of the PSD avoids high velocity of air movement in the station induced by the piston effect due to the train movement. The ventilation shafts are kept open, such that the train piston effect will bring outdoor air into the tunnel, while hot tunnel air will leave through the ventilation shafts of the next station. This will significantly reduce (1) the infiltration of hot tunnel air into the platform area and (2) the infiltration of hot and humid outdoor air from station entrance into the station area. Thus, the station air-conditioning cooling load can be reduced by approximately 30% compared to the Closed System.

Even during the summer seasons in temperate regions (e.g. when outdoor temperature is 32 degrees Celsius), the station temperature can be maintained at around 24 to 28 degrees Celsius. This provide greater comfort to the passengers. There will be an improvement in the station air quality as well, as there will be less infiltration of dust from tunnel and the outdoor into the station. Another advantage in the installation of PSD is that it can reduce the number of accidents where passengers fall onto the tracks.

Therefore, the PSD has been widely used in metro systems in the South-East Asia region in recent years (such as metro system in Shenzhen, Hong Kong, Singapore, etc.).

Figure 2: Station with Platform Screen Door

2.4 Comparison Among Various ECS Schemes

The table below summarizes the comparison among the three systems:

Table 1: Comparison of Various ECS Schemes

	Open System	Closed System	PSD
Air-conditioning system required in station	No	Yes	Yes
Station temperature during summer months	Higher than outdoors (Without air-conditioning system in station)	3~5°C lower than outdoors	24~28°C

3. SUSTAINABLE FEATURES

3.1 Free Cooling During Cooler Months

During the cooler months, the concept of "Free Cooling" with the aid of mechanical ventilation can be adopted to relief energy consumption. Cool outside air is induced into the station by mechanical ventilation to maintain the station temperature and humidity within comfort level. The station return air is discharged to ambient. The "Free Cooling" operation can be automated simply by installing sensors to keep track of the enthalpy difference between indoors and outdoors.

3.2 Energy Saving Lighting System

Lighting system is the second highest source of energy consumption in a metro station, taking up to 15 to 20% of the overall energy consumption [5]. Through appropriate selection of lighting lamps, the lighting energy consumption can be reduced significantly:

Table 2: Comparisons between Traditional Incandescent, Halogen Incandescent, CFLs, and LEDs based on same light levels provided [8]

	60W Traditional Incandescent	43W Energy-saving Incandescent	15W CFL	12W LED
Energy Saved (%)	-	25%	75%	75~80%
Annual Energy Cost *	$4.80	$3.50	$1.20	$1.0
Bulb Life (hours)	1,000	1,000~3,000	10,000	25,000

* Based on 2 hours/day of usage, an electricity rate of 11 cents per kilowatt-hour, shown in U.S. dollars.

It can be seen from above that by replacing the station lighting system with LED lights, the lighting energy efficiency can be increased by about 75~80% as compared to the traditional incandescent lights. Furthermore, the bulb life of LED is 25 times of the traditional incandescent lights, and 2.5 times of Compact Fluorescent Lights (CFL).

On the other hand, selection of lighting lamps also has an impact on the cooling load for the HVAC system. Based on project experience, heat dissipation rate is reduced by selecting LED over T5 Fluorescent tube, and thus the cooling load of the HVAC system can be reduced by at least 25%.

The popular use of LED lighting was brought about not only by its energy saving and durability benefits, but also its flexibility for integration with the design of the space. The concept of "Light without Fitting (LwF)" is a new lighting concept to integrate the lighting system with other facilities or even as a display of art. For example, LED lighting can be integrated into the escalator to avoid having light fittings mounted above the escalator. This also have the benefit of avoiding work at height during maintenance of the light fittings at ceiling level.

Another example of using LED lighting for functional and decorative use is the upcoming station in Singapore. The LED lights are fitted into the edges of the walls or inside part of a wall. From Figure 3, LED lights are also embedded into the escalator balustrades.

Figure 3: Functional LED in upcoming Singapore Metro Station.

3.3 Adoption of Central Chiller Plant (CCP)

The conventional design approach is to provide a self-contained, stand-alone chilled water plant comprising of an underground chiller plant and an at-grade cooling tower plant for each station. As the chiller plant and cooling tower plant require large plant rooms to accommodate them, this poses a challenge in the design for some stations, especially for station located in urban built up areas where large at grade space for cooling tower may not be available.

To adopt a centralized chilled water plant concept, the centralized chilled water plant can be located at one station. The chilled water will be supplied to other adjacent stations through supply and return pipes that run along the tunnels connecting the stations.

The advantage of the CCP scheme is that the use of larger chiller instead of a few smaller ones is more energy efficient. There is also reduction in station footprint for stations that do not have a chiller plant. The freed up above and underground space can be used for other developments. However, the longer pipe run in tunnels may introduce additional pump head and increase piping cost. Hence, careful analysis should be made to determine whether there is justifiable saving for the CCP scheme over the conventional scheme.

ORIGINAL SCHEME	STATION A	STATION B	STATION C
CHILLER PLANT	290m²	290m²	290m²
COOLING TOWER PLANT INCLUDING MAKEUP WATER TANK	190m²	190m²	190m²

CCP SCHEME	STATION A	STATION B	STATION C
CHILLER PLANT	-	360m²	-
COOLING TOWER PLANT INCLUDING MAKEUP WATER TANK	-	530m²	-

Figure 4: Comparison of conventional (Original Scheme) approach and by adopting a central chiller plant (CCP Scheme) [3]

This concept can be further developed for implementation to not only several adjacent metro stations, but also on a metro station together with their surrounding development to achieve greater overall energy efficiency.

4. PLANNING OF THE METRO SYSTEM

In many major cities, the challenge posed in finding available above-ground space for new development is increasing rapidly. Alongside the population boom in densely populated cities such as New York, Hong Kong, etc., the housing availability cannot meet the demand. Thus, the concept of integrating commercial or residential development above metro stations has become a trend in recent years. This approach is widely adopted in many major cities such as Hong Kong, Shenzhen, Shanghai, Beijing, etc. This approach demonstrates a well-designed sustainable development for both the station and the building to achieve optimization in the use of resources, environmental-friendliness, technological innovation, safety and convenience, etc. The advantages are elaborated in details as follows:

4.1 Integrated Development

With commercial development, such as shopping malls above a metro station, it can increase patronage for both the building and the metro station. Such approach has been widely adopted in Hong Kong. For example, Kowloon

Bay station which has a shopping mall and residential development on top.

As such, the HVAC system design of the entire integrated development can be considered as a whole to achieve a more energy efficient design. For example, the concept of centralized chiller plant mentioned before can be considered to be adopted for the station and the development on top of it.

4.2 Transit Orientated Development (TOD)

For a metro station where topside development is not possible, a better connection between the station and other nearby development is crucial as it can help to mitigate impact to the above ground traffic. For example, in Singapore, the City Hall Station and Esplanade Station are connected by long linkway to transfer passengers travelling from the North South East West Line to and from the Circle Line, and some other nearby buildings, without another means of transport above ground.

Singapore MRT has also adopted another environmental friendly approach. In the promotion of better public health while aiming to reduce the number of on-road vehicles, the government encourages the public to get to the metro stations by bicycles. As such, underground bicycle parks have been designed in some metro stations. Having the bicycle park underground saves above-ground space while also protects the bicycles from different weather conditions. The underground bicycle parking area is fire compartmented away from station main concourse area and has its independent air-conditioning and smoke control system.

4.3 Making Use of Incidental Space

During the construction of cut & cover tunnels, the space between the tunnel and ground needs to be back-filled. This space when incidentally built but not on purpose may be used for commercial activities instead of backfilling it. This is a good demonstration of a better use of land and serves as a solution to the problem of depletion of available land for new developments.

4.4 Open Top Vent Shafts

The ventilation of underground stations and tunnels rely on ventilation shafts extended above ground. In conventional design, ventilation outlets are provided with weather proof louvres. To avoid excessive pressure loss through the louvre and regenerated noise, the face velocity at the louvres is designed at 2-3m/s. This normally will result in big louvre areas and plenums which would take up large land space and appears to be aesthetically unpleasant.

To mitigate this, some metro systems, e.g. Singapore MRT Thomson-East Coast Line Stages 4 & 5, which is under construction, used open top vent shaft arrangement. It uses gratings to allow higher face velocity (e.g. 7-8m/s), in lieu of louvres. This eliminated the use of large louvres and plenums, hence reduces the total vent shaft height and associated above ground structure by up to 40%.

5. OTHER SUSTAINABLE DESIGN

Apart from the topics discussed above, other sustainable concepts that are implemented in the metro system are discussed in detail in the paper "Energy Efficiency in Singapore's Rapid Transit System" [9]. Items below are some brief description of the measures:

1. Variable Speed Drive (VSD)

A VSD is a device that controls the rotational speed of motor-driven equipment. When in use with fans, chilled water pumps and cooling towers, it allows the equipment to run at part load condition efficiently to meet the design requirements. It can reduce the overall station power consumption by 0.4%.

2. Outdoor Air Control

Carbon dioxide (CO_2) sensors are installed to automatically adjust the fresh air supply rates to the station public areas while ensuring CO_2 level is below 1000ppm. This ensures that the energy consumption of the HVAC system is reduced without compromising the in-station air quality. The estimated reduction in energy is up to 0.36% of a typical station's power consumption.

3. Regenerative Energy

During the braking of trains, energy can be regenerated instead of dissipated as heat to the tunnel. A suitably equipped train has the potential for feeding energy back into the "line" during braking. This regenerated energy could be used to power onboard equipment as well as to satisfy the power needs of a train operating in the vicinity of the regenerating train. Other methods have been proposed for handling the regenerated energy. One feasible design consists of recycling the regenerated energy to be fed-back to the stations or utility feeders, and yet another proposes storing the energy in a flywheel device making the energy available for future use. The amount of energy regenerated varies from one metro system to another. Among the variables affecting regeneration are: track profile, train headway, the type of traction motor used, electrical distribution network circuitry, and the type of regeneration scheme. Because of the complex dependence on various system parameters, the results obtained for a particular system cannot be universally applied. In addition, the sparse measured data available is not always presented in a usable way. In some projects, regeneration was found to be more than 30% [10]. The percentage of regeneration was defined as:

$$Wr/Wtotal \times 100\% \qquad (1)$$

where, Wr was the energy regenerated to the line during braking,

Wtotal is the total energy supplied from the line during the whole run without regeneration.

6. CONCLUSION

With the metro system being one of the most commonly used means of transportation in many major cities, a smart and energy-efficient design for the metro system is crucial in reducing the cities' carbon footprints. For large scale underground metro stations, especially in warmer and humid regions such as in the South-East Asia, the HVAC system is one main contributor to the overall energy consumption of an underground metro station. Thus, designing an energy-efficient HVAC system can not only have a positive impact on the environment, but also greatly saves the running cost of the station. Unconventional approach such as installation of platform screen door or the adoption of central chiller plant has been proven to be more energy efficient.

As many cities are growing rapidly, the issue of the depletion of land has come to the attention of many governments. The concept of integrated development and transit orientated development has become a trend. Not only does this save land space, but it will also reduce the overall carbon footprint of the city, as many passengers can avoid driving or taking the bus to and from the metro station. Engineers should participate in the overall planning of the metro with other stakeholders to reduce the site footprint and achieve better energy-efficiency in the design of the metro systems and its surrounding development. To promote public health and a more active lifestyle, the construction of underground bicycle park at metro stations can be considered as well.

7. ACKNOWLEDGMENTS

The authors gratefully acknowledge the Land Transport Authority of Singapore and WSP Hong Kong and Singapore offices for their assistance in providing project information and data for reference in this paper.

8. NOMENCLATURE

HVAC = Heat, Ventilation and Air-Conditioning
PSD = Platform Screen Door
CCP = Central Chiller Plant
LED = Light Emitting Diode
CFL = Compact Fluorescent Lamps
LwF = Light without Fitting
TOD = Transit Orientated Development
VSD = Variable Speed Drive
CO_2 = Carbon dioxide

9. Subscripts

N.A.

10. REFERENCES

[1] ASHRAE. 2017 *ASHRAE Handbook—Fundamentals*. Atlanta: ASHRAE.
[2] MTR Corporation Limited. (2016). MTR Sustainability Report 2016. Retrieved from http://www.mtr.com.hk/en/corporate/sustainability/2016rpt/download.php
[3] Xie, W., and Koo, S.L. (2016, September). Centralized chilled water plant for underground metro stations. Retrieved from https://network.wsp-pb.com/article/centralised-chilled-water-plant-for-underground-metro-stations
[4] Lu, M., He, T., Chen., Z.Q., and Pei, X.H. (2010). Analysis of the Electric Power Consumption of Beijing's Urban Rail Transit. *2010 The Second China Energy Scientist Forum*, 1985-1989.
[5] Wijaya, H., Chiam, B.H., Ang, K.W., Xie, W., and Lai, S. (2017). Smart green underground metro station in Singapore, *HKIE Transactions*, 24:2, 113-120, DOI: 10.1080/1023697X.2017.1313136
[6] Lai, S., Xu, E., and Xie, W. (2009). Sustainable development of an integrated shopping mall with a hub station serving three subway lines in Shanghai. *Joint Symposium 2009: Design for Sustainable Performance*, 2.1-2.5.
[7] Wang, Y., Feng, H., Song, L., Hao, X., Xi, X. (2017, August). On energy saving of subway HVAC system: investigation and autonomous control, Energy for Sustainable Development, 39, 1-12.
[8] U.S. Department of Energy. How Energy-Efficient Light Bulbs Compare with Traditional Incandescent. Retrieved from https://energy.gov/energysaver/save-electricity-and-fuel/lighting-choices-save-you-money/how-energy-efficient-light
[9] Thong, M., Cheong, A. (2012, May). Energy Efficiency in Singapore's Rapid Transit System, *JOURNEYS, May 2012 Issue*, 38-47.
[10] Parsons Brinckerhoff. (2014, November). SES Version 6 User's Manual, page 8-13 & 8-14.

Energy Performance and Occupant Comfort in an Office Building: Co-simulation of an Agent-based Behavior Model with EnergyPlus

Mohamad Awada **Mohamad Hajj-Hassan** **Daoud Kiomjian**

Issam Srour **Hiam Khoury**

ABSTRACT

There is a lack of consensus in the literature regarding the interplay between energy efficiency and occupant comfort. Some sources argue that energy-efficient buildings are congruent with improved occupant comfort, while others state that energy efficiency efforts have a dampening effect on occupant comfort. This lack of agreement is rooted in the fragmented nature of the building industry that requires conducting energy simulations during the design phase when little is known about the building envelope and the behavior of the occupants. Although design parameters are considered central for energy modeling, behavioral aspects are of paramount importance as well and their impact on energy performance should also be considered. Therefore, this paper co-simulates an Anylogic agent-based model of occupant comfort and behavior with an EnergyPlus building energy model. To that end, a hypothetical case study of an office building is developed to identify the optimal position to place the office furniture that can balance comfort levels and energy consumption. Provided with three control features (thermostat, Clo level and internal shades), the thermal comfort level of an occupant is assessed according to the employees' Predicted Mean Vote (PMV). The findings generated by modeling multiple scenarios indicate that comfort levels are not the sole triggers of behavior and that the agent behavior emerges as a result of the autonomy agents have over a set of control parameters within their environment.

BACKGROUND

Energy and Comfort Interplay

In the past decades, the primary energy consumption has been increasing annually and the building sector plays a substantial role in the energy consumption on both; the national and the global level (Pérez-Lombard et al., 2008). In developed countries, buildings tend to consume around 20-40% of their primary energy. Typically, the average energy consumption by buildings in Europeans countries is around 37%, which is slightly exceeded in the United Kingdom (39%). Nevertheless, in the United States, 41% of the total energy consumption and 74% of the energy used for electricity in 2014 were from residential and commercial building sectors (EIA, 2015). Consequently, numerous studies have been conducted to investigate methods that can be used to improve energy efficiency and to reduce energy wastes in buildings.

In addition to the conservation of energy studies, occupants' comfort and well-being is considered a crucial building performance metric. There are extensive studies on occupants' comfort in buildings because it is estimated

Mohamad Awada is a Master's student in the Department of Civil Engineering, American University of Beirut. **Mohamad Hajj-Hassan** is a PhD student in the Department of Civil Engineering, American University of Beirut. **Daoud Kiomjian** is a PhD student in the Department of Civil Engineering, Amercican University of Beirut. **Issam Srour** is an associate professor in the Department of Civil Engineering, American University of Beirut. **Hiam Khoury** is an associate professor in the Department of Civil Engineering, Amercican University of Beirut.

that on average people tend to spend between 80-90% of their lives indoors (Dounis and Caraiscos, 2009). This implies that the environmental conditions of a building are key drivers to occupants' well-being, health, and productivity (Fisk, 2000, Wong et al., 2008).

Recent studies in commercial and residential buildings indicate that occupant behaviors within an indoor environment have a dual effect on both: the overall building energy use and satisfaction with the indoor environmental quality (IEQ) (Langevin et al., 2015). Growing out of efforts to understand the interdependency between these two factors, the process of simulating occupant behavior has gained an increasing attention in recent simulation research. Yet, there is still a lack of total agreement regarding the interplay between energy and comfort in buildings (Helbing, 2012).

Several methods have been implemented to mimic occupants' behaviors in buildings. The traditional method used to simulate occupants in buildings is the hourly schedule or the deterministic approach. However, new emerging approaches have been recently implemented: stochastic approach, random walk approach and the Agent-Based Modeling (ABM) approach. ABM is a powerful simulation paradigm mimicking real life systems by looking into its constituent units and testing what-if scenarios. These units are called agents whereby every agent has its own properties that trigger replicating certain real-life acts. Recently, ABM has been applied in the context of simulating building occupants' behavior. Most of these studies focused on the effect of adaptive behaviors of individual agents on energy consumption. For instance, Azar and Menassa (2011) proposed an ABM model to estimate energy consumption in commercial buildings. The model predicts more realistic energy consumption during the building design phase, by simualtig multiple energy consumption patterns among the occupants.

A key contributor to the lack of understanding of the interplay between occupants comfort and their energy consumption is the fact that Building Energy Modeling Programs (BEMPs) are not able to grasp the way people interact with indoor built environment. BEMPs have developed rapidly in the recent years with several programs adopted worldwide. One prominent programs that is recognized gloaby for its accuracy is EnergyPlus(Crawley et al., 2001, Ellis et al., 2008). Therefore, this engine will be used in the energy modeling process to betterr predict the energy consumption implications that result from thermal comfort triggered actions; mainly changing the Clo level, the thermostat and the internal shades.

Thermal Comfort

The condition of comfort within a certain environment is heavily dependent on the interaction between several physical, physiological and psychological aspects.ASHRAE (2013) standard 55 defines thermal comfort as: "*that condition of mind which expresses satisfaction with the thermal environment and is assessed by subjective evaluation*".

The Predicted Mean Vote (PMV) model, proposed by Fanger (1970), is considered to be one of the most recognized thermal comfort models. It permits researchers to express quantitatively the degree of thermal comfort level within an environment, from the measurement of the physical parameters of the environment; air temperature, mean radiant temperature, air speed and humidity, and personal parameters of the subject itself: metabolic rate and clothing level.

The PMV model is mainly used in air conditioned spaces. It is based on a 7 point scale ranging from "Very Cold" (-3) to "Very Hot" (+3) to investigate the thermal comfort and sensation of people within an indoor environment. Each level on the scale is calculated based on Fanger's equations for the PMV, given a certain combination of personal and environmental parameters. ASHRAE standard 55 identifies one's judgment equal to the value of zero as the optimal situation for thermal comfort. This situation is usually referred to as "Thermal Neutrality". Also, ASHRAE defines the comfort zone within a PMV range [-0.5, 0.5].

STUDY OBJECTIVES

This paper co-simulates an Agent-Based modeling of occupant comfort and behavior with an energy simulation engine, EnergyPlus. As such, the paper builds towards two major objectives:

1. Mimic the behavior of a single occupant in its workplace environment while considering the effect of thermal discomfort on the behavioral decision making.
2. Identify the optimal position within a predefined hypothetical office that balances between the occupant thermal comfort and energy consumption.

METHODOLOGY

General Description

The proposed model simulates the behavior of an employee in a 20 m^2 rectangular office (5m x 4m). Four different positions will be studied, to identify the optimal position that will grant the employee the maximum number of thermal comfort hours while minimizing the energy consumption. Figure 1 represents the layout of the office under study. The glazed surface is oriented towards the North with a window to wall ratio (WWR) of 0.9 and no external shading device (overhang). Internal shades are modelled, as required by ASHRAE std. 90.1-2010, with a solar transmittance of 0.21, and a solar reflectance of 0.23.

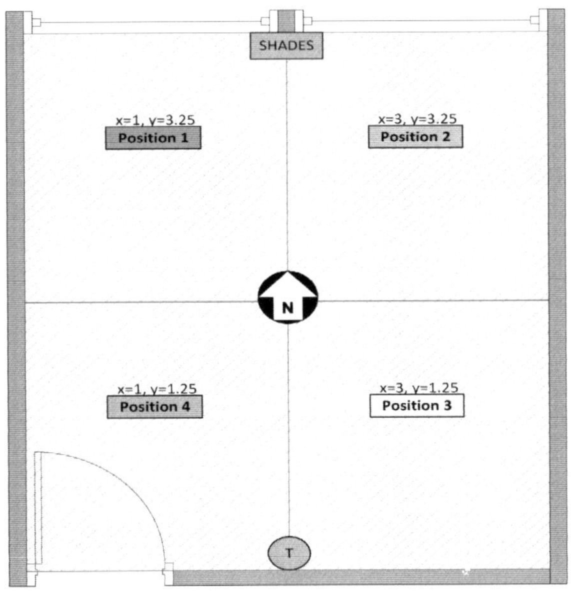

Figure 1 Layout of the office with four studied positions

A human behavior model is usually based on three main processes that are considered vital in the field of behavior research: behavior definition, trigger recognition and measurement of the effects associated with the accomplished behavior. The proposed model in this paper focuses on behaviors that are triggered by thermal discomfort and can only affect the thermal satisfaction of the occupants. Given that the focus of the paper is on air conditioned spaces, a list of predefined behaviors, along with the thermal comfort factor affected by each option is shown in Table 1.

Table 1. Behavior Options

Behavior Option	Affected Fctors
Changing the Clo Level	Clo Level
Pulling up/down Shades	Mean Radiant Temperature
Thermostat Set-point	Air Temperature

The simulation will take place over a whole year, taking into consideration the effect of the different seasons over the mean radiant temperature and the occupants' choice for their initial clo level at the beginning of each day. Each day starts at 8:00 AM and ends at 6:00 PM. Saturdays and Sundays are considered as vacations so they will not be considered in the simulation process.

Mathematical Model

The proposed methodology for describing the human behavior is based on a psychological behavioral model, proposed by Brian Fogg. The Fogg Behavioral Model (FBM) states that for any person to complete a behavior, he needs to be motivated, have the ability to perform the behavior and needs a trigger to perform this behavior. Applying this context, into the proposed model:

- The thermally dissatisfied occupant is motivated to perform an action that will make him comfortable.
- The thermally dissatisfied occupant considers different behavioral actions, and selects the easiest option to perform.
- An occupant is triggered to behave in the context of thermal comfort if he was thermally dissatisfied.

Sensation Motivation

An occupant feeling thermally dissatisfied tries to alter the settings of the environment in hope of feeling comfortable, thereby the occupant will choose the option that enhances the thermal conditions, and grant him comfortable thermal sensation. The mathematical equation reflecting the sensation motivation is given by equation 1.

$$\alpha = \begin{cases} 1, & if\ \ option\ grants\ the\ occupant\ thermal\ comfort \\ 0, & if\ \ option\ fails\ to\ grant\ the\ occupant\ thermal\ comfort \end{cases} \quad (1)$$

Ability to Perform

The ability to perform a certain behavior described in FBM is linked with the distance separating the occupant from the necessary tool to conduct an action. The longer the distance is the harder it is to perform the corresponding option and as such reduces the ability of the agent toward it and increases the cost associated with that behavior. The mathematical equation governing the cost for the "Ability to Perform" notion is given by equation 2.

$$\beta = \frac{Distance\ of\ option\ under\ consideration}{Longest\ distance\ associated\ with\ an\ option} \quad (2)$$

Learning

During the first week no learning occurs and workers behave randomly. Afterwards, the agents observe the impact of different decisions on their comfort. This impact is calculated as follows:

$$\Delta = PMV_t - PMV_{t-1} \quad (3)$$

Overall Utility Function

The utility function governing the process of behavior selection is exemplified in equation (4).

$$Utility\ function = k_1 \times \alpha + k_2 \times \beta + \gamma \quad (4)$$

The weighting constants k_i should be determined using a survey. In this paper, both factors are assumed to be equal to 0.5.

Agent Based Model

An agent based model was developed with the aim of evaluating the impact of the occupant's position in a 5m x 4m space on their thermal comfort and eventually their behavior. Since the studied space is an office environment, naturally the time unit for the model was selected to be hours. The model includes two agent types, Employee and Room. The Employee agent represents the occupant of the office space, whereas the Room agent is the space itself. The Employee receives information from the Room and adjusts its parameters and the Room parameters accordingly. In the proposed model the Employee can manipulate two building components to become more comfortable. The first component is the set-point temperature of the HVAC system and the second is the position of the shading devices. The proposed system has 6 inputs and 11 outputs and these are outlined in Tables 2 and 3 respectively.

Table 2. ABM System Inputs

Parameter	Definition	Agent
Tr	Mean Radiant Temperature (°C)	Room
Clo	The clo level Extracted from a triangular distribution Limits vary according to season	Employee
Humidity	Humidity level set at 50%	Room
Met	Metabolic rate of employee Depending on the activity level Exctracted from a triangular distribution	Employee
Velocity	Air velocity set at 0.1m/s	Room
T	Room air temperature (°C)	Room

Model Inputs

The first input to the developed system is the mean radiant temperature which is the temperature felt as a result of the weather conditions and the location of the employee relative to a fenestration system. And since this is a characteristic of the environment it is passed to the Room agent. The second input is the Clo, which is the level of clothing for the Employee agent. In order to accommodate for multiple clothing levels this parameter is drawn from a triangular distribution, whose limits vary according to the season. The third and fourth inputs are humidity and air velocity which were set according ASHRAE (2013) recommendations. The fifth input is the metabolic rate of the employee and this varies according to the Employee's activity level, i.e. whether they are walking sitting down or running. The final input required by the ABM is the air temperature which is the default set point of the HVAC system in the room. This input was set to 24 degrees during spring, summer and fall, and to 20 degrees during winter.

Model Outputs

The developed system generates 11 outputs and these are described in Table 3.

Table 3. ABM System Outputs

Parameter	Definition	Agent
PMV_Final	Potential PMV after employee take action	Employee
Satisfied_Season	Total number of hours employee agent was satisfied during a given season	Employee
NonSatisfied_Season	Total number of hours employee agent was dissatisfied during a given season	Employee
Ta	HVAC system set point temperature set by the employee	Room
Shades	A binary variable describing the status of the shading device in the room If value is set to true, the shades are on If value is set tofalse, the shades are off	Room

The PMV_Final is the potential mean vote for the Employee agent after it changes their clo level and takes an action (changes the set point temperature or changes the setting of the shading device) . The second and third outputs are a set of four variables each. The second set is the total number of hours the Employee agent was satisfied during each of the four seasons (i.e. the total number of hours the following is satisfied: -0.5<PMV<0.5), the third set is the total number of hours the Employee agent was dissatisfied during each of the four seasons. The fourth output is the set point temperature of the HVAC system in the room as set by the Employee agent after it takes an action. The final output is a binary output and it is the status of the shading device after the Employee agent takes action. If the agent puts the shading device in the ON position then the shades parameter is set to True and False otherwise.

Agent Logic

Since the Employee agent is the agent taking action it is the active agent and since the Room agent's parameters are modified by the Employee agent exclusively it is the passive agent. As such, the Employee agent is the one with a statechart which is shown in Figure 2.

The Employee agent has two states which are Home and Office. When in the Home in the office the Employee agent is assumed to be off duty and thus takes no action. The agent stays in that state for 14 hours every day. Whereas while in the Office state the agent is at work and thus takes actions affecting its clo level, and modifies the ambient temperature or the status of the shading devices. All the transitions are time triggered transitions except for ET3 which is a condition triggered transition. A time transition is one which occurs after the passage of time, whereas a condition transition occurs when a certain condition is satisfied. A transition labeled ET is an external transition which means, it changes the state of the agent, whereas a transition labeled IT is an internal transition. Internal transitions are ones where agents take action without changing their state.

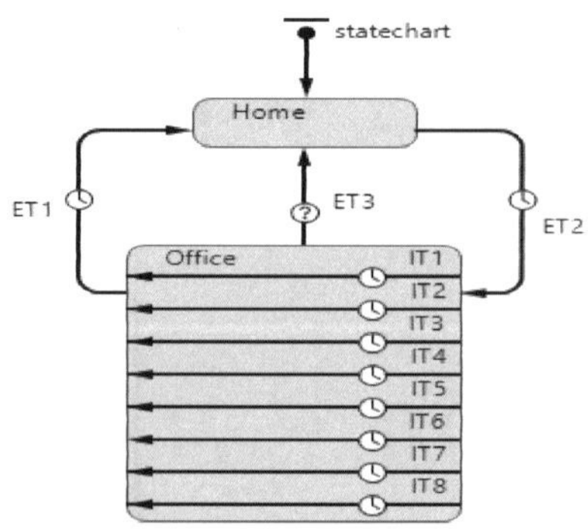

Figure 2 Employee Agent state-chart

External Transitions

Three external transitions are involved in the Agent Logic. ET1 changes the state of the agent from to Home after the passage of 10 hours and thus we are assuming a 10 hour work shift. The same happens in ET3 yet this transition occurs immediately if the day of the week is Saturday or Sunday (A weekend). On a week day the Employee stays in the Home state for 14 hours and then ET2 is triggered and the agent goes back to the office state.

Internal Transition

Although all the internal transitions are time triggered, these transitions could be divided into two categories learning transitions (IT1, IT3, IT5, and IT7) and regular transitions (IT2, IT4, IT6, IT8). The first set of transitions occurs every hour during the first week of every season. In this model we are assuming that learning does not carry forward from one season to another. During these transitions the Employee agent starts by checking it's PMV, if it is not within the comfort range, the agent changes its clo level (increments it during cold weather and decrements it during hot weather). After changing the clo level the agent checks its PMV again, if it's still uncomfortable, the agent changes the shades or the set temperature with a probability of 50% each. After taking the decision the Employee calculates the benefit it obtained from taking such a decision. Benefit in this case is defined as the improvement in the agent's PMV (how close to 0 did it become?).

The second set of transitions occurs during the rest of the season in a similar fashion to the ones above, however in these transitions the agent does not select changing the temperature or the shading device with equal probabilities, it calculates the utility of each and selects the one with the higher utility. For more information on utility calculations please refer to the "Mathematical Model" section.

Statistical Analysis

A statistical analysis was conducted to look at the variation of the number of thermal comfort hours over the course of the simulation. Figure 3 shows a box plot representing the effect of the occupant's position within the office and the building exposure on the number of comfort hours.

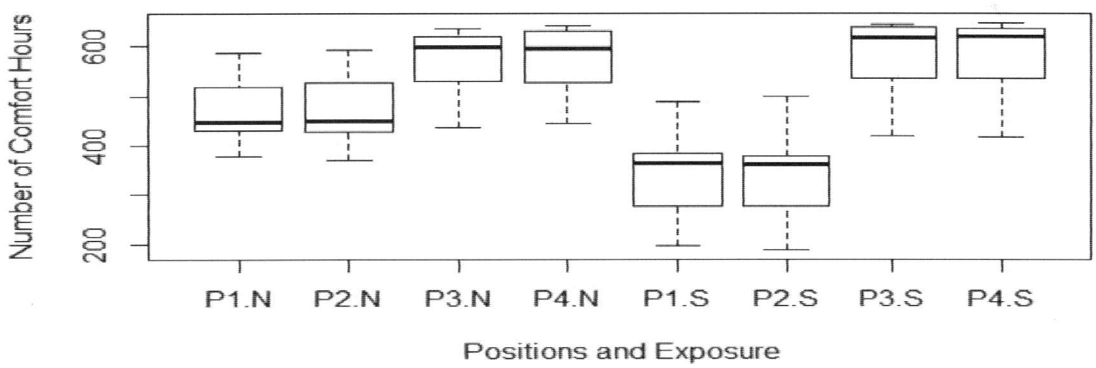

Figure 3 Comfort hours vs positions and exposure

Energy Model

An energy model was developed with the aim of evaluating the impact of the occupant's beahavior on energy consumption, given the four proposed positions and two different orientations; north and south. The values for the diffenet envelope components are summarized in Table 4. These are the technical or static features of our model, whereas the dynamic ones include the behavioral patterns; which for this study include thermostat set point and status of internal shades – as set by the emplyee/agent.

Table 4. Envelope Modeling Values

Opaque Elements		Fenestrataion Element	
Roof (w/m2k)	0.49	Assembly U(w/m2k)	3.04
Wall (w/m2k)	0.45	Assembly SHGC	0.55
Floors (w/m2k)	0.6	Assembly VT/SHGC	1.2

The dynamic behavioral patterns were obtained from the ABM model in the form of logged data for 1960 time-steps and converted into typical daily schedules over four seasons, as shown in Figure 4. The figure illustrates the case of the thermostat setpoint with a north exposure; the same was done for the south exposure as well as the internal shades status. Morever, given these schedules, a base energy model is constructed to compare the energy numbers generated from the simulated schedules. As such, ASHRAE (2016) standard 90.1 is used to model the same building while alterning the thermostat set point as required in the building trade-off schedules and loads. Similarly, the shades are modeled to be lowered when the solar transmitted energy exceeds 100 watt/m2 and then remain lowered for rest of the day. Accordingly, four different energy models are construcred; 2 base models for north and south exposure and 2 AMB models for the latter exposure. Results of the energy simulation are presented in Figure 4. Clearly the south exposed models consume more energy, but what is intriguing is the energy consumption for the base models at positions 3 and 4. Rationally, consumption should increase at positions 1 and 2 and this was established in the ABM model as shown in Figure 5. The reason behind this invalid outcome is the criteria used to model the shades irrespective of the daylight sensor position.

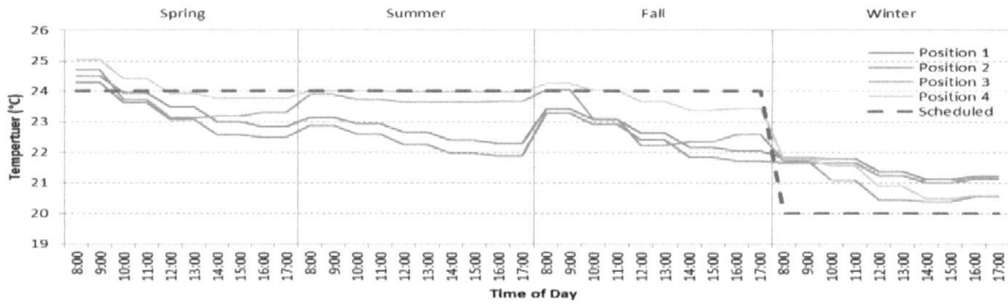

Figure 4 Behavioral impacts on Thermostat set-point for a North Exposure

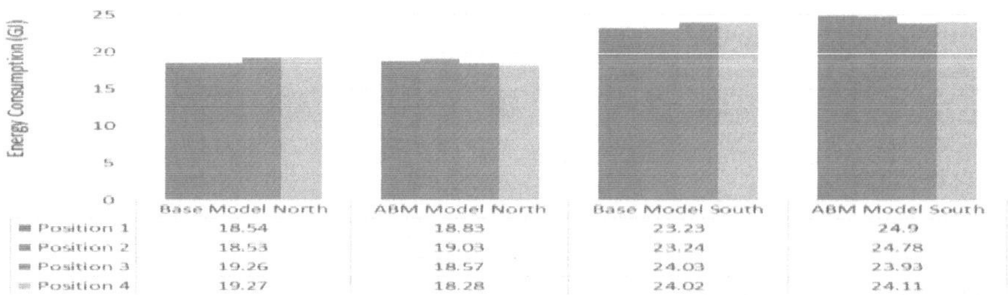

Figure 5 Behavioral impacts on Thermostat set-point for a North Exposure

CONCLUSION

The paper at hand presents a model that simulates a single occupant behavior in an operating office. The model employs a co-simulation between ABM and EnergyPlus to identify the optimal position to place the employee's desk in a way that maximizes its thermal comfort hours while minimizing the energy consumption over a year. The results of the ABM and EnergyPlus showed that Position 4 is the optimal position that maximizes thermal comfort hours and minimizes energy consumption. Also, it was evident that the building's north exposure boosts the thermal comfort of the occupant and helps reduce energy consumption in comparison with the south exposure.

NOMENCLATURE

PMVt = PMV at time t

PMVt-1= PMV at time t-1.

γ= tmean of Δ factor over time t shown in Equation (3).

α = Sensation Motivation

β = Ability to Perform

γ = Learning acquired

k_i = Weighting constants for attribute *i* (range is 0-1 with $\sum k_i = 1$)

REFERENCES

ASHRAE 2013. *ANSI/ASHRAE Standard 55: Thermal environmental conditions for human occupancy,* Atlanta, American Society of Heating, Refrigerating and Air-Conditioning Engineers.

ASHRAE 2016. ANSI/ASHRAE/IES Standard 90.1: Energy Standard for Buildings Except Low-Rise Residential Buildings. Atlanta: American Society of Heating, Refrigerating and Air-Conditioning Engineers.

AZAR, E. & MENASSA, C. C. 2011. Agent-based modeling of occupants and their impact on energy use in commercial buildings. *Journal of Computing in Civil Engineering,* 26, 506-518.

CRAWLEY, D. B., LAWRIE, L. K., WINKELMANN, F. C., BUHL, W. F., HUANG, Y. J., PEDERSEN, C. O., STRAND, R. K., LIESEN, R. J., FISHER, D. E. & WITTE, M. J. 2001. EnergyPlus: creating a new-generation building energy simulation program. *Energy and buildings,* 33, 319-331.

DOUNIS, A. I. & CARAISCOS, C. 2009. Advanced control systems engineering for energy and comfort management in a building environment—A review. *Renewable and Sustainable Energy Reviews,* 13, 1246-1261.

EIA, U. 2015. Energy Information Administration: Annual Energy Outlook 2015 with projections to 2040. *US Energy Information Administration, Office of Integrated and International Energy Analysis, US Department of Energy, Washington, DC,* 20585.

ELLIS, P. G., TORCELLINI, P. A. & CRAWLEY, D. 2008. Simulation of energy management systems in EnergyPlus. National Renewable Energy Lab.(NREL), Golden, CO (United States).

FANGER, P. O. 1970. Thermal comfort. Analysis and applications in environmental engineering. *Thermal comfort. Analysis and applications in environmental engineering.*

FISK, W. J. 2000. Health and productivity gains from better indoor environments and their relationship with building energy efficiency. *Annual Review of Energy and the Environment,* 25, 537-566.

HELBING, D. 2012. Agent-based modeling. *Social self-organization.* Springer.

LANGEVIN, J., WEN, J. & GURIAN, P. L. 2015. Simulating the human-building interaction: Development and validation of an agent-based model of office occupant behaviors. *Building and Environment,* 88, 27-45.

PÉREZ-LOMBARD, L., ORTIZ, J. & POUT, C. 2008. A review on buildings energy consumption information. *Energy and buildings,* 40, 394-398.

WONG, L., MUI, K. & HUI, P. 2008. A multivariate-logistic model for acceptance of indoor environmental quality (IEQ) in offices. *Building and Environment,* 43, 1-6.

Principles of Split Mass Flow and Heat Shifting Psychrometrics toward Efficient Comfort Management

Peter Phillips
Member ASHRAE

ABSTRACT

When designing energy efficient buildings much attention is devoted to reducing building fabric, internal sensible heat loads and ventilation heat loads. Latent heat loads from people and ventilation are beginning to dominate our thinking toward satisfying our concepts of occupant comfort and mold control building wellness.

The paper discusses the merits of split mass flow psychrometrics (SMFP) and cold cooling heat exchangers to disconnect the sensible and latent cooling processes. In association with SMFP, heat shifting within either air mass stream can provide further disconnection of the cold cooling process from the saturation curve to enhance occupant comfort management and control of the potential for mold growth within ducting systems.

INTRODUCTION

Air conditioning system design has followed several generational conceptualization trends. Each challenge is assessed and with varying successes solved toward providing superior outcomes. Energy efficiency perhaps is the greatest challenge presented to modern designers. This drive toward greater efficiency has several subsets of complexity to resolve. Designing a better building could seek to reduce sensible heat loads. Improving façade efficiencies, reducing internal equipment loads and reducing pressurization and occupancy ventilation, all work toward minimizing sensible heat loads (refer ASHRAE 2017, page 36.12). Certainly there will consequentially be a reduction in latent loads from reduced façade leakage and lower outdoor ventilation rates. But what is the implication of these efficiencies upon psychrometric processes? One could image the perfect building envelope and no internal heat loads, with the only load presented to heat exchangers emitted by occupants. What would our system design endeavor to solve, perhaps only the comfort solution for the occupant? Of course this narrowing of heat flow analysis within buildings at this stage is potentially naïve, however working toward a net zero energy building will have contemplations and complexities beyond our current thinking.

This paper will discuss principally opportunities to disconnect our psychrometric processes from the saturation curve and energy efficiency justifications to afford enhancements of the concepts presented.

RESOLVING THE LATENT COOLING EQUATION

If we consider our standard person, ASHRAE 2017, page 18.4, Table 1 suggests an office worker would require a heat exchanger to deliver a sensible heat ratio of 0.58. Further, if we consider comfort could be achieved for the occupant with a space condition of 24°C (75.2°F), 55%RH, inside the comfort zone indicated by ASHRAE 2017, page 9.12, Figure 5, will a simple cold cooling heat exchanger provide an appropriate supply air condition to offset the heat load?

Peter Phillips is the Principal Consultant of PCES Consulting, a private practice HVAC engineering consultancy based in Australia.

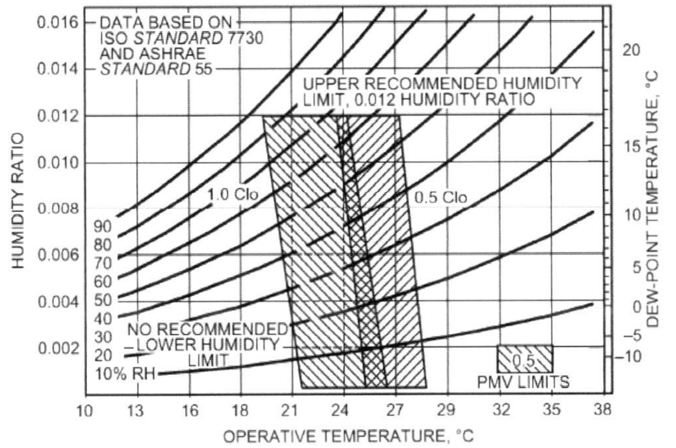

ASHRAE 2017, page 9.12, Figure 5, Summer and Winter Comfort Zones

All too often as designers we conceptualize heat exchanger performance akin to ASHRAE 2017, page 1.13, Figure 5, a process line projected to intersect the saturation curve. However a SHR of 0.58 does not intersect the saturation curve for the stated comfort condition. Our attention would then be drawn to ASHRAE 2016 page 25.2, Figure 2 suggesting adding sensible heating to achieve a dew point driven comfort solution.

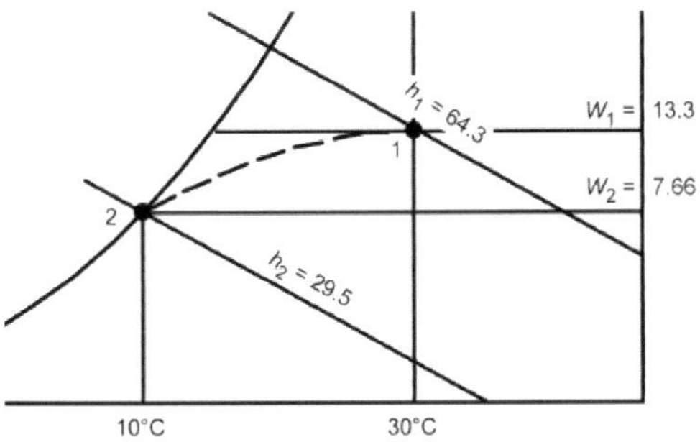

ASHRAE 2017, page 1.13, Figure 5

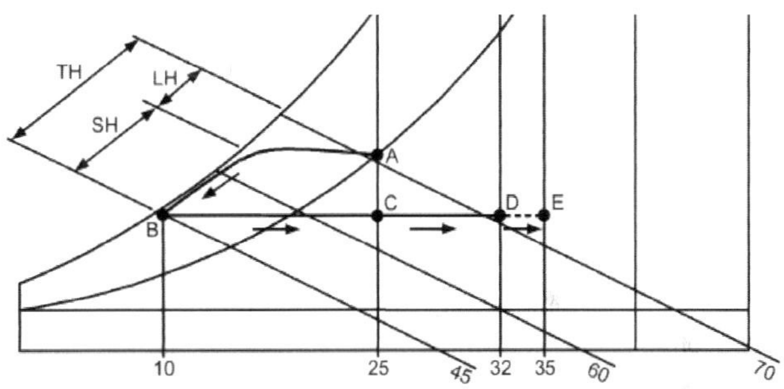

ASHRAE 2016, page 25.2, Figure 2

The addition of wrap around coils indicated ASHRAE 2016, page 25.10, Figure 16 and 17 suggests how a psychrometric process incorporating three heat exchangers can with no (minimal) additional energy (by means of heat shifting), separate the leaving air condition away from the saturation curve. This suggests that air conditioning designers may not be tied to the saturation curve to deliver cold cooling comfort management.

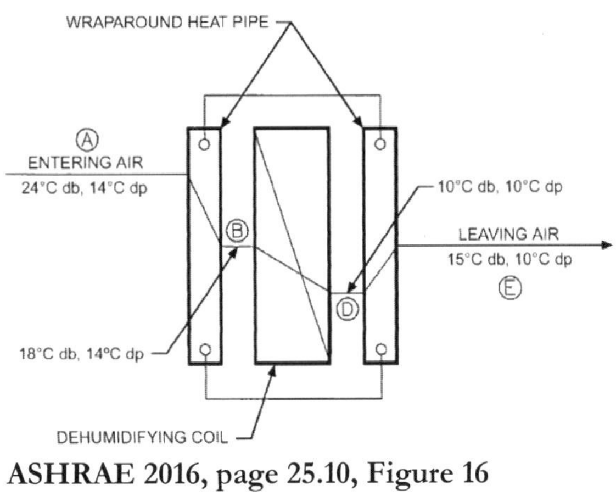

ASHRAE 2016, page 25.10, Figure 16

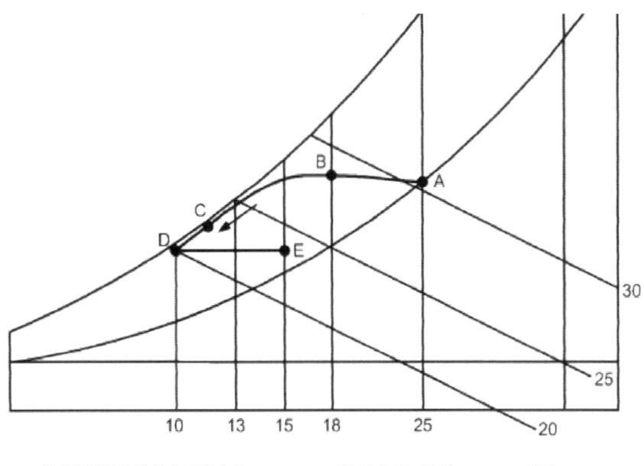

ASHRAE 2016, page 25.10, Figure 17

If we consider now the ventilation requirement for our occupant, the use of dedicated outdoor air systems to pre-treat the ventilation air supply as indicated through ASHRAE 2016, page 4.14, Figure 14, suggests an opportunity to manage separately air mass flows. Although Figure 14 suggests a series configuration psychrometric process, parallel split mass flow processes are also valid.

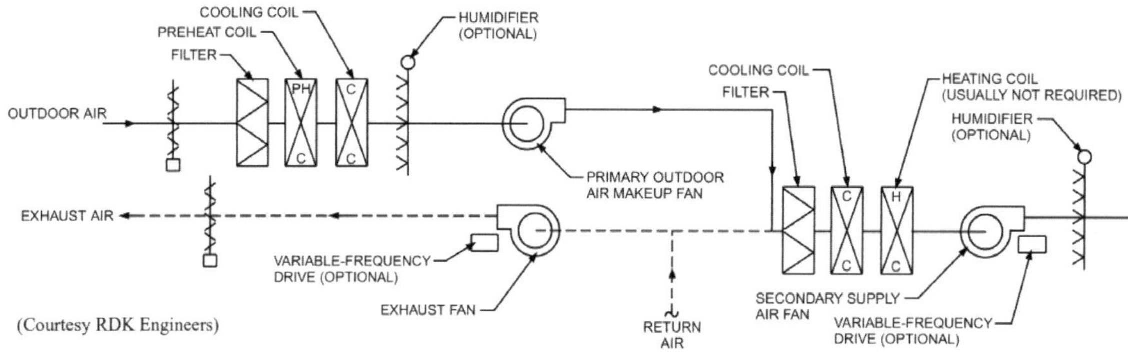

(Courtesy RDK Engineers)

ASHRAE 2016, page 4.14, Figure 14

Figure 1 indicates a variant on the ASHRAE Figure 14 schematic. The outdoor air stream and the return air stream are separately psychrometrically treated prior to mixing within the supply air fan suction chamber (heat shifting coils are indicated). This configuration maintains psychrometric independence and reduces load on the primary outdoor air make up fan indicated.

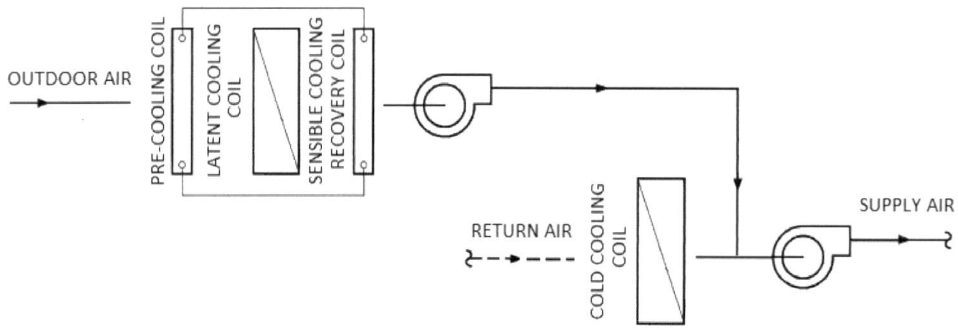

Figure 1, Figure 14 modified to a parallel split mass flow psychrometric solution incorporating heat shifting

If we then consider that there is an opportunity to separately process the ventilation air mass flow from the recirculating air mass flow through split mass flow psychrometrics (SMFP), we can as indicated by ASHRAE 2017, page 36.12, use outdoor air heat exchangers to deliver moisture content related comfort, and the recirculating air heat exchangers to deliver dry bulb temperature related comfort.

If we consider an outdoor ventilation rate of 7.5 L/s (15.3 cfm) per person and an office worker emitting 55 Watts (188 BTU/hr) of latent load, depressing the outdoor air leaving air condition by 2.5g/kg (17.5 g/lb) below the desired room moisture content would achieve a comfort balance. This would require lowering the cold heat exchanger leaving dew point by approximately 5°C (9°F). Alternatively it may be more energy efficient to increase the dehumidification mass flow. This can be achieved through the introduction of recirculating airflow into the outdoor air stream. Increasing the dehumidification mass flow to 12.5 L/s (26.5 cfm) can reduce the leaving dew point differential from 5°C to 3°C (9°F to 5.5°F).

MOLD MANAGEMENT

If we then consider the other moisture management issue within our air conditioning system environment, mold.

ASHRAE 2017, page 25.16 states, 'mold in buildings is of concern to occupants', and further suggests, 'mold growth can be usually avoided by surface relative humidity over 80% for only a short period time periods'. Then if we can separate duct humidity away from the saturation curve through use of heat shifting, perhaps mold can be substantially reduced from systems without the use of additional energy resources.

PSYCHROMETRIC PLOT

If we consider a simple psychrometric plot of the split mass flow and heat shifting (wrap around coil) processes as depicted in Figure 1, the psychrometrics of each airstream and the end points will follow Figure 2. Figure 2 depicts Heat exchanger A pre-cooling the dehumidification air stream, this should generally be the ventilation requirement outdoor air flow. Heat exchanger B then depresses moisture content from the dehumidification air stream to a level that satisfies occupant comfort. Heat exchanger C recovers sensible cooling associated with the dehumidification process, the extent to which this process shifts energy will be dependent on either the duct relative humidity set point or the occupied space dry bulb set point. Heat exchanger D sensibly cools the recirculating air stream, its leaving air condition can be limited by either the duct relative humidity set point or the occupied space dry bulb set point.

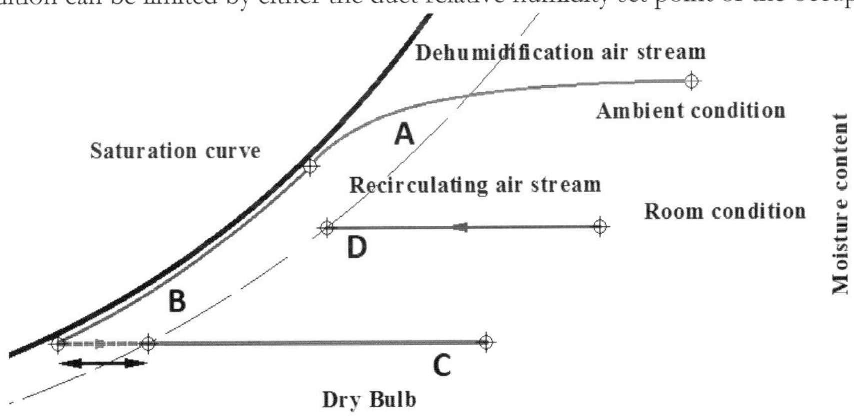

Figure 2, General psychrometric representation of split mass flow and heat shifting processes

Of course variants on the depicted psychrometric processes indicated will depend on whether recirculating air flow is introduced to the outdoor air flow and whether heat shifting is required within the return air process as part of the mold management requirements.

LOAD PROFILES

If we expand our thinking back toward real building scenarios, we understand that other heat load influencers are very real. Increased sensible heat loads will change the conditioned space supply airflow rates of delivery and fabric related latent heat loads will depress leaving heat exchanger dew point requirements. Managing acceptability criteria within the schematic design process may lead us to adjusting airflow rates of delivery and psychrometric split mass flows to suit load extremes. If we consider a simple psychrometric plot (Figure 3) of the region of 'outdoor air' heat exchanger leaving air conditions (G) to satisfy variable latent cooling scenarios. Leaving border (a) satisfies the maximum duct relative humidity set point (A). Leaving border (b) satisfies the room condition dew point (comfort) set point (B). Leaving border (c) delivers dehumidification capacity (C) to the conditioned space based on conditioned space latent load. Leaving border (d) delivers sensible cooling capacity (D) to the conditioned space based on conditioned space sensible load and the conditioned space below dry bulb temperature control range (E). The leaving air condition (F) therefore satisfies the conditioned space latent control comfort set point, in part the conditioned space sensible control comfort set point and the outdoor air duct mold management set point. The heat exchanger line (H) is a generalization of the cold cooling, heat shifting psychrometric process.

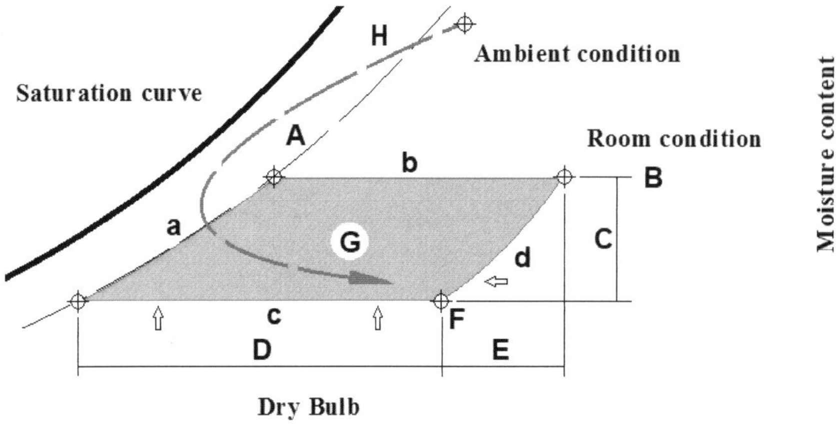

Figure 3, Dehumidification leaving air condition psychrometric representation

If we consider a simple psychrometric plot (Figure 4) of the 'recirculating air' heat exchanger leaving air conditions (G). Leaving border (a) satisfies the maximum duct relative humidity set point (A). Leaving border (b) is a consequence of the dehumidification strategy which satisfies the room condition dew point (comfort) set point (B). Leaving border (d) delivers sensible cooling capacity (D) and potentially dehumidification capacity (C) to the conditioned space based on conditioned space sensible load (the extent to which latent cooling being provided will be determined by mass flow proportioning and cold cooling medium entering temperature). The leaving air condition (F) therefore satisfies the conditioned space dry bulb control comfort set point, in part the conditioned space latent control comfort set point and the supply air duct mold management set point. The heat exchanger line (H) is a generalization of the cold cooling, heat shifting psychrometric process.

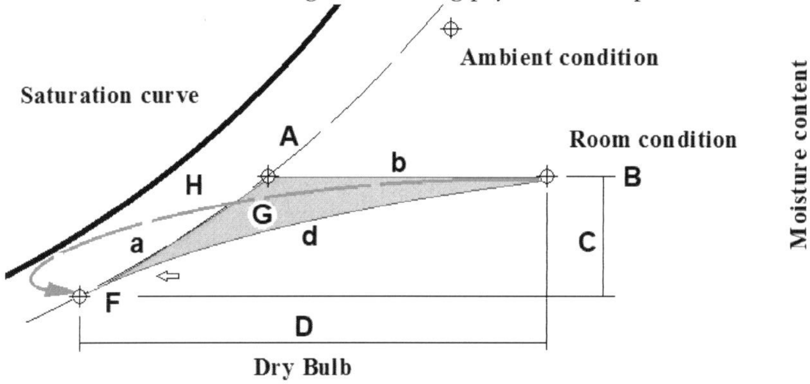

Figure 4, Sensible cooling leaving air condition psychrometric representation

We can then consider that the supply air condition based on variable cooling load scenarios would be a mass flow mix of the two heat exchange processes depicted in Figure 5. The mixed air, supply air condition (M) satisfies variable load scenarios with a greater degree of independence relative to the saturation curve. The room sensible heat load (D) and room latent heat load (C) are offset with the supply air condition (M) ostensibly without a requirement for sensible re-heating.

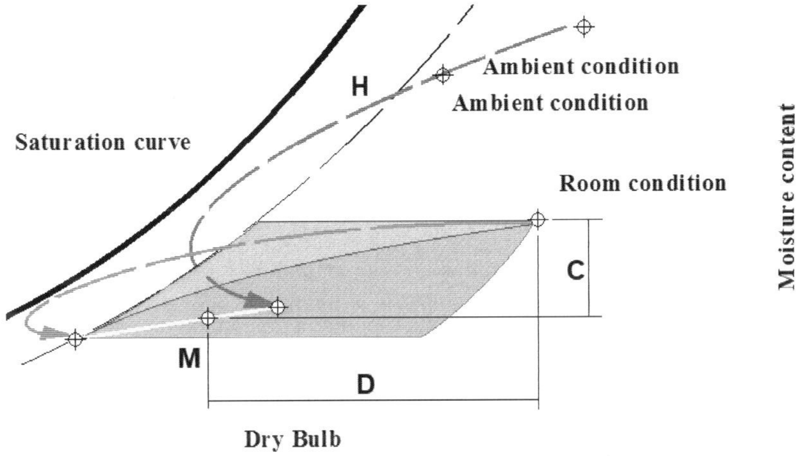

Figure 5, Parallel split mass flow mixing psychrometric representation

Although total independence cannot be created with this method, any requirement for re-heat within the process should be significantly reduced. Peak load and part load performance of heat exchangers utilizing SMFP and heat shifting can create separation from the saturation curve to achieve low non intersecting sensible heat ratios to satisfy comfort solutions for occupants.

RETROFIT POTENTIAL

Where an existing air handling plant is not satisfying occupant comfort objectives, adding a heat exchanger to the outdoor air stream and introducing a SMFP strategy managing thermal capacity delivery to offset moisture stress, should improve comfort management at peak and part load. A parallel arrangement mixing the split air stream post treatment should not result in additional fan power. Selecting the outdoor air heat exchanger and its filtration arrangement with the same or less pressure drop than the existing arrangement should be accommodated by the existing motive arrangement. Typically the return air heat exchanger will required fewer rows within its configuration than the outdoor air heat exchanger. Lowering air flow face velocity for the outdoor air heat exchanger will overcome the air resistance disparity and the potential for moisture carry over.

Adding equipment into available space will always be a challenge for the retrofit designer. Introducing only the outdoor air dehumidification heat exchanger and not the heat shifting mold control strategy may be the solution for the system being evaluated. Adjustment of the desired moisture content set point controlling the dehumidification strategy may be introduced to manage supply air duct relative humidity.

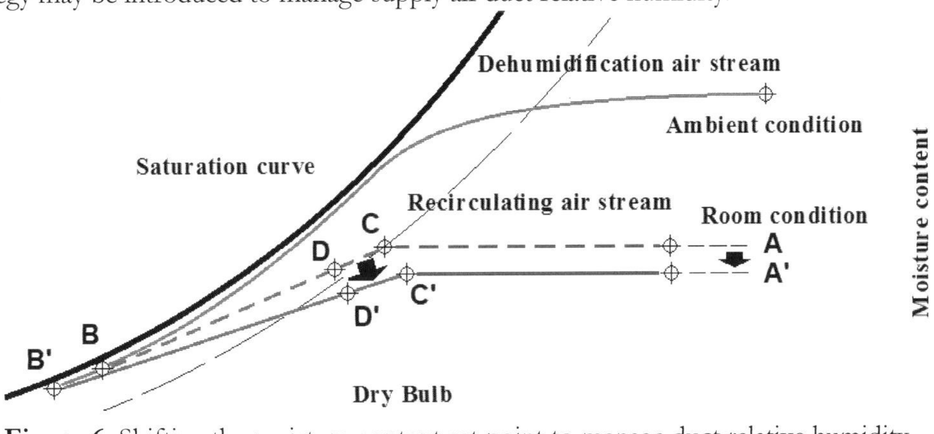

Figure 6, Shifting the moisture content set point to manage duct relative humidity

Figure 6 indicates the room condition moisture content controlling set point shifting from A to A'. This will result in the dehumidification heat exchanger leaving air condition to shift from B to B', the recirculating air steam heat exchanger will then consequentially reduce its capacity with a resulting process curve moving from A to C to A' to C', thereby effectively drying the required supply air duct condition away from the required relative humidity set point.

CLIMATE CHANGE

Whether the long term consensus of the impact of climate change is real or inconclusive, changing expectations upon system performance in many instances is exposing marginal system designs and plant performance. Additional thermal capacity and enhanced psychrometric management can be achieved through the addition of heat exchangers configured to a SMFP solution.

ENERGY EFFICIENCY

Air handling heat exchanger performance ratings and airflow resistances are discussed on page 23.6, ASHRAE 2016. Several precautionary statements are offered including designing to maximum limits. These precautionary statements provide sound engineering advice potentially at the expense of reduced recurrent energy consumption. Although it is acknowledged that the creation of plantroom space does impose a capital cost impost on new build projects, reducing space to engineering limits does have the potential to increase recurrent energy costs.

If we consider a conventional six row single cold cooling heat exchanger providing an air off condition to accommodate a room sensible heat ratio of 0.8, with a face velocity of 2.5m/s (492 f/m), if face velocity is reduced to 1.5m/s (295 fpm), pressure drop should reduce notionally to 36% of the initially proposed pressure drop allowance. This suggests that the six row could be expanded to 16 rows. Lowering the face velocity to 1.0m/s (197 f/m) would suggest that 37 rows could be afforded. Notional selections suggest a 20 row heat shifting series heat exchanger operating at 1.25m/s (246 f/m) provides an effective psychrometric performance selection with a reduction in fan power of approximately 15%. An assessment of opposing acceptability criteria should be undertaken. On one hand we have a six row heat exchanger squeezed into a plant space delivering a marginally acceptable comfort solution and an industry metric energy consumption. On the other hand with SMFP's we have a deeper heat exchanger configuration accommodated in a larger plant space performing a superior comfort solution potentially consuming less fan power.

Cold cooling heat exchangers have conventionally been designed with an approach temperature differential (entering cooling medium to leaving air temperature) of approximately 5 to 7°C (9 to 12.6°F). Increasing available heat exchange surface area has the potential to reduce the approach temperature differential. This should permit elevation of compressor saturated suction temperatures permitting a reduction in energy consumed by the vapor compression cycle (refer ASHRAE 2016, page 43.9, and ASHRAE 2015, page 42.12). Actively managing elevated chilled water temperature to improve coolth generation energy efficiency will require additional chilled water and/or airside mass flow rates to satisfy load/capacity equilibrium (refer ASHRAE 2015, page 42.14). Understanding the relationships of compressor power, pumping power and fan power within a multi-dimensional energy consumption model requires skilled assessment (refer ASHRAE 2015, Section 42).

Figures 7 and 8 are a representation of HVAC plant energy efficiency depicted psychrometrically for a generic building located in Dubai. Figure 7 assumes a conventionally designed cooling process incorporating industry practice part load energy efficiency management. Figure 8 assumes the design enhanced with split mass flow psychrometrics with low approach temperature heat exchanger selections supported by load feedback management of the chilled water plant. A 38% improvement in energy efficiency is possible, occupant comfort management should be improved and mold development within ducting systems should be reduced.

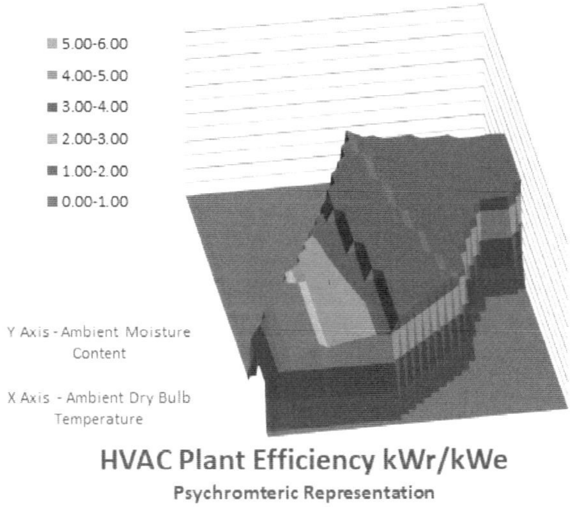

Figure 7, Generic system design, plant efficiency

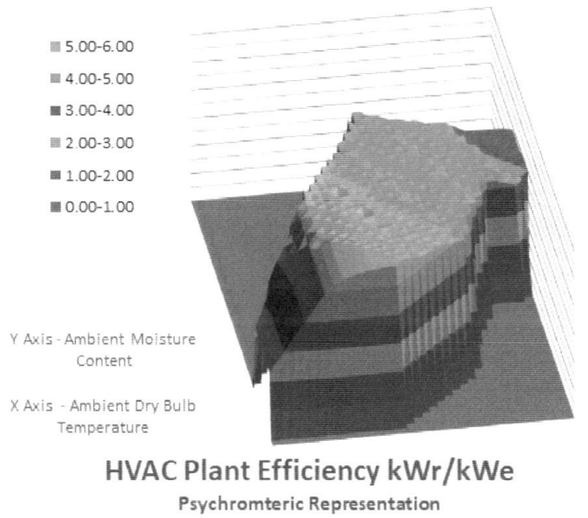

Figure 8, Split mass flow system design, plant efficiency

It is recognized that assertions of energy efficiency gain potential should commence with baseline concepts followed by sustainability enhancements to balance economic, environmental and societal values. Each technical opportunity should be tested beyond merely capital cost affordability.

CONCLUSION

It is acknowledged that the upper limit moisture content management of the occupant comfort equation had little audience within the post 1970's energy crisis HVAC world. However our collective drive toward reducing energy consumption through the creation of superior energy utilization strategies resulting from the global warming debate, has heightened a desire to effectively understand the occupant comfort equation (refer ASHRAE 2017, Section 9). Reducing sensible loads have created an increased reliance upon effective latent cooling strategies to drive the comfort solution. Development of the ASHRAE Summer Winter Comfort Zones depicted in ASHRAE 2017, page 9.12, Figure 5, suggests a maximum conditioned space dew point of less than 17°C (62.6 °F). Thermal zoning now should not merely consider effective sensible load profile management, latent load zoning should also be considered.

The application of parallel split mass flow and sensible heat shifting psychrometrics has the potential to provide effective sensible heat ratios whereby separation of leaving air conditions away from the saturation curve makes possible energy efficient occupant comfort management whilst minimizing the potential for mold growth within ducting systems. Effective control of the multiple heat exchangers employed maintains the occupant comfort equation through the range of part to full load scenarios. Effective selection of heat exchangers can manage fan power requirements within general acceptance criteria and the application of energy efficiency part load strategies within the chiller plant can enhance the HVAC system efficiency.

Split mass flow psychrometrics incorporating heat shifting can enhance; societal acceptance of our modern enclosed built environments through improved comfort management and mold minimization, economic acceptance through reduced recurrent expenditure, and environmental acceptance through the application of energy reduction strategies associated with its application.

REFERENCES

ASHRAE. 2015. *ASHRAE Handbook—HVAC Applications.* Atlanta: ASHRAE.
ASHRAE. 2016. *ASHRAE Handbook—HVAC Systems and Equipment.* Atlanta: ASHRAE.
ASHRAE. 2017. *ASHRAE Handbook—Fundamentals.* Atlanta: ASHRAE.

World Class Energy Efficient HVAC System for New 'Twisty Tower' in South Africa

Pieter J. de Bod, PE
Member ASHRAE

ABSTRACT HEADING

The international auditing firm PricewaterhouseCoopers (PwC) in South African has taken occupation of the brand new modern office building in November 2016, a project that deserves global recognition. The new iconic 26 storey PwC head office is showcasing on the global prospectus with a beautiful "twisty" high rise tower architectural design which gently twists through in height with the top and ground floor rotated by 30 degrees. Keeping South Africa's environmental challenges of water scarcity and limited electricity supply in mind, the new energy efficient air-conditioning system is a low water and electricity consumption 4 pipe chilled and hot water system. Using two new state of the art water cooled chillers plus three new multifunction 4-pipe units', the carbon footprint of the building is substantially reduced compared to most other buildings in the area. At the same time the annual on-site water consumption is also reduced. The efficient building design boasts numerous innovative and energy saving technologies which contributes to the measured low energy consumption as well as the excellent occupancy comfort and environmental quality.

INTRODUCTION

The developer and tenant were very clear on the overall design brief: The new PwC HQ had to be an iconic building that would be unique to the development, and which would mark the property as a new top tier destination for business. It also had to conform to the internationally recognized LEED BD+C v3.0 (2009) Silver rating criteria. On completion, this 516,667ft² (48,000m²) building boasts 25 levels prime office space and meeting rooms, and an adjacent annexure building with sound studios, training rooms, large commercial kitchens, a staff canteen, forensic rooms, computer rooms, archive rooms, triple volume entrance foyer and a structured parking basement. A future tri-generation plant could easily be accommodated and integrated into the current HVAC system. This article outlines the HVAC design approach that was followed, including the thermal zoning strategy, fresh air supply strategy as well as equipment selection. The incorporation of various advanced technologies in the field of air-conditioning in combination with the innovative designs of the architect, façade engineer, mechanical engineer and the other professionals have delivered a world class energy efficient building.

P. J. de Bod is Technical Director in the Department of Mechanical Engineering at WSP Consulting, Johannesburg, South Africa.

DESIGN APPROACH

The air-conditioning for this building has been designed to meet the following criteria: commercially economical, high operational efficiency, excellent occupancy comfort, good thermal zoning, simple and easy to maintain, simple to operate, good service access to all HVAC equipment, relative high fresh air rates to occupants, good system reliability, long service lifespan, high filtration efficiency, integrated heating and cooling recovery, high performance glazing, automated internal blinds system, extensive carbon monoxide (CO) and carbon dioxide (CO_2) control, and advanced controls. During the initial stages of the project, various chiller technologies were researched to enable the delivery of the best solution that complies to the client brief. During detail design stages, various sun studies and the energy modelling was done to assist the designers to deliver excellent occupancy comfort.

Figure 1 The new 26 floor "twisty" high rise building, displaying its concave façade and fire escape staircase, with the annexure building at the base of the tower.

Figure 2 The typical triangular concave floor plate and the central core is observed during the construction phase.

SYSTEM DESCRIPTION

The air-conditioning system comprises a central cooling and heating 4 pipe system. Chilled water is generated on site by a combination of (a) two state-of-the-art high efficiency water cooled screw chillers plus (b) three multifunctional 4 pipe air-cooled units. Hot water is generated on site by means of the three multifunctional 4 pipe units. These five chillers work in combination to provide the best energy efficiency, lowest water consumption and lowest energy cost, all depending on the cooling demand, the ambient temperature conditions and the electrical time of use tariff.

The chilled water is pumped by means of constant primary and variable secondary pumps to various fan coil units (FCU's), air handling units (AHU's) and Computer Room Air Handling Units (CRAH's) in insulated steel piping. Chilled and hot water are supplied at a temperature of 46.4°F (8.0°C) and 95.0°F (35.0°C) respectively to FCU's or AHU's can cool or heat the internal space to meet design conditions.

THERMAL ZONING STRATEGY

To ensure the occupants are comfortable, the correct thermal zoning was of utmost importance. The typical shape of the tower floor plates is in principle a triangle, but the three straight sides areconcave. In addition to this, the tower floor plates rotate 1.2 degrees clockwise every floor, so much that the top floor and the bottom floor are rotated 30 degrees clockwise compared to the ground floor. Refer to Figure 1 and Figure 2 that illustrates the 'twisty' effect and a typical office floor plate during construction.

Perimter Zone:

The collective effect of the concave shape and floor rotation resulted in multiple thermal zones per floor, each to be controlled carefully to ensure best occupancy comfort. Following a long design review process, it was decided to install various FCU's in the ceiling of perimeter zone 12ft (3.65m), each with individual temperature control, for the following main reasons:

1. Optimal zone control: Compared to a central variable air volume AHU system, FCU's provides superior independent zone control for each zone along this building's façade, keeping in mind that the facade direction varies at each level as the building twists upwards and each office level the facade faces three different directions.
2. Excellent Flexibility: It is easier to deal with tenant changes on a FCU compared to a central AHU system. To change, add, and remove FCU's to accommodate major office churn is straight forward. The FCU can serve a meeting room, executive office or canteen, without affecting the surrounding air-conditioning zone or systems.
3. Future proofing: Chilled and hot water piping is provided in the ceiling with additional pipe T-offs with isolating valves for future FCU's.

Internal Zone:

The internal zone of the 'triangle' floor plate is served by means of a variable air volume air handling unit with variable flow diffusers. Each floor is served by its own AHU.

Each FCU contains the necessary washable removable filters, insulation, cooling and heating coil and three-speed supply air fans. In most cases return air are drawn into the ceiling voids and mixed with a regulated amount of filtered outside air. No electrical heating is installed anywhere in the HVAC system, as this is very inefficient and environmentally unfriendly.

The internal design temperature for the office area is 72.5°F (22.5°C) in summer and 70.7°F (21.5°C) in winter. The above temperatures are those at which the majority of people are considered comfortable, and which are accepted as standard indoor design conditions in South Africa. The occupancy comfort modelling that was done have proved the same result of good occupancy comfort.

OUTSIDE AIR SUPPLY STRATEGY

The requirement of high fresh air rates was also considered a very important design aspect of the HVAC system. In the tower building, two dedicated outside AHU's are located on the roof of the tower, providing treated fresh air to all the FCU's in the perimeter zone. Air is supplied at a neutral temperature of 64.4°F (18.0°C). Provision is made to fit rotating heat recovery wheels in the future to the dedicated outside AHU's. The riser ducting for the central toilet extract system for the tower is linked to a backward curved centrifugal fan that is located close to the dedicated outside AHU's, which is ideal for heat recovery. The toilet extract air is exhausted to the atmosphere at high velocity.

The variable air volume air handling units on each typical tower floor draw in fresh air and mixes it with return air before supply into the internal space. Each AHU is linked with a CO_2 meter to constantly monitor and control the fresh air. The same design approach was followed the rooms in the training workshops, staff canteen and circulation areas.

All the AHU's are capable to perform a night time flushing cycle, and an economy cycle, both functions to conserve energy. During the summer season, the night time flushing cycle supplies outside air to the office space, typically in the early hours of the morning. The objective is to remove the inertia heat buildup (heat that is present in the slab and other materials in the building), thereby reducing the cooling demand at start up the next day, saving energy.

WATER COOLED CHILLERS

The heart of the cooling system is considered to be the two ultra-efficient water cooled chillers, each producing 300 RT (1055kW) cooling. Each water cooled chiller is piped to a dedicated closed circuit cooling tower (with VFD on the fans) to cool the condenser water. The chiller manufacturer claimed it to be the world's first integrated variable speed, water-cooled, screw chiller, which incorporates significant breakthroughs in water-cooled chiller technology to provide excellent reliability and achieve superior efficiencies at true operating conditions all without compromising the environment. The chiller provides high energy efficiency using variable speed, positive displacement screw compressor. The chiller has the ability to reduce speed and optimise operation independent of ambient conditions. This capability permits the chiller to precisely match building load and conserve energy for:

1. High part-load performance. Table 1 indicates the high part load conditions which contribute to the low energy cost in providing chilled water.
2. Extreme High Reliability using the advanced tri-rotor compressor design, featuring a balanced rotor geometry and shorter screw lengths, resulting in vastly reduced compressor bearing loads and a minimum L10 compressor bearing life in excess of 500 000 hours when operated at Air-Conditioning and Refrigeration Institute (ARI) standard conditions.
3. Superior oil management where the chillers regulate oil temperature, viscosity and pressure via a positive displacement pump compared to a differential system pressure. This allows continuous operation with cold condenser water at all loads. Should the input power to the chiller be lost, the system is designed to assure proper lubrication of the bearings during coast down.
4. Cooler tube expansion: The cooler tube is fitted with expansion values at center support sheets to prevent unwanted tube movement and vibration, thereby reducing the possibility of premature tube failure.
5. Low harmonic distortion: The chiller generates less than 5% total harmonic distortion at the input to the Variable Frequency Drive (VFD) without the use of any external filters or line reactors. The integrated VFD provides a soft start, further reducing stress on the compressor and inrush current at start-up. Drive suitable for name plate voltage plus or minus 10% and provide 100% of name plate amps continuously and 150% of name plate amps for 5 seconds.
6. Positive pressure design: This ensures that air, moisture and other performance-degrading contaminants are not sucked inside the chiller. This eliminates purge units and their associated maintenance.
7. Product Integrated Control: direct digital electronic controls provide great flexibility and functionality. Each unit integrates directly with the Comfort Network system, providing a solution to controls applications.

Table 1. Water Cooled Chiller Performance

Chiller Performance	Load %	kWc	Condenser Leaving Temp [°C]	Condenser Entering Temp [°C]	C.O.P.
Full Load	100%	1050	33.00	25.00	6.39
Part Load	90%	945	30.78	23.67	7.15
Part Load	80%	840	28.57	22.33	8.02
Part Load	70%	735	26.39	21.00	8.99
Part Load	60%	630	24.24	19.67	9.88
Part Load	50%	525	22.11	18.33	10.89
Part Load	40%	420	21.35	18.33	11.04
Part Load	30%	315	20.59	18.33	11.06
Part Load	20%	210	19.94	18.33	6.19
Part Load	10%	105	19.19	18.33	4.16

MULTIFUNCTIONAL 4-PIPE CHILLED AND HOT WATER GENERATORS

The heat recovery multifunctional 4-pipe unit is mainly sized to meeting the space heat demand for the building. Each chiller can operate in 3 different modes namely (a) heating only (b) cooling only (c) simultaneous cooling and heating. The maximum efficiency of the chillers is reached with simultaneous loads, during opposite overlapping thermal loads. Other benefits of these 4-pipe units include the chiller plant areas that are reduced, hydronic circuits that are simplified, and maintenance that is reduced by half. Using traditional ratings such as EER and COP to measure efficiency of 4-pipe units would be limiting, but to objectively measure performance under simultaneous load conditions, the rating Total Efficiency Ratio (TER) is used.. The TER is calculated as the ratio between the sum of the delivered heating and cooling power and electrical power input. In part cool, part heat mode, the unit operates like a warm-water unit, as all the evaporating and condensing energy is used for the system, and only requires part of the total energy, and in this particular state, the exchangers are "oversized", thus achieving an even higher efficiency.

The multifunction 4-pipe unit is sized based on ambient summer condition 95.0°F (35°C) dry bulb and ambient winter condition 32.0°F (0°C).

AUTOMATED INTERNAL BLINDS

This building is equipped with a sophisticated automated blind system linked to numerous solar sensors located on the external façade of the building. Appropriate sun studies revealed that the same façade can have sun and shade, a result from the concave triangle design and rotating floor levels. A few minutes before the sun's rays hit the glass façade, the blinds will close. This system automatically opens and closes blinds on various sectors of the facades depending on the angle of the sun and the time of the day. This system is incorporated to control glare and minimize solar influx into the building. Two different types of blinds were used: (a) perforated roller blind (for the office floors in the tower), and (b) horizontal venetian blinds with a drop down function and auto tilting function.

RECEPTION DESK AND CAST IN SLAB HEATING

Johannesburg experiences cold dry winters, and the comfort of receptionists in the foyer is ensured by designing and installing an innovative heating system integrated into their desks. An un-insulated hot water pipe is located in the foot well position of the desk. Receptionists may want to rest their feet on the hot water pipe to provide heating during cold days. Each receptionist desk has their own heater and thermostat. Low pressure hot water circulates through the pipes. The hot water system is hydraulically separated from the main buildings hot water system by means of a plate heat exchanger. A small circulation pump circulates hot water to the reception desk heating and in slab underfloor system.

Underfloor hot water heating are also installed in the main lobby in the ground floor slab. This heating system is designed to combat the cooling effect of the large single glazed glass in the lobby area, to make the space more comfortable for occupants.

ENERGY AND ENVIRONMENTAL IMPACT MONITORING SYSTEM (EMS)

The purpose of the environmental impact monitoring system (EMS) is to display, monitor and record the electrical and domestic water consumption and to make the occupants aware of the environmental impact that the building and its systems have on the planet. Further functionality of the EMS is to create and distribute custom made reports and graphics which assists in the understanding and maanaging of the energy and consumption pattern of the large electrical and water consumption components.

BASEMENT EXTRACT SYSTEM

Most of the car parking is in an underground basement. The entire parking basement is mechanically ventilated with large axial fans, connected to the smoked detection system and a CO sensor system. The same system double up for a fire smoke extraction system in line with the fire engineers rational fire design. The operation of the axial fans is controlled by the CO monitoring system. If the CO levels are less than a predetermined value, the fans switches off, and will start and ramp up based on the deteced CO levels. In case of a smoke or fire signal, all the fans on the level will switch on.

COOKING HOODS WITH CAPTURE JETSTREAM AND ULTRA VIOLET LIGHT

The staff canteen is capable to provide meals for up to 4000 employees. Stainless steel canopy hoods with Ultra Violet lights and capturing jetstream device are installed in heavy cooking areas in the commercial kitchens to break down grease. (Ultra-Violent producing lamps have proven to be beneficial in destroying grease.)

Most of the grease exhausted from the hood can be classified as fatty acids which consist of long chains of molecules connected by double bonds, which are chemically reactive. One of the chemical reaction that occurs is called photolysis and takes place when UV-light hits the long molecular chains and breaks the double bonds. Ozonolysis also takes place which is facilitated by ozone generated to react with the smaller chains of grease and grease vapors to chemically alter them. The resulting substance reduces the particles that stick to the ductwork or fans which lowers the fire risk. Special low noise, high pressure backward curved fans is located on the roof of the canteen building and extract the canopy smoke. Fire suppression system is also installed in the canopies.

The smoke capturing jetstream device decreases the airflow, whilst increasing the smoke capturing efficiency and reducing the fan energy consumption. The induced jetstream minimizes the heat and impurities produced by cooking appliances spreading to the work area. Compared to conventional exhaust-only hoods, this capture jetstream is claimed to be 28% more efficient.

BUILDING AUTOMATION SYSTEM (BAS)

The building is equipped with a new Building Automation System (BAS), and in South Africa this is known as a Building Monitoring System (BMS). The BAS is a localized computer-based monitoring system that monitors the building's HVAC system, generators, UPS, main electrical meters, main water meters, and fire detection system alarms. All the information is transmitted to a central computer in the control room on level 1 where it is displayed in graphical format. The information on the BMS can also be assessed remotely and can also be set to trend and store data for troubleshooting purposes.

The BAS is also be used for alarm management. Critical alarms are sent via text messages to the facility manager. Water and electricity consumption are also tarcked and the information used for reporting purposes. The BMS network uses dedicated LAN cable and is separate from the PwC network.

Pressure independent control valves (PICVs) have been installed to all the chilled and hot water supplies to FCU's, AHU's and CRAH's to automatically compensate for chilled and hot water pressure variations. This enables continual balancing functions to maintain system performance at varying loads, and to prevent cooling and heating at the same time. These PICVs maximizes energy savings of variable flow pumping systems.

STAIR AND LIFT PRESSURISATION

The lift shafts and fire escape staircases in the tower are pressurized by means of axial fans which are located on the roof of the tower. To ensure that temperature do not build up in the tower staircases, a motorised fire damper on each floor have been installed at ceiling level , connected to a thermostat. The damper will open when the temperature in the staircase exceeds 77.0°F (25.0°C) allowing secondary air from the office space to flush the staircase. This was tested during commissioning phase and proved to be very effective.

24 HOUR COMPUTER ROOM SYSTEM COOLING

A dedicated chilled cooling system provides cooling to the important server room, patch rooms and other similar rooms. Either FCU's or CRAH's are connected to the chilled water system to ensure these rooms are cooled uninterruptedly. A separate air cooled chiller provides the cooling, but in case the chiller fails, an automatic changeover is made to the central cooling system. On the basis that the cooling demands is not higher than 30TR (105kW) it was decided to design a dedicated system to avoid running the main chilled water plant continuously during after hours.

BUILDING FLUSHING

In order to comply with the minimum requirements of the Indoor Environmental quality section of the LEED rating system, the building have been flushed out in accordance with the LEED requirement Path 1, which is to flush the building out after construction, prior to occupancy, with all interior finishes installed, with new filter media, supplying a total air volume of 14,000ft^3/ft^2 (2400m^3/m^2) of outdoor air while maintaining an internal temperature of at least 60°F (15°C) and a relative humidity less than 60%.

This equated to a total of 6,356 million ft^3 (180 million m^3) of outside air, equivalent of 72,200 Olympic sized swimming pools. This was done over a period of between 10 and 21 days, using the 34 air handling units and accompanied fresh air fans. The purpose of the flushing is to evacuate airborne contaminants.

GLAZING SELECTION AND SOLAR REFLECTANCE OF CONCAVE FACADES

The glazing selection for this building was one of the most important design elements. The glass that was selected was a toughenable double silver coating with high solar protection and superb light and energy performance. This tinted grey glass is well appreciated by glass processors and architects for its low interior and exterior light reflection On top of this, its coating can withstand the toughening process, enabling glass processors the flexibility to process the glass as they want so they can offer customers' options as glass can be curved or enameled. The angles of the solar reflection on the uniquely warped concave shape of the façade of the PwC building was carefully modelled upfront to prevent any high levels of solar reflections on the ground and neighboring buildings. Concentration contours on the ground have been modelled by the façade engineers, using modeling programmes.

The modelling process revealed that the initial solar concentrations were up to 0.55kW/ft^2 (6kW/m^2) in some places, which is about six times stronger than normal. The spandrel panels were acid-etched glass with the outside surface acid etched to create a non-reflective surface. This resulted in a reduction of the reflective area with roughly a third. The sun's rays are also dispersed before hitting the ground because glass is not perfectly flat. The final solar concentrations where slightly more than 0.18kW/ft^2 (2kW/m^2) and it was agreed to hard and soft landscape the surrounding areas. The team has considered installing pipes in the affected areas to heat up water, but this was not viable because the amount of effective time that the pipes received solar reflectance was minimal. Examples of solar reflectance of concave buildings is the Walkie Talkie building in London (20 Fenchurch Street), which have been blamed for melting car parts in streets below.

CONCLUSION

The incorporation of various advanced technologies in the field of air-conditioning in combination with the innovative designs of the architect, façade engineer, mechanical engineer and the other professionals have delivered a world class energy efficient building design that have inspired and continue to inspire many young engineers and architects, and provide a comfortable and pleasant new office for an international firms like PwC.

Energy Conservation for an Office Building in a Hot Climate

Prof. Hari S. Dalal
Member ASHRAE

ABSTRACT:

Kuwait is located in Middle east region. The weather is very hot.During summer the Ambient temp becomes 52 • C db, and relative humidity varies between 40 % to 60%. Kuwait area is also very dusty due to sand dust .

HVAC design is a very challenging today. ASHRAE 90.1 – is a standard guidelines related to energy saving.Year after year the power requirement becoming less to save energy as well carbon di oxide generation to control the environment. Ministry of electricity and water (MEW) have formulated some guidelines in this aspect. Façade glazing, Thermal Energy storage (TES) , Energy Recovery Unit (ERU) etc will play related to Energy savings. Manufacturer also trying to improve Energy Efficiency Ratio (EER) of products like chillers, fans etc and they are concentrating on development of product by research and development. Noise reductions also Energy Conservation for an Office Building in A Hot Climate , Kuwait

In this paper, An office building have been considered. Office area is air conditioned by Underfloor system. To reduce HVAC plant capacity , Thermal storage – Chilled water by Indirect interface system have been considered.

The influence of following parameters like UFAD (Underfloor air distribution system}, Facade glazing factor like "U" (Overall Heat Transfer co efficient}, SHGF (solar Heat gain factor} , Thermal Energy Storage (TES) for Partial loading , ERU (Energy Recovery Unit) etc. will be discussed and the contribution as percentage of energy saving will be presented.

An approx 10-15 % energy have been saved .

Keywords, UFAD,SHGF,MEW,ERU, TES.

DESCRIPTION

An office building consists of Auditorium, Big waiting area , Offices , Atrium area etc. Auditorium area is airconditioned with conventional system supported by Energy Recovery System. The air distribution is made via under seat with small type jet diffuser suitable placed.

Offices area is air conditioned by UFAD system considering return air via ceiling and supply air under-floor plenum with active fan mounted underfloor diffuser and its controller.

Big waiting area, Atrium area is air conditioned is done by conventional system like from ceiling diffuser.

UFAD SYSTEM

ASHRAE 90.1 2010 , are recommending . Based on USGBC and also MEW related to energy consumption saving target is kept approx 25 % . ASHRAE 62.1 , 2010 also have been recommend for Indoor air quality.

Fig 1 describes the UFAD of a typical floor for one fire Zone. This Fire Zone have 4 air conditioned Zone. Each Zone have separate Hivar unit / Air Cabinet Unit . As Floor diffuser is an active which is fan assisted , the plenum pressure have been kept at @ 12 PA . To maintain privacy some Zones have been divided in two or three

Author HARI S. DALAL is a HVAC Consultant working in a consultancy firm, The Associated Engineering Partnership, KIPCO Tower, Kuwait

sub zones.

To combine both smoke & air conditioning , the system is designing in such a way that supply duct can be considered as fresh air for smoke ventilation and return air will be connected with smoke exhaust system. As explained above in Fig 1, One Fire zone have 4 air conditioned zone, suitable fire rated Dampers are located so that during fire or smoke ventilation, the fire damper will open and the ceiling will operate as one fire zone which is related to smoke exhaust. To compensate with fresh air for proper smoke evacuation, the door will be automatic operated by fire alarm panel.

Fig 2 describes the operation of UFAD system where supply air in underfloor via plenum and return air via ceiling as in conventional way.

BMS plays an important role related to control during un-occupancy period by sensor and also providing VFD for control of air flow in Hivar unit - supply air fan.

FENESTRATION

In Kuwait, the glazing area and wall area ratio is approx 90 -95 percent which carries 30 – 35 % of cooling load for the building for standard double glazing glass . Day light is also important related to hygiene and also energy saving . Visible Light Transmission (VLT) , Over all heat transfer co efficient, (U) , and solar heat gain co efficient (SHGC) are decision factor for glazing selection. LSG is an important parameter = VLT / SHGC. Related to selection of glazing type in particular area.

As Kuwait climate is hot , dry in summer Low solar gain is preferable and Low E coating is placed inside of outer layer of double glazing tinted glass.

Based on prescriptive requirements as mentioned in MEW latest edition and also ASHRAE 90.1,2010 better glass have been suggested . and the benefit related to cooling load becomes approx 100 TR (approx 6 % saving) is obtained.

The following table have been considered.

PERFORMANCE OF GLASS

Sl	Description of Glass, Glazing Type	VLT	SHGC	LSG	U
01	6 mm double Tinted	90	0.75	1.2	3.62
02	6 mm double reflective	35	0.5	0.7	3.33
03	Spectrally selective – uncoated	79	0.7	1.13	2.72
04	Spectrally selective- 1 silver layer	73	0.52	1.4	1.75

THERMAL ENERGY STORAGE

Chilled water thermal energy storage (TES) in naturally stratified tanks is considered to reduce operating cost and main Chiller plant capacity. In a naturally stratified chilled water storage tank, warm and cool volumes of water are stored together without an intervening physical barrier. A stable density difference between hot and cold water is maintained by varying the direction flow through the tank.. Fig 5 explains the operation . plate type heat exchanger keeps the two circuits separate , An Indirect Transfer System between thermal storage and main chiller plant.

The Capacity is decided based on cooling load profile refer Fig 4 ,July, August (summer month) .The figure of merit (FOM) is considered approx. 0.9. the formula considered are as below :

FOM = (Ratio of Available Cooling Capacity during Discharge) / (Theoretical Maximum Cooling Capacity of Fully charge Tank) . or (Load Ton Hours x 12000) / (Volume, Gallons x 8.36 x sp gr x ΔT, deg F) .

Here, Load = 2000 Ton Hr, ΔT, deg F = 12 . , the tank size becomes = 240,000 Gallons = 930 M3. Size considered = 120 m2 x 8 M.

Importance of UFAD system

UFAD system have an important advantage related to building volume if return air is considered in

floor . A lot of architectural constraint like leak proof partition and a planned furniture layout is required. Practically it becomes impossible .

Basic advantage of UFAD system:

1) Improved comfort & indoor air quality
   a) Due to fresh air induction by removing stale air .
   b) Floor swirl diffuser since they strongly mix the warm supply air say 16.67 deg C into the occupancy microclimate area.
   c) The ventilation effectiveness is greater for air supply from floor level than for air distribution from the ceiling.
   d) Since they operate at lower pressure and velocity, underfloor systems generally produce less noise .
   e) Operation flexibility of FTU.
2) Energy Impacts :
   a) As supply air temperature is approx 17 degC , chilled water temperature can be kept at 7.25 deg C instead of 4.5 deg C. Chiller Co efficient of performance (C.O.P) will increase.
   b) As Static pressure loss is minimal, Fan power requirement is minimum.
3) Savings in Capital & Operating cost :
   - Capital cost may be lower due to followings:
   a) Reduction in power, data/ voice networking installation cost.
   b) Reduced ductwork, lighter duct materials.
   c) Lower horsepower fans.
   d) Lower capacity chiller say approx 15-20 % lower.
   e) Construction time and material cost savings.
   - Operating Cost will be lower due to followings
   a) KW/ TR will be lower .
   b) Control will play an important role for savings.

Benefits of TES system

The importance of energy management like TES system depends on Cooling Load profile . Ref Fig 4-- Chiller design day load for July which is maximum requirements in Kuwait. Discharge requirement from TES = 2200 Ton-HR which is less than Charging 3800 Ton- HR .

Factors discourage for TES :

1) Utility rates do not differ between Night and Day .
2) Space is not Available.
3) Cooling Load profile is flat.

Benefit : Energy savings is approx 15-18 %. Capital cost due to space, tank etc can be overcome within 5 years.

Discussion

In this project , the total cooling load is approx 1750TR . Plant capacity= 1140 TR.
Effect of cooling load reduction due to Fenestration , like SHGC, U factors is approx 245 TR.
Comparing with base load . the glass is spectrally selective layer 1, approx glass area- 7000 m2.
Basically due to fenestration advantage of 245 TR, the cooling load becomes approx. 1500 TR.
Based on base load , the power demand will be @ 1.5 KW/ TR 1750 x 1.5 = 2625 KW.
Power requirement due to Chiller , pump, cooling tower etc == @ 1.00 KW/ TR= 1140 KW.
UFAD system cooling demand = 600 TR @ 0.2 Kw/TR= 120 KW.
Conventional system = 900 TR , @ 0.3 KW/ TR = 270 KW. ERU fan == 30 KW.
TES pumps Power == 18 KW. UFAD electric power == 50 KW.
Total power becomes = 1140+120+270+30+18 +50=1628 KW. Savings = (2625 - 1625) =1000 KW .
which is equal to 1000 / 2625 = 38 %. Saving. A great saving.

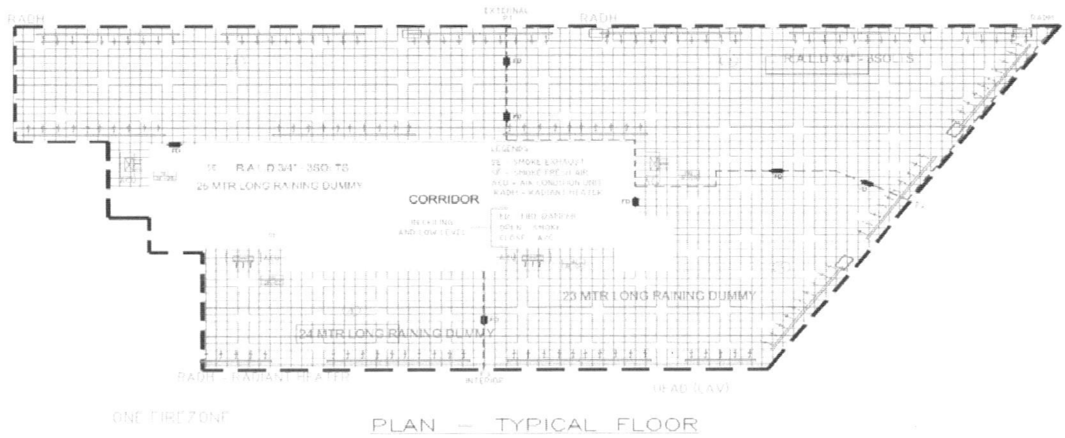

Fig 1: A Typical Floor for one fire Zone and 4 no UFAD system.

One Fire zone have 4 Hivar zone which means 4 hivar unit which have separate supply and return air.

During smoke the partition between zone in ceiling level will be opened by Fire damper to make ceiling as smoke exhaust. Fresh air will be ducted at low level by automatic door.

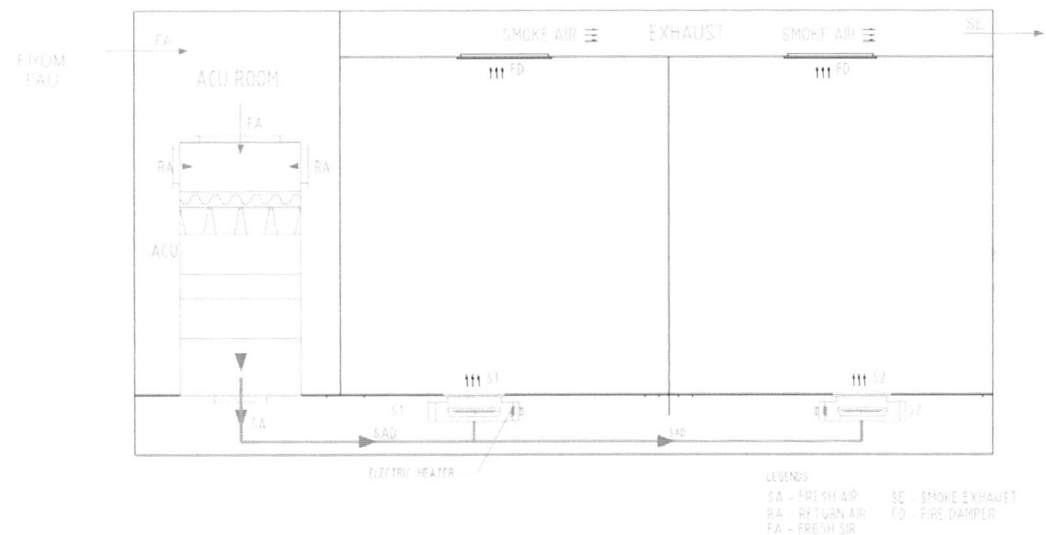

Fig 2: One HIVAR Zone. supply through underfloor. Return via ceiling

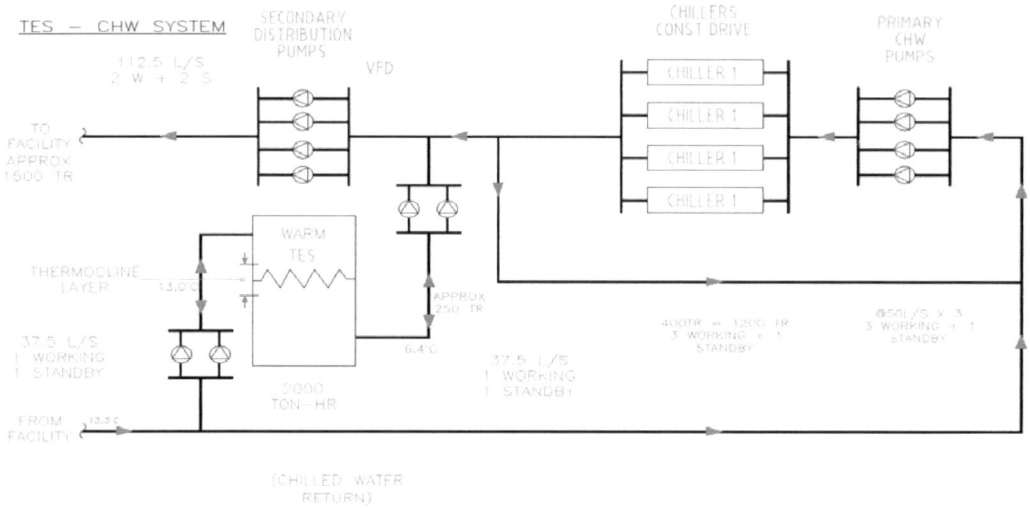

Fig 3: Thermal Energy Storage (Partial Type)

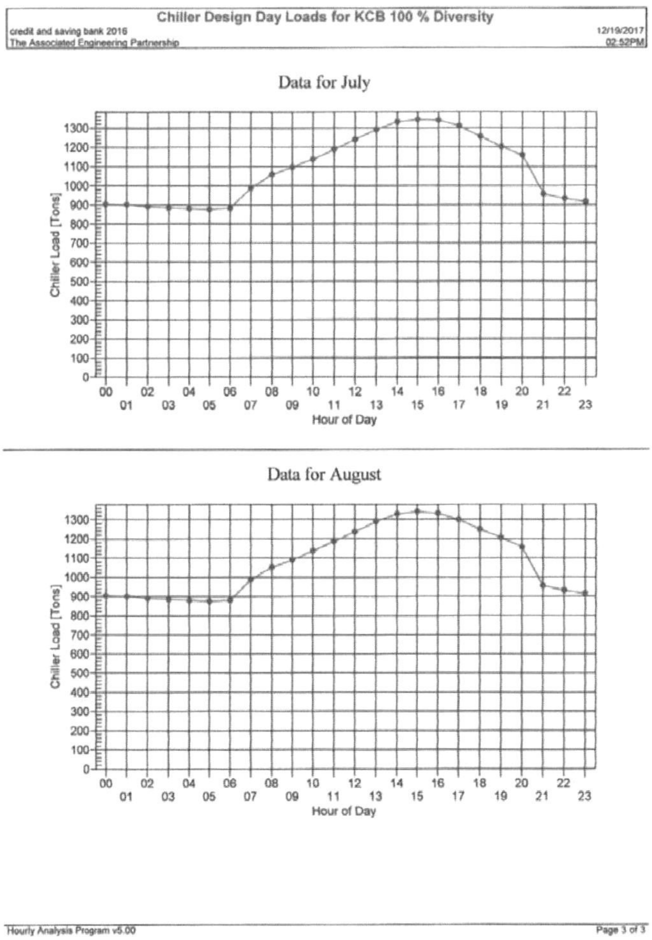

Fig 4: Cooling Load Profile

NOMENCLATURE

ASHRAE-- American Society of Heating Refrigeration and air conditioning Engineers	C.O.P - Coefficient of performance.
ERU- Energy Recovery Unit.	F.O.M -- Figure of Merit
FTU-- Floor Terminal Unit	HAP-- Hourly Analysis Program.
HSG – High Solar Gain (Coating) .	KW-- Kilo-watt.
KCB-- Kuwait Credit Bank.	LSG – Light to Solar Gain Ratio. (VLT/ SHGF)
PA-- Pascal.	MEW – Ministry of Electricity and Water
TES – Thermal Energy Storage.	SHGF - Sensible Heat Gain Factor.
U – Overall Heat Transfer Co-efficient.	TR - Tonnage of Refrigeration.
VLT- Visible light transmittance	UFAD – Underfloor Air Distribution .

REMARKS

Energy conservation in hot climate is very difficult related to ASHRAE 90.1 and ASHRAE 62.1 . A proper co-ordination is required from design to execution and also during commissioning. In this project , Fenestration, TES have played very important role to get Energy benefit of Apprpx 30 %.

ACKNOWLEDGMENTS

This paper is related to design of Kuwait credit Bank in Kuwait. As author is designer in a consultant company , I would like to thank this office. (The Associated Engg Partnership, Kuwait).

REFERENCES

1. 2008 ASHRAE handbook , Systems and Equipment's , Chapter 50 -- Thermal storage.
2. A. Bhatia – PDH online course M 217, HVAC- overview of Underfloor Air conditioning system.
3. National research Council , Canada -- Selecting window glazing for optimum energy performance. Transactions 100(1):1065–72.

A Simplified Personalized and Displacement Ventilation Model for Predicting Passive Contaminant Spread in Office Spaces

Douaa Al Assaad, MS **Carine Habchi, PhD** **Kamel Ghali, PhD**
Member ASHRAE

Nesreene Ghaddar, PhD
Member ASHRAE

ABSTRACT

The purpose of this work is to develop a mathematical multi-layer, multi-plume model of passive contaminant spread for office spaces equipped with displacement ventilation (DV) coupled with a personalized ventilation system (PV). The model studies the transport of species and therefore predicts occupant risk of infection in typical office spaces. The particle concentrations in the thermal plumes and the surrounding air are separately evaluated in each air layer in the space. The PV is taken into account in the occupant breathing zone and affects the species' concentration in the rising thermal plume. The temperatures and concentrations at each layer are computed by solving the energy and mass balance equations. On the other hand, the temperature and concentration at the breathing zone are calculated by discretizing the PV jet into vertical control volumes and performing energy and mass balance at each PV layer. The model is then validated experimentally in a climatic chamber equipped with a DV+PV system and a heated cylinder representing an occupant in a typical office space. Validation of the model is performed for a DV flow rate of 60 L/s (127.13 CFM) and a supply temperature of 19°C (66.2°F) along with a typical PV flow rate of 3.5 L/s (7.41 CFM) and supply temperature of 22°C (71.6°F). Good agreement is obtained between measured and predicted values by the model with maximum errors of 7.44% for temperature and 4.47% for concentration found in the macroclimate air.

Key-words: *personalized ventilation, displacement ventilation, mathematical model, indoor air quality*

INTRODUCTION

Nowadays, people are spending most of their times in buildings (office spaces, homes…). Therefore, Heating ventilation and air conditioning (HVAC) systems are concerned with providing thermally comfortable and clean indoor environments for occupant health, work efficiency and welfare (ASHRAE, 1991). To achieve thermal comfort, HVAC systems lower the space temperature to suitable levels by supplying cool air. More critically, HVAC systems should assure good indoor air quality (IAQ) and protect occupants from the different types of passive contaminants present in the space. These contaminants can compromise human health by inflicting nausea, irritations and headaches (Redlich et al., 1997). Passive indoor pollutants include gases emitted by human breathing (such as carbon dioxide), or office appliances (such as printers, computers). Other contaminants result from office furniture or flooring materials (rugs, paints, varnishes) and are considered as volatile organic compounds (VOCs) (Mølhave et al., 1997). Therefore, the ventilation systems should also supply clean fresh air to dilute these pollutants and meet with IAQ requirements to protect occupants' health. Typical HVAC systems include the displacement ventilation system (DV). This system is know to provide both IAQ and thermal comfort. It is also superior to the conventional mixing ventilation system (MV) where cool high-momentum recirculated air is supplied overhead. MV systems can provide

Douaa Al Assaad is a graduate student in the Department of Mechanical Engineering, American University of Beirut, Beirut, Lebanon. **Kamel Ghali** is a professor and chair in the Department of Mechanical Engineering, American University of Beirut, Beirut, Lebanon.**Nesreen Ghaddar** is a professor in the Department of Mechanical Engineering, American University of Beirut, Beirut, Lebanon and the corresponding author, email: farah@aub.edu.lb.

uniform thermal conditions, but thermal comfort is not always met due to risks of thermal draft (Cho et al., 2002). Moreveor, due to recirculation and mixing, MVsystems can reduce IAQ. For this reason, DV systems are favored and considered as superior alternative systems (Yuan et al., 1998).

Unlike MV, DV systems reverse the location of the supply grill by placing it at low levels near the floor. The cool fresh supplied has low velocities (0.2 m/s (0.66 ft/s)) and relatively higher temperatures than MV (18°C (64.4°F)) (Lin et al., 2005) to prevent thermal draft. In DV systems, the air heats up gradually as it encounters heat sources in the space (occupants, walls…). Due to this buoyant effect, a stratified flow is formed (Bjørn et al., 2002) which divides the space into a cool clean lower zone and a warm contaminated upper zone.

Despite being preferred to MV, standalone DV systems have some limitations in terms of efficiency. In fact, the low momentum and relatively warmer supply air makes it unable to remove large heat loads (maximum of 40 W/m^2 (0.00352 Btu/s.ft^2)) (Behne et al., 1999). In addition, when the heat source does not constitute a pollution source, high amounts of DV fresh air are needed to dilute the contaminants present in the room making the DV an unsuitable system (Brohus et al., 1996). Hence, when large heat loads are present; and when the heat source is different from the contamination source, the DV performance is compromised and it needs to be assisted by another HVAC system that can restitute thermal comfort and IAQ without compromising energy costs.

A viable system that can assist DV is the personalized ventilation (PV) (Cao et al., 2014). PV system are compact HVAC systems that can be coupled to the present background ventilation mode. It delivers cool and clean fresh air targeting the occupant's breathing zone. This system can enhance thermal comfort and provide high breathable air quality at minimal energy costs (Al Assaad et al., 2017). Makhoul et al., (2012a, 2012b) developed and validated a simple mathematical model for DV+PV systems and showed that PV was able to provide comfort compared to a standalone DV system. However, the model of Makhoul et al. (2012a, 2012b) was limited, as it did not consider the development and decay of the impinging PV jet. Moreover, their model did not study IAQ and the distribution of contaminants in the space in the case of DV+PV.

This work studies the performance of a constant PV+DV system with respect to IAQ through simplified mathematical modeling based on first principles. Simplified modeling is cheaper and chronologically economic and allows for fast system optimization compared to computational fluid dynamics simulations. The developped model will be used to understand the interaction between the DV flow, thermal plumes as well as the PV jet. The model will also calculate the temperatures and species' concentration in the space. To prove its robustness, the model will be experimentally validated in a climatic chamber equipped with a DV+PV system and tracer gas representing species.

PROBLEM DESCRIPTION AND MATHEMATICAL FORMULATION

A typical office space is considered as seen in Fig. 1(a). The space is conditioned by a DV system and an occupant workspace equipped with a personalized ventilator delivering clean fresh air towards the seated person. The DV supplies cool fresh air at low velocities and higher supply temperatures than MV. The PV is supplied horizontally towards the occupant at a flow rate m_{PV}. To efficiently filter the PV and DV air, high efficiency particulate air (HEPA) filters are used. A contaminating source is considered in the space at a critical location near the floor as seen in Fig. 1(a). Ths is to study the ability of the PV+DV system is occupant protection against floor contaminants.

In DV systems, the cool air is heated by present heat sources in the room. Due to density differences also known as buoyancy forces, the air starts to rise which creates temperature gradients in the vertical direction. Rising plumes are also formed (Fig. 1(a) & (b)) when the cool air comes into contact with a heat source. A vertical species' concentration gradient is also established in the space. This is presented in Fig. 1(b) which shows the DV space and the stratified flow along with the different transport phenomena (diffusion, convection and entrainment). Four zones are identified in the space: (a) the macroclimate air, (b) the thermal plume air, (c) PV jet region at the breathing height and (d) wall plumes. Each region is characterized by its temperature and species' concentration respectively: (a) (T_a, C_a), (b) (T_{sp}, C_{sp}) (c) (T_{PV}, C_{PV}) and (d) (T_{wp}, C_{wp}).

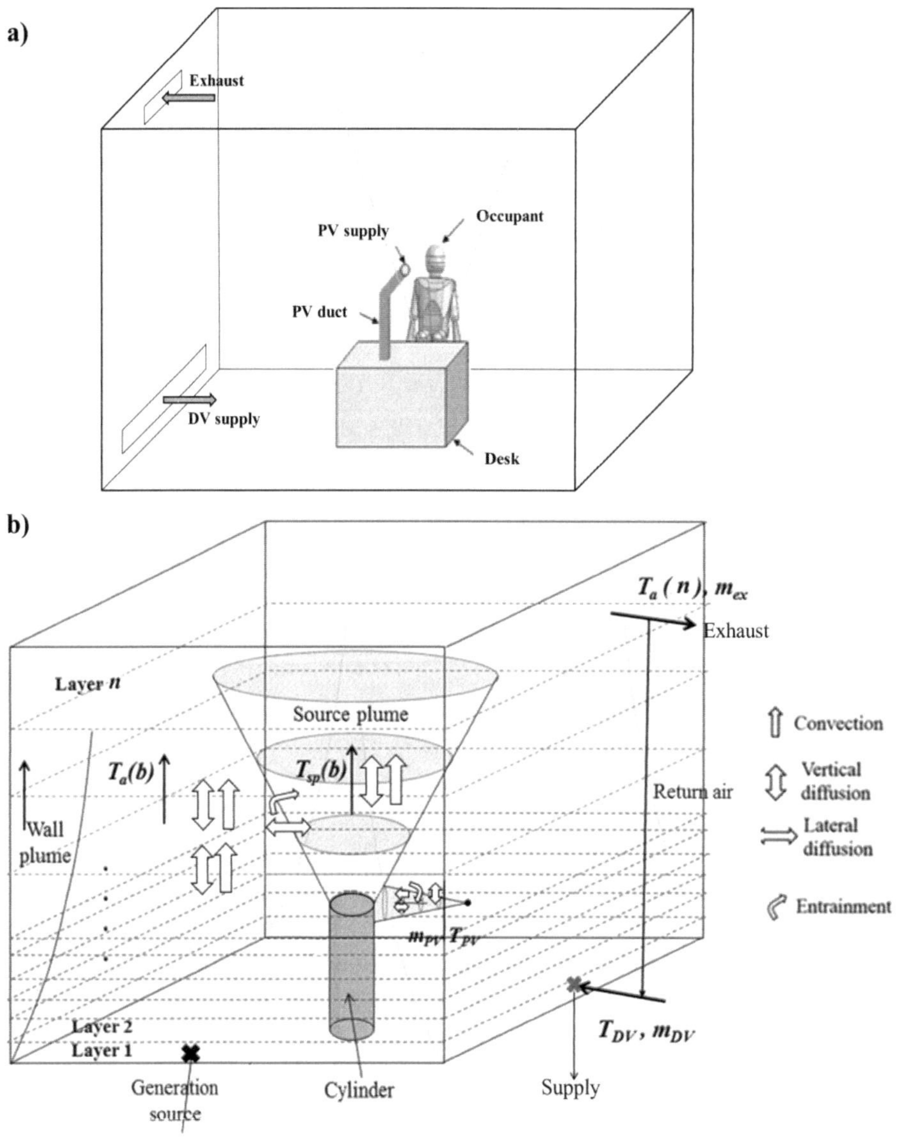

Figure 1: Configuration of: (a) the office space conditioned by a DV and assisted by a PV system, (b) the discretized space with the different physical phenomena.

The PV jet is situated at the breathing height (1.1 m) and is supplied at a specific temperature and species' concentration. The jet entrains air from the macroclimate and is heated up as it reaches the occupant. Moreover, its species' concentration increases horizontally. Thus, at each height, two regions are present: the macroclimate air and thermal plume, and at the breathing height, the PV jet zone is added exclusively. Since there are temperature and concentration variations in the vertical direction, the space is discretized vertically into several layers (Fig. 1(b)). The layers at the beginning contain the seated occupant modeled by a cylinder (diameter of 0.47 m (1.54 ft) and a height of 1.1 m (3.61 ft), surface are similar to human being: 1.8 m^2 (19.375 ft^2)). These layers contain the thermal plume (T_{up}, C_{up}) and the macroclimate air (T_a, C_a). At the breathing layer (Fig. 1(b)), the PV jet is discretized into several layers in the horizontal direction with each layer (T_{PV}, C_{PV}) interacting with the adjacent layers and macroclimate air. To compute temperatures and concentrations, the source plumes' mass flow rate m_{sp} will be calculated using the model of Mundt et al., (1995) which was previously used in the study of Makhoul et al., (2012a) and yielded good results. As for the wall plumes, the model of Jaluria et al., (1980) was used to calculate their corresponding mass flow rates m_{wp}. It is noteworthy that for the plume model to be valid, the rising plume profile should be conserved. Therefore, limitations should be met on the PV maximum supply flow rate and minimum supply temperatures. A value of 7 L/s (14.83 CFM) and 20°C (68°F) were found to distort the plume in the model of Makhoul et al. (2012a).

The macroclimate air mass flow rate m_c is concluded by performing mass balance at each layer between wall and source plumes' mass flow rates and the DV supply flow rate m_{DV}. When m_{DV} exceeds m_{sp} and m_{wp}, circulated air m_c moves upwards. However, when the opposite is true, circulated mass moves downwards. The PV jet supply mass flow rate $m_{PV,supply}$ is determined based on typical values found in literature. The impinging PV jet as seen in Fig. 2(a) has a vritual origin and an opening angle θ having a value of 24° (Shujan et al., 2006). The PV nozzle has a diameter d and situated at close proximity to the occupant such that the occupant lies in the potential core region (distance ranging from $4d$ to $6d$); cleanest zone of the jet. The PV jet velocity V_{PV} decays in the x direction as seen in equation (1):

$$V_{PV}(i) = V_C \cdot \left[1 - \left(\frac{r}{R_{PV}(i)}\right)^{\alpha}\right]^2 \quad (1)$$

Where i is the index of x coordinate of the PV, V_C is the centerline velocity equal to the supply velocity and R_{PV} is the expanding radius of the jet. The exponent α is a constant upon which the shape of the jet profile is determined (Musculus et al. 2009). The PV mass flow rate m_{PV} is calculated using equation (2). It increases horizonally due to macroclimate air entrainment.

$$m_{PV}(i) = \int_{0}^{R_{PV}(i)} 2\pi \rho . V_{PV}(i) . r . dr \quad (2)$$

The different temperatures and species' concentrations in the space are calculated by solving the energy and mass balance equations. This requires proper understanding of the different physical phenomena occurring among the different regions in the space. Fig. 2(b) illustrates a cross sectional area of the different regions in the space and the different occurring phenomena. In fact, the macroclimate air interacts with its adjacent layers through diffusion due to temperature and concentration vertical gradients. Convection also occurs due to rising air momentum. These phenomena also occur in the thermal plume as well as the PV jet in the horizontal direction. The macroclimate air interacts with the thermal plume and PV jet through entrainment and lateral diffusion. In addition, the macroclimate air exchanges heat with the walls, floor and ceiling through convection. Equations (3), (4) and (5) illustrate the mass balance equations in the macroclimate air, PV jet and thermal plume respectively. It is noteworthy that the temperature balance equations are similar to the mass balance equations and can be obtained from the latter by replacing the concentration variable with temperature. Since this study focuses on assessing IAQ, these equations are not showcased in this work.

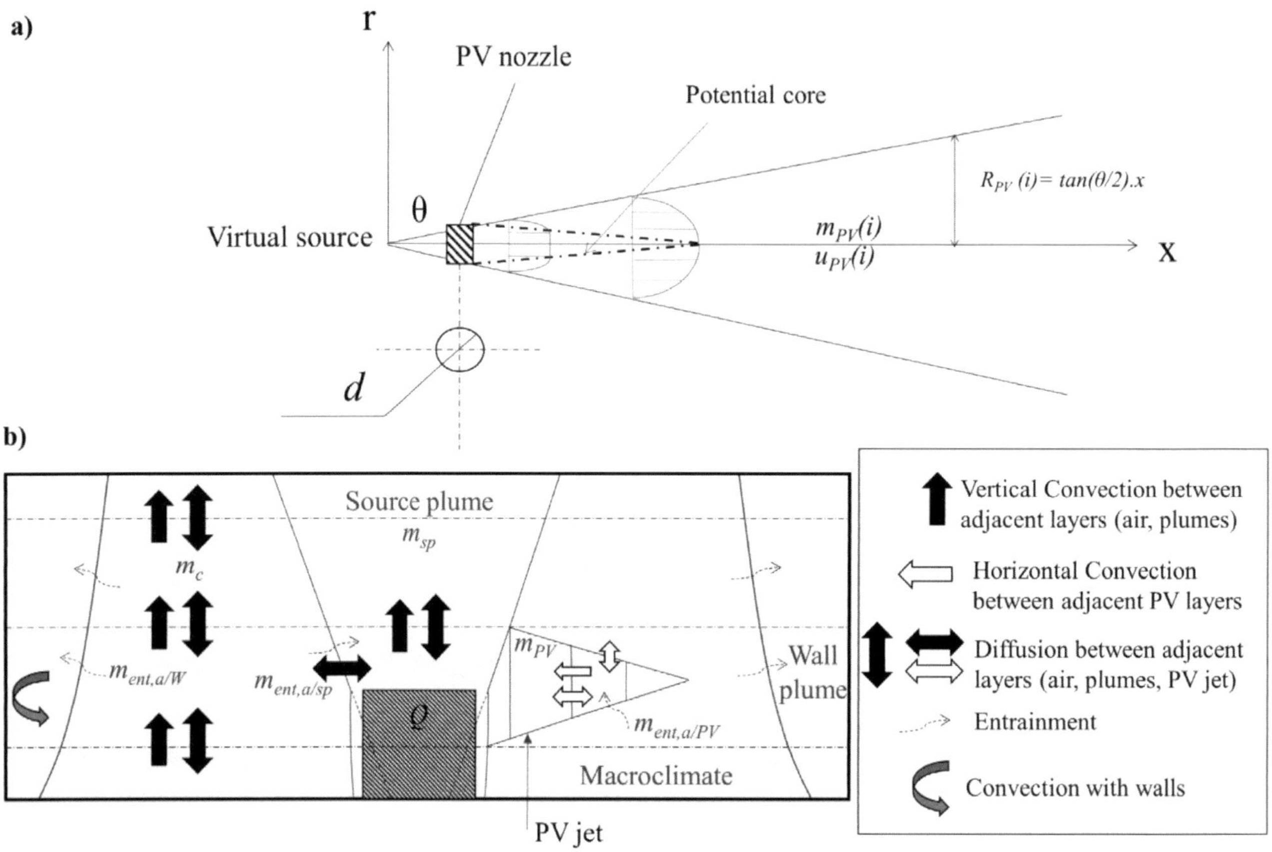

Figure 2: Illustration of: (a) the impinging PV jet, (b)cross sectional area of the discretized space with the different physical phenomena.

$$
\begin{aligned}
&\left[\underbrace{\max(m_c(j-1),0).C_a(j-1)-(\max(m_c(j),0)+\max(-m_c(j-1),0)).C_a(j)+\max(-m_c(j),0).C_a(j+1)}_{Net\ convective\ fluxes\ between\ two\ consecutive\ layers}\right] \\
&+\underbrace{\left[-S_a(j-1).D.\rho\left(\frac{\partial C_a}{\partial z}\right)_{j-1}+S_a(j).D.\rho\left(\frac{\partial C_a}{\partial z}\right)_j\right]}_{Net\ Brownian\ diffusion\ between\ two\ consecutive\ layers}+\underbrace{\left[S_{int}(j).(D+D_t).\rho\left(\frac{\partial C_{sp}}{\partial r}\right)_j\right]}_{Net\ turbulent\ diffusion\ air-plume} \\
&-\underbrace{\left[m_{ent,a/sp}(j).C_a(j)\right]}_{Entrained\ mass\ air\ to\ plume}-\delta*\underbrace{\left[m_{ent,a/PV}(j).C_a(j_{PV})\right]}_{Entrained\ mass\ air\ to\ PV}+\underbrace{\left[M_{gen}(1).C_{gen}(1)\right]}_{Sepcies\ generation\ source}+\delta \\
&*\underbrace{\sum_{j=1}^{nPV}\left[S_{int,PV}(j_{PV}).(D+D_t).\rho\left(\frac{\partial C_{PV}}{\partial r}\right)_{jPV}\right]}_{Net\ turbulent\ diffusion\ air-PV}=0
\end{aligned} \tag{3}
$$

Where M_{gen} and C_{gen} are the species' mass flow rate and concentration of generation at floor level respectively, D and D_t are the molecular and turbulent diffusion coefficients of species into the air respectively. D is equal to 1.39×10^{-5} m^2/s (1.5×10^{-4} ft^2/s), and the value of D_t was calculated by the correlations of Chen et al. (1998) that relay this number to the jet speed and specific length. These relations were used in the work of Kanaan et al. (2010) and gave good results. S_a is the area between adjacent air layers. S_{int} and $S_{int,PV}$ are the interfacial areas between the macroclimate air and thermal plume and PV jet respectively while $m_{ent,\ a/sp}$ and $m_{ent,\ a/PV}$ are the entrained masses from the macroclimate air into the thermal plume and PV jet respectively. The symbol δ is the Kronecker symbol, equal to 1 in the layer containing the PV and zero elsewhere.

$$\underbrace{[m_{PV}(i-1).C_{PV}(i-1)-m_{PV}(i).C_{PV}(i)]}_{\text{Net convection between consecutive PV layers}}+\underbrace{[m_{ent,a/PV}(i).C_a(i)]}_{\text{Entrained species from air to PV}}$$
$$+\underbrace{\left[-S_{PV}(i-1).D.\rho\left(\frac{\partial C_{PV}}{\partial x}\right)_{i-1}+S_{PV}(i).D.\rho\left(\frac{\partial C_{PV}}{\partial x}\right)_i\right]}_{\text{Net Brownian diffusion between consecutive PV layers}}+\underbrace{\left[S_{int,PV}(i).(D+D_t).\rho\left(\frac{\partial C_{PV}}{\partial r}\right)_i\right]}_{\text{Lateral turbulent diffusion air-PV}}=0 \quad (4)$$

Where C_{PV} follows a profile similar to that of velocity (equation (1)) and S_{PV} is the area between adjacent PV layers.

$$\underbrace{[m_{sp}(j-1).C_{sp}(j-1)-m_{sp}(j).C_{sp}(j)]}_{\text{Net convection between consecutive plume layers}}+\underbrace{[m_{ent,sp}(j).C_a(j)]}_{\text{Entrained species from air to plume}}$$
$$+\underbrace{\left[-S_{sp}(j-1).D.\rho\left(\frac{\partial C_{sp}}{\partial z}\right)_{j-1}+S_{sp}(i).D.\rho\left(\frac{\partial C_{sp}}{\partial z}\right)_j\right]}_{\text{Net Brownian diffusion between consecutive plume layers}}+\underbrace{\left[S_{int}(i).(D+D_t).\rho\left(\frac{\partial C_{sp}}{\partial r}\right)_j\right]}_{\text{Lateral turbulent diffusion air-PV}}+\delta$$
$$*\underbrace{[m_{PV}(n_{PV}).C_{PV}(n_{PV})]}_{\text{Entrained mass from PV to the plume}}=0 \quad (5)$$

To solve for temperature and concentration in the difference zones in the space as well as mass flow rates, the model requires several inputs. The inputs include the air properties, the DV and PV supply temperatures and flow rates, the room geometry, species concentration at the inlets and at floor level. The outdoor conditions are also required to compute the indoor wall temperatures (Makhoul et al., 2012a). The space is discretized into n layers in the vertical direction. These layers are the macroclimate air and thermal plumes. The PV jet is discretized horizontally into n_{PV} layers. An iterative method is used to solve for the different variable since all layers are interdependent and since the species' equations depend on solving the energy balance equations. The solution is initialized by making proper assumptions for macroclimate air, wall temperatures as well as vertical gradient. First, the energy balance is solved for each layer to compute temperatures and update the corresponding mass flow rates accordingly. Once the mass flow rates are known, the mass balance equations at each layer are solved to compute the species' concentration in the space. Convergence is considered reached when the residuals are lower than 10^{-6}.

EXPERIMENTAL VALIDATION

In order to enusre that the PV+DV model is able to predict temperature and species' concentration in the different zones in the space, experiments were conducted in a climatic chamber to validate it. The chamber can be seen in Fig. 3. It has dimensions of 2.5 m (8.2 ft) × 2.75 m (9 ft) × 2.8 m (9.18 ft) and is equipped with a DV system (supply: 0.582 m (1.91 ft) × 0.24 m (0.787 ft), return: 0.44 m (1.44 ft) × 0.19 m (0.623 ft)) coupled with a PV nozzle. The PV has an outlet diameter of 10 cm (0.033 ft) and separated from the occupant represented by a heated cylinder by a distance of 40 cm (0.131 ft). The cylinder has a height of 1.1 m (3.6 ft) and a diameter of 0.47 m (1.54 ft), totaling to a surface area of 1.8 m^2 (19.38 ft^2) similar to the surface area of a human. To generate heat flux, a heat source of 100 W (341.21 Btu/hr) was installed inside the cylinder. To represent species, CO_2 is used as a tracer gas (Al Assaad et al., 2017). Four CO_2 tubings are placed at low levels in the room as seen in Fig. 3 and supply each, 0.5 L/min (0.0176 ft^3/s) of CO_2 (Abou Hweij et al., 2015). The DV and PV supplied air have typical CO_2 concentrations of 450±10 ppm. Temperature measurements were perfomed using T-type thermocouples (OMEGA) having a measurement accuracy of ± 0.1°C (32.18°F). They are connecting to an OMEGA data logger to record the data. CO_2 concentrations were measured using FIGARO CDM7160 CO_2 sensors with a detection range of [300 – 5000] ppm. They are also connected to an OMEGA data logger. To start experiments, the lights (100 W) and cylinder heat source (100 W (341.21 Btu/hr)) are turned on. The HVAC systems are also turned on. The walls in the room are adiabatic with small U-value of 0.3 W/m^2.K (0.0528 Btu/hr.ft^2.°F). The DV and PV supply temperatures are specified at 19°C (66.2°F) and 22°C (71.6°F) respectively while the DV and PV supply flow rates are specified at 60 L/s (127.13 CFM) and 3.5 L/s (7.41 CFM). The CO_2 sources are set to deliver 0.5 L/min (0.0176 ft^3/s) of CO_2 each. The values of temperature and concentration were measured by the thermocouples and CO_2 sensors respectively. The experiment was repeated five times for accuracy and repeatability of the predictions.

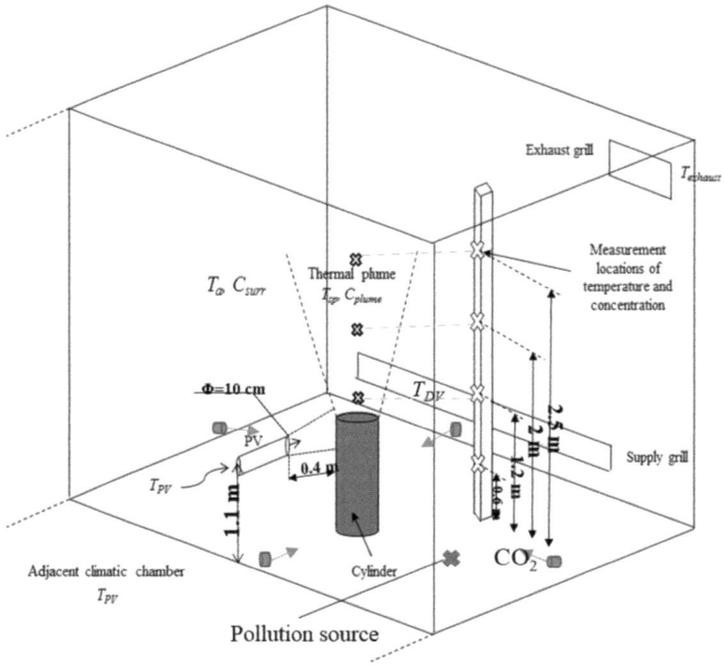

Figure 3: Climatic chamber setup equipped with DV+PV systems and different sensors for temperature and concentration measurements.

RESULTS AND DISCUSSION

Figures 4 and 5 present the experimental and predicted values of temperature and CO_2 concentrations variations with height for the macroclimate air and thermal plume respectively. The comparison is done for a typical load in the room of 40 W/m² (0.00352 Btu/s.ft²). It is shown according to Fig. 4(a) that the temperature in the macroclimate air increases with height. This is due to the heat exchanged with the different heat sources through entrainment and diffusion. The plume temperature shows a decreasing trend (Fig. 4(b)). This is due to the plume entraining cooler DV air. Also, the plume is further cooled by the presence of the PV jet by 0.25°C (32.45°F). There was good agreement between the experiment and predicted values by the model with a maximum relative error of 7.44% in the macroclimate air.

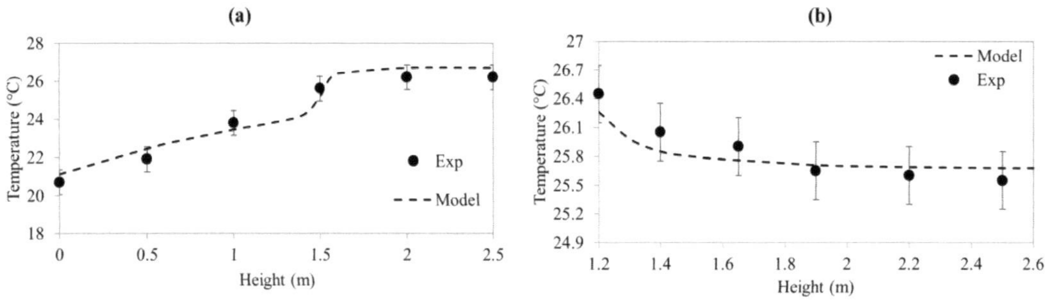

Figure 4: Comparison of experimental and predicted values of temperature gradients with height for: (a) macroclimate air, (b) thermal plume.

According to Fig. 5(a), the CO_2 concentration in the macroclimate air was very high at low levels due to the presence of the pollution source. It decreases gradually due to diffusion and entrainment with the plume (source plume and wall plumes) as well as the PV jet at the breathing height of 1.1 m (3.6 ft). Above 1.1 m (3.6 ft), the concentration continues its decreasing trend, however more slowly. A peak observed at a height of 1.46 m (4.8 ft), which is the stratification height where accumulation of species is observed. On the other hand, the plume species' concentration increases due to entrainment of the polluted DV air reaching a maximum at the stratification height and decreasing above it due to the descending macroclimate air. However, the presence of the PV jet decreased the CO_2 concentration in the thermal plume compared to standalone DV system by almost 40 ppm justifying the use of PV to improve the breathable air quality. Good agreement was found between experimental and predicted values with a maximum relative error of 4.47% in the macroclimate air.

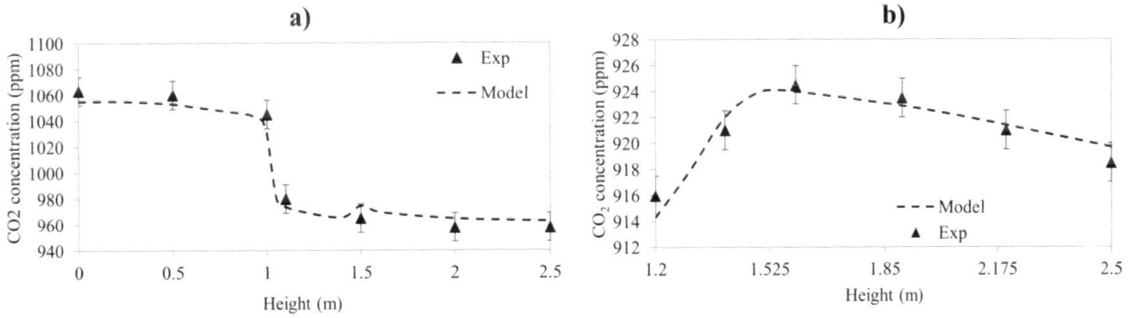

Figure 5: Comparison of experimental and predicted values of CO_2 concentration gradients with height for: (a) macroclimate air, (b) thermal plume.

CONCLUSION

A PV+DV simplified mathematic model was developed. This model is able to predict temperature and species' concentration in the different regions in the space: macroclimate air, thermal plumes as well as the PV jet. To validate the model's ability in predicting temperatures and concentrations, experiments were conducted in a climatic chamber. Typical DV and PV conditons were chosen and a tracer gas was selected to assess the distribution of species in the space. Results showed good agreement between experimental and model values and that the model is robust in predicting thermal and concentration fields in DV+PV spaces. The model can be coupled with a bio-heat model to determine thermal comfort and to assess the effects of different PV operating parameters on comfort and IAQ.

NOMENCLATURE

S_a = Macroclimate air layer area (m^2)
S_{int} = Interfacial area between macroclimate air and source plume (m^2)
$S_{int,PV}$ = Interfacial area between macroclimate air and PV jet (m^2)
S_{PV} = PV jet layer area (m^2)
S_{sp} = Source plume layer area (m^2)
C_a = Macroclimate air species' concentration (ppm)
C_{BZ} = Species' concentration at the breathing zone (ppm)
C_{gen} = Species' generation concentration (ppm)

C_{PV} = PV jet species' concentration (ppm)
C_{sp} = Thermal source plume species' concentration (ppm)
d = PV nozzle diameter (m)
D = Mass diffusion coefficient (m^2/s)
D_t = Turbulent diffusion coefficient (m^2/s)
DV = Displacement ventilation
HEPA= High particle efficiency air
HVAC= Heating ventilation and air conditioning
i = PV jet layers index
IAQ = Indoor air quality
j = Macroclimate air layers index
j_{PV} = Macroclimate air layer containing the PV jet
max = maximum operator
m_c = Circulated mass (kg/s)
m_{DV} = DV supply flow rate (kg/s)
$m_{ent,a/PV}$= Entrained mass from the macroclimate air into the PV jet (kg/s)
$m_{ent,a/sp}$ = Entrained mass from the macroclimate air into the source plume (kg/s)
$m_{ent,a/W}$= Entrained mass from the macroclimate air into the wall plume (kg/s)
M_{gen} = Species' generation mass flow rate (kg/s)
m_{PV} = PV mass flow rate (kg/s)
$m_{PV,\ supply}$= PV supply flow rate (kg/s)
m_{sp} = Source plume mass flow rate (kg/s)
MV = Mixing ventilation
m_{wp} = Wall plumes mass flow rate (kg/s)
n_{pv} = Number of PV layers
PV = Personalized ventilation
Q = Heat source strength (W)
r = Radial coordinate
R_{PV} = PV jet radius (m)
T_a = Macroclimate air temperature (°C, K)
T_{PV} = PV jet temperature (°C)
T_{sp} = Thermal source plume temperature (°C, K)
V_c = PV centerline velocity (m/s)
VOC = Volatile organic compound
V_{PV} = Average PV velocity (m/s)
x = Cartesian coordinate
z = Room height (m)
a = Empirical constant
θ = PV jet opening angle
ϱ = Density (kg/m^3)
δ = Kronecker symbol

REFERENCES

ASHRAE Handbook. Heating, Ventilating, and Air-conditioning Applications. American Society of Heating Refrigerating and Air-Conditioning Engineers, (1991).

C. A. Redlich, J. Sparer, M. R. Cullen, Sick-building syndrome. The Lancet, 349(1997): 1013-1016.

L. Mølhave, L., G. Clausen, B. Berglund, J. de Ceaurriz, A. Kettrup, T. Lindvall, Total volatile organic compounds (TVOC) in indoor air quality investigations, Indoor Air 7, 4 (1997): 225-240.

Y. Cho, H. B. Awbi, T. Karimipanah, A comparison between four different ventilation systems, In Proceedings of the 8th International Conference on Air Distribution in Rooms (2002): 181 -184.

X. Yuan, Q. Chen, Q., L.R. Glicksman, A critical review of displacement ventilation. ASHRAE transactions 104(1998): 78

Z. Lin, T.T. Chow, K.F. Fong, C.F. Tsang, C.F. Q. Wang, Comparison of performances of displacement and mixing ventilations. Part II: indoor air quality. International Journal of Refrigeration, 28(2005): 288-305.

E. Bjørn, P.V. Nielsen, Dispersal of exhaled air and personal exposure in displacement ventilated rooms. Indoor air, 12(2002): 147-164.

M. Behne, Indoor air quality in rooms with cooled ceilings: Mixing ventilation or rather displacement ventilation? Energy and Buildings, 30(1999): 155-166.

H. Brohus, H., P.V. Nielsen, Personal exposure in displacement ventilated rooms. Indoor Air, 6(1996): 157-167.

G. Cao, H. Awbi, R.Yao, Y. Fan, K.Sirén, R. Kosonen, A review of the performance of different ventilation and airflow distribution systems in buildings. Building and Environment, 73(2014): 171-186.

D. Al Assaad, Douaa, K. Ghali, N. Ghaddar, C. Habchi, Mixing ventilation coupled with personalized sinusoidal ventilation: Optimal frequency and flow rate for acceptable air quality, Energy and Buildings 154 (2017): 569-580.

A. Makhoul, K. Ghali, N. Ghaddar, A simplified combined displacement and personalized ventilation model. HVAC&R Research 18(2012a): 737-749.

A. Makhoul, K. Ghali, N. Ghaddar, The Energy Saving Potential and the Associated Thermal Comfort of Displacement Ventilation Systems Assisted by Personalised Ventilation. Indoor and Built Environment 22(2012): 508–519.

E.Mundt, Displacement ventilation systems—Convection flows and temperature gradients. Building and Environment 30(1995): 129–133.

Y.Jaluria, Natural convection: heat and mass transfer, 5(1980)

A. Shujan, B. Thurow, Axisymmetric turbulent jets: Development and Structure, American Institute of Aeronautics and Astronautics 2006.

M.P. Musculus, Entrainment waves in decelerating transient turbulent jets, Journal of Fluid Mechanics, 638(2009): 117-140.

Q. Chen, W. Xu, A zero-equation turbulence model for indoor airflow simulation. Energy and buildings, 28(1998): 137-144.

M. Kanaan, N. Ghaddar, K. Ghali, Simplified model of contaminant dispersion in rooms conditioned by chilled-ceiling displacement ventilation system. HVAC&R Research, 16(2010): 765-783.

W. Abou Hweij, N. Ghaddar, K. Ghali, C. Habchi. Optimized performance of displacement ventilation aided with chair fans for comfort and indoor air quality, Energy and Buildings 127 (2016): 907-919.

Feasibility Assessment for Retrofitting an Energy-efficient Hospital Building through Energy Modelling and Field Investigation

Fu-Jen Wang, Ph.D., P.E.
Fellow ASHRAE

Cheng-Shu Kuo, Ph.D.

Hung-Wen Lin, Ph.D.
Member ASHRAE

Pei-Yu Yu, Ph.D.

Chen-Pu Wang
Student Member ASHRAE

ABSTRACT

The objective of this study, including field investigations and energy modelling, is to evaluate the feasibility of retrofit for energy performance of a hospital in Taiwan. This hospital consumes around 8500000 kWh annually with 37490 m^2 floor area including twelve stories and two floors of basements. The current facility systems of this hospital present relevant signs of degradation and obsolescence with high energy consumption. The retrofitting plan of facility systems for this hospital is assessed and conducted to improve energy performance accordingly. Based on ASHRAE procedures for commercial building energy audits (PCBEA), field study and energy audits were conducted to collect information of building geometry, building materials, space types, density and activities of people, capacity of HVAC system, loads and operating schedules of lighting and other facilities. The energy consumption baseline can be derived through PCBEA and BEMS (building energy management system) data. Energy modelling using Energy Plus code has been conducted extensively to evaluate the energy saving approaches. The results reveal that the energy saving strategy of increasing chiller water supply temperature 1 °C could obtain 1% of annual energy saving, while increasing room set point 1 °C may reach 2% saving of annual energy consumption. Besides, applying the insulation film for glass also presents about 0.8% of annual energy saving through energy modelling. The retrofit feasibility has also been evaluated in terms of annual savings and pay-back period of the investment. The procedural investigation of building energy consumption should provide valuable information for facility engineers facing energy saving concerns.

INTRODUCTION

Taiwan is a small island with rare indigenous energy, 97% of energy imported from overseas. Recently, scales of building in Taiwan such as office buildings, department stores, hospitals and hotels were increasing, consuming large amount of electricity. From the current situation and future trend of electricity demands and percent reserve margin of capacity of Taiwan, shown in the Figure 1(Taipower company, 2011), it can be found that electricity demands of Taiwan increase with years while percent reserve margin of capacity decreases with years. Therefore, how to reduce the electricity consumption and improve energy efficiency has become a key issue for energy management of Taiwan. According to investigations (Bureau of Energy of Taiwan, 2015) from Bureau of Energy of Taiwan, there are 145 hospital buildings with contract capacity above 800 kW, consuming 2.1 billion kWh per year, about 14.9 % of electricity consumption of non-productive industries in Taiwan. In addition, average electricity consumption per unit area of medical centers and regional hospitals are 245 kWh/m^2.yr and 205 kWh/m^2.yr,

Fu-Jen Wang is professor and **Chen-Pu Wang** is graduate student in the Department of Refrigeration, Air Conditioning and Energy Engineering, National Chin-Yi University of Technology, Taiwan. **Cheng-Shu Kuo**, **Hung-Wen Lin** and **Pei-Yu Yu** are senior researchers in Green Energy and Environment Lab, Industrial Technology Research Institute, Taiwan.

respectively, about 1.47 to 1.75 times of the one of office buildings. Hospitals provide patients with 24 hours of lighting, air conditioning and healthcare facilities to meet the requirements of medical care, thus consuming large amount of electricity. The Taiwan government has set the energy saving target that Mega-Watt energy users would have 1% of energy reduction per year in the following 5 years (Bureau of Energy of Taiwan, 2015). Concerning this matter, hospital buildings would have to implement energy-saving measures in order to reduce electricity consumption and to reach the Taiwan government's energy saving target.

Energy simulation is a useful method to evaluate influence factors on building energy performance such as building materials, power densities of lighting and internal electrical equipment, air conditioning system, and operation schedules etc., especially useful for the early designing or retrofit planning stage of such large scale hospital buildings. Energy simulations have been applied to analyze energy performance of hospital buildings in several previous works. The building energy simulations were employed to evaluate the energy-saving efficiency of HVAC system by studying two green hospital buildings (Chen and Kan, 2014), to examine thermal performance of hospital buildings with various geometries (Shchuchenko et al, 2013), to calculate heating and cooling loads of the hospital (Saipi et al, 2014), to assess the feasibility of 50% energy savings for large hospitals in various climate zones of USA (Bonnema et al. 2010), and to evaluate impacts of both envelope renovation and HVAC system selection for a hospital building (Ascione et al, 2016) etc. It is indicated that the building energy software can help to obtain a better understanding of energy consumption throughout the whole building. Therefore, EnergyPlus building energy simulation software developed by Department of Energy of USA was used in this report to calculate energy consumptions of energy saving measures applied on the hospital.

In this report, we will evaluate the feasibility of energy saving measures applied for the retrofit of a hospital building, located in the midst of Taiwan. In order to obtain completed information of the hospital building, ASHRAE procedures for commercial building energy audits (PCBEA) (ASHRAE, 2011) were followed to collect information of building geometry, building materials, space types, density and activities of people, capacity of HVAC system, loads and operating schedules of lighting and internal equipment. Based on collected information, an energy model for this hospital was established using EnergyPlus software, while this energy model was calibrated by actual monthly electricity as a retrofit baseline. With reference to required operation conditions of this hospital, energy saving potentials of several energy saving measures for air conditioning system, lighting devices, internal equipment, building materials and energy management resulting from EnergyPlus simulations have been examined. It will compare annual energy saving and pay-back period of the investment among the existing baseline and energy saving measures to evaluate retrofit feasibility.

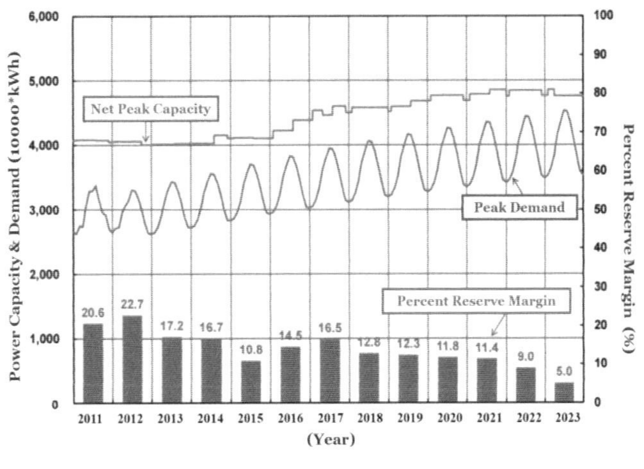

Figure 1. Trend of net peak capacity, peak demand and percent reserve margin of Taiwan.

ENERGY MODEL OF THIS HOSPITAL BUILDING

Specification of Building

This hospital building is located in the midst of Taiwan, built in twelve floors above ground and two floors underground, with base area 64 meter by 58.6 meter. The front of this hospital building faces northwest. The first floor is the lobby and emergency rooms. From the 5th floor to the 11th floor, all spaces are wards except some spaces in 10th floor are used for negative pressure isolation rooms. The 12th floor contains an auditorium, meeting rooms and negative pressure isolation rooms. The detailed information of area and space types of each floor of this building is shown in Table 1.

Table 1. Specification of building

Floor	Floor area (m^2)	Space property
B1	3749	**Radiology department; Morgue; Parking lot**
B2	3749	**Restaurant; Pharmacy; Medical records department**
1F	3749	**Lobby; Emergency room**
2F	3749	**Outpatient Clinic**
3F	3749	**Intensive care unit; Operating room**
4F	3749	**Air conditioning machine room**
5F	1874.5	**Ward**
6F	1874.5	**Ward**
7F	1874.5	**Ward**
8F	1874.5	**Ward**
9F	1874.5	**Ward**
10F	1874.5	**Ward; Negative pressure isolation room**
11F	1874.5	**Ward**
12F	**1874.5**	**Negative pressure isolation room; Auditorium; Meeting room**

(a) (b)

Figure 2. (a) Actual appearance of the hospital and (b) 3D model of the hospital.

Construction of The Building

According to the architectural drawing of this building, the construction, appearance and interior partition of the building are established. The roof of this building is reinforced concrete with regular insulation (overall heat transfer coefficient U = 0.99 W/m^2.K). Exterior wall is 15 cm reinforced concrete with tiles (overall heat transfer coefficient U = 3.49 W/m^2.K). Window glass is 8 mm brown colour glass (overall heat transfer coefficient U = 6.07 W/m^2.K, solar heat gain coefficient SHGC = 0.82). The building model is built up following the geometric dimensions of the building in the architectural drawing. The complicated geometric dimensions of the building are simplified without influencing the simulations. The actual appearance and 3D model of the building are shown in Figure 2 and 3. The interior partitions of each floor of this building are independently established according to the architectural drawing. Some spaces without air conditioning or with the same property are simply combined into a space to reduce the complexity of building model.

Interior Load Parameters

According to field investigation, the average people, lighting density and internal electrical equipment density of each floor of this hospital are shown in the Table 2.

Table 2. Average interior load of building.

Floor	People number	Lighting density (W/m^2)	Internal electrical equipment density (W/m^2)
B2	100	15	15
B1	200	30	15
1F	300	25	15
2F	160	25	20
3F	160	20	10
4F	5	5	10
5F	160	25	10
6F	160	25	10
7F	160	25	10
8F	160	25	10
9F	160	25	10
10F	160	25	15
11F	160	25	10
12F	100	15	10

Schedule

Figure 3 shows the people activity in the hospital building. Schedule 1 is weekday schedule of occupancy while schedule 2 is for weekend and holiday. Activity factor of occupancy schedule for special spaces such as emergency rooms and intensive care unit remains one for 24 hours throughout the whole year. Figure 4 plots the lighting schedules in the hospital building. Schedule 1 is weekday schedule of lighting while schedule 2 is for weekend and holiday. Operation factor of lighting schedule for special spaces such as emergency rooms and intensive care unit remains one for 24 hours throughout the whole year. Figure 5 illustrates the internal electrical equipment schedule in the hospital building. Schedule 1 is weekday schedule of internal electrical equipment while schedule 2 is for weekend and holiday. Operation factor of internal electrical equipment schedule for special spaces such as emergency rooms and intensive care unit remains one for 24 hours throughout the whole year. Figure 6 plots the air conditioning system schedule in the hospital building. Schedule 1 is weekday schedule of air conditioning system while schedule 2 is for weekend and holiday. Operation factor of air conditioning system for special spaces such as emergency rooms and intensive care unit remains one for 24 hours throughout the whole year.

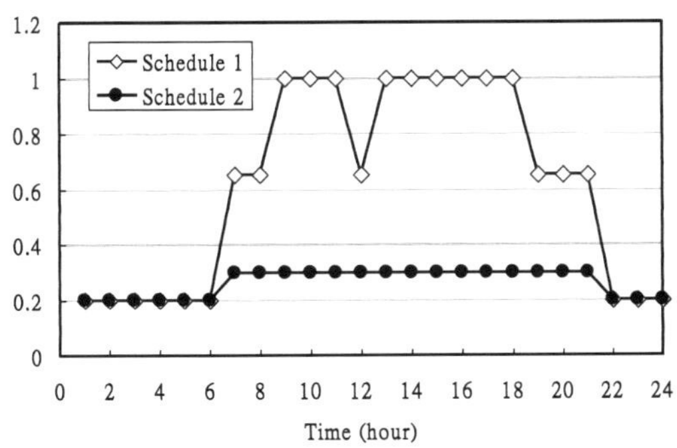

Figure 3. Occupancy Schedule.

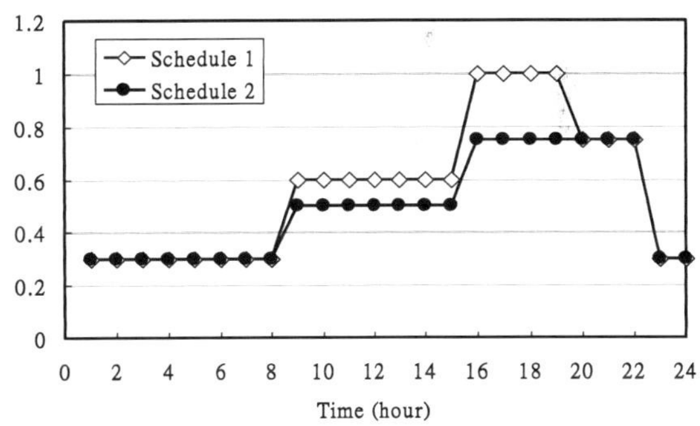

Figure 4. Lighting schedule.

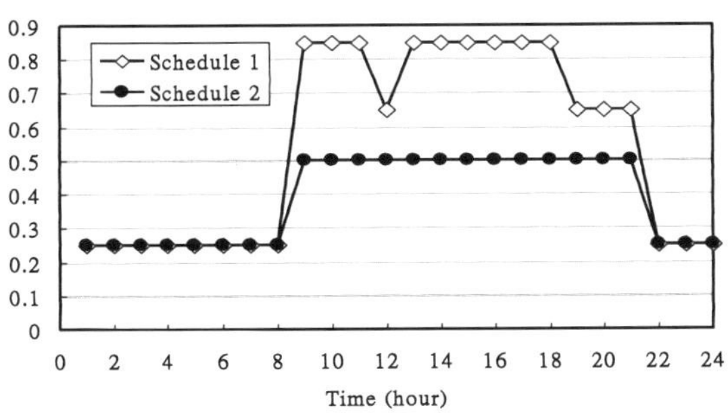

Figure 5. Internal electrical equipment schedule.

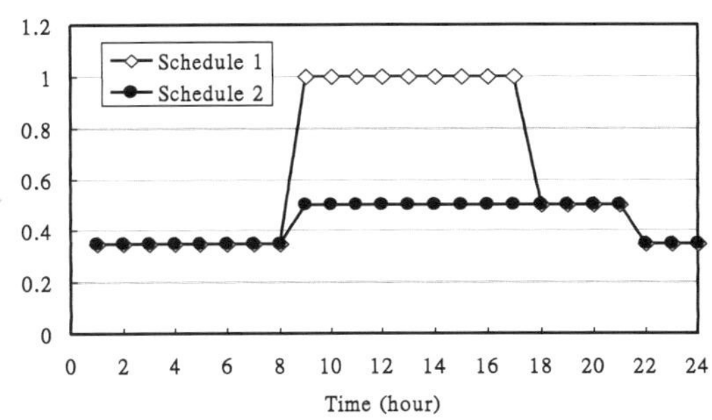

Figure 6. Air conditioning system schedule.

Air Conditioning System

This air conditioning system of this hospital contains two 540 RT chillers in a parallel configuration and a 250 RT chiller, with rated COP of 5.0, 5.0 and 4.0, respectively. These chillers work using 5 chilled water pumps and 7 regional chilled water pumps to cool the rooms inside the building, illustrated in Figure 7. Also, these chillers employ 5 cooling water pumps and 5 cooling towers to release heat to the environment, shown in Figure 8. Specifications of chillers are indicated in Table 3. The chiller of 250 RT works 24 hours whole year to meet the requirements of cooling loads of special spaces such as emergency rooms and intensive care unit. Two 540 RT chillers satisfy the cooling demands for all regular rooms, operating with the air conditioning system schedule indicated in Figure 6. The efficiency of fans used in air conditioning system is 0.6. The room temperature of regular space is set at 25 ℃.

Table 3. Specifications of chillers.

No.	CH1	CH2	CH3
Capacity (RT)	540	540	250
Power consumption (kW)	384	384	198
Inlet/outlet chilled water temp. (℃)	12/7	12/7	12/7
Inlet/outlet cooling water temp. (℃)	32/37	32/37	32/37

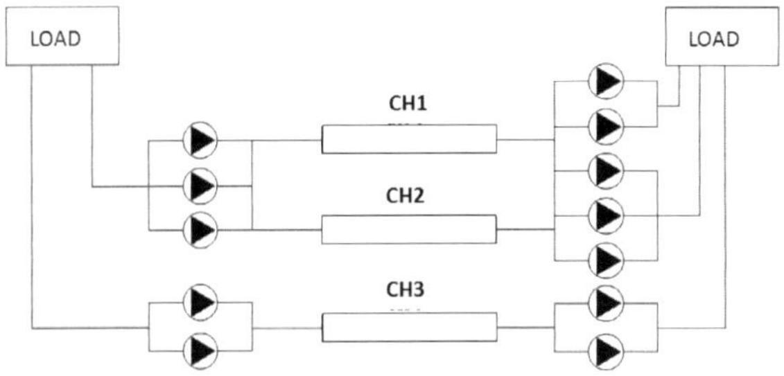

Figure 7. Chillers, chilled water pumps and zone chilled water pumps.

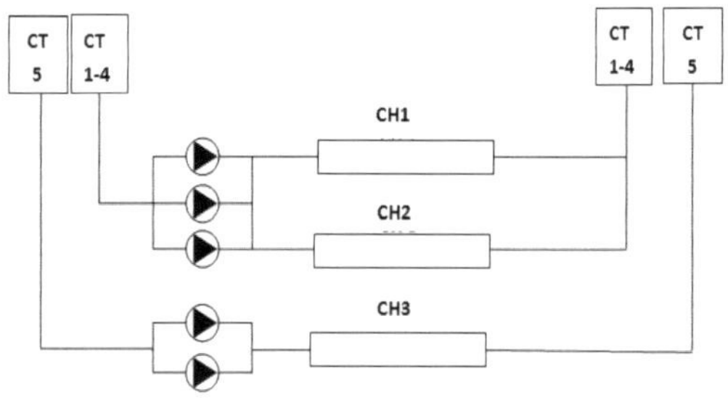

Figure 8. Chillers, cooling water pumps and cooling towers.

Baseline Model

A building model with above settings including construction and materials, internal loads, schedules and air conditioning system is simulated using EnergyPlus software. The simulation result of electricity consumption of this building is compared with measured electricity consumption in 2015, as shown in Figure 9. The deviation of total annual electricity consumption between simulation and actual measured data is about 3.64% and the trends of monthly electricity usages of both are similar and closed. The allowable difference between predicted and measured data is that the annual simulated energy usage should be within 10% of metered data, while a difference of less than 25% is acceptable on a seasonal basis (Rahman et al, 2010). From the simulation result, it is indicated that this building model can be the baseline for retrofit evaluations in terms of total annual electricity consumption.

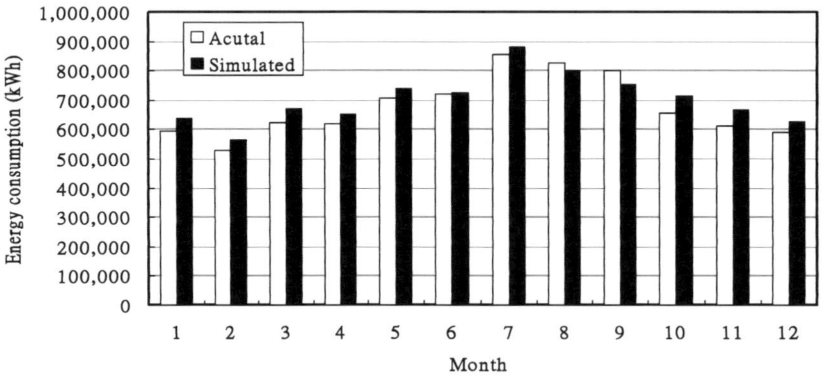

Figure 9. Simulation result and comparison with measured data.

EVALUATION OF ENERGY SAVING MEASURES

Based on baseline energy model of this hospital building, calculations of energy saving and pay-back period of investment of energy saving measures applied on air conditioning system, ventilation system, lighting system, internal electrical equipment, building materials and energy management are conducted as follows. Different measure are shown in Table 4, 5, 6 respectively

Baseline: The EnergyPlus building model with current internal system, which has been calibrated with the monthly electricity consumptions of the hospital building in 2015.

Measure 1: Original 250 RT chiller with COP = 4.0 is replaced by a new 250 RT chiller with COP = 5.0. After this improvement, total annual energy consumption is 8,115,158 kWh, saving 3.8% of electricity about reduction of 961,671 NTD (New Taiwan Dollar). If cost of this improvement is 2,500,000 NTD, the return on investment (ROI) will be 2.6 years.

Measure 2: This measure is to increase outlet chilled water temperature of chillers from 7°C to 8°C. After this setting, total annual energy consumption is 8,345,617 kWh, saving 2.1% of electricity about reduction of 540,591 NTD. Since cost of this improvement is 0 NTD, the return on investment (ROI) is 0 year.

Measure 3: Decreasing inlet cooling water temperature of chillers from 32°C to 31°C is applied to save energy. After this change, total annual energy consumption is 8,365,667 kWh, saving 1.1% of electricity about reduction of 270,294 NTD. Since cost of this improvement is 0 NTD, the return on investment (ROI) is 0 year.

Measure 4: Employing the tinted window film on window glass of the building as thermal insulation, the solar heat gain coefficient (SHGC) of window glass with tinted film is reduced from 0.81 to 0.51. After this improvement, total annual energy consumption is 8,365,667 kWh, saving 0.8% of electricity about reduction of 210,143 NTD. Since cost of this improvement is 2,875,128 NTD (total glass area is 2395.94 m^2 and cost is 1200 NTD/m^2), the return on investment (ROI) is 13.7 years.

Measure 5: The original lighting systems for existing building using T-8 fluorescent lamps are replaced by T-5 fluorescent lamps. Therefore, by this energy saving measure, the average lighting density in all space will be 0.85 times of the original lighting density. After this improvement, total annual energy consumption is 8,112,018 kWh, saving 3.8% of electricity about reduction of 971,092 NTD. Since cost of this improvement is 6,887,000 NTD, the return on investment (ROI) is 7.1 years.

Measure 6: High-performance appliances are applied to reduce energy consumption of internal electrical equipment. By this measure, energy consumption of internal electrical equipment is estimated to be 0.94 times of original one. If this improvement is conducted, total annual energy consumption will be 8,300,822 kWh, saving 1.6 % of electricity about reduction of 404,679 NTD. Since cost of this improvement is 2,000,000 NTD, the return on investment (ROI) is 4.9 years.

Measure 7: This measure is to improve the fan efficiency in the ventilation systems from 0.6 to 0.7. If this improvement is conducted, total annual energy consumption will be 8,417,819 kWh, saving 0.2% of electricity about reduction of 53,689 NTD. Since cost of this improvement is 2,000,000 NTD, the return on investment (ROI) is 37.3 years.

Measure 8: The temperature of regular spaces inside the hospital building is increased from 25°C to 26°C. Therefore, by this energy saving measure, energy consumption will be 8,342,484 kWh, saving 1.11% of electricity about reduction of 279,690 NTD. Since cost of this improvement is 0 NTD, the return on investment (ROI) is 0 years.

Table 4. Evaluation for measure 1, 2 and 3.

Month	Baseline (kWh)	Measure 1 (kWh)	Measure 2 (kWh)	Measure 3 (kWh)
1	638,165	605,065	622,475	628,378
2	566,942	544,168	598,826	558,894
3	669,196	636,645	554,790	660,159
4	653,532	624,917	632,630	646,134
5	739,281	716,105	627,204	732,909
6	723,314	701,414	736,843	716,678
7	881,724	859,645	725,095	874,862
8	799,616	775,209	889,260	792,462
9	752,193	722,164	800,329	743,800
10	715,404	686,169	733,259	709,064
11	668,029	637,066	699,360	659,729
12	628,319	606,591	635,447	622,548
Total	8,435,715	8,115,158	8,255,518	8,345,617
Energy saving (%)	--	3.8	2.1	1.1
ROI (year)	--	2.6	0	0

Table 6. Evaluation for measure 4, 5 and 6.

Month	Baseline (kWh)	Measure 4 (kWh)	Measure 5 (kWh)	Measure 6 (kWh)
1	638,165	630,553	611,381	624,654
2	566,942	554,136	540,879	553,547
3	669,196	667,345	647,416	659,457
4	653,532	651,027	633,432	643,968
5	739,281	736,112	713,282	728,766
6	723,314	719,197	697,673	713,000
7	881,724	877,679	840,544	871,286
8	799,616	794,211	769,419	788,709
9	752,193	745,460	724,794	741,659
10	715,404	711,665	691,725	705,129
11	668,029	666,202	645,876	658,567
12	628,319	612,080	595,597	612,080
Total	8,435,715	8,365,667	8,112,018	8,300,822
Energy saving (%)	--	0.8	3.8	1.6
ROI (year)	--	13.7	7.1	4.9

Table 7. Evaluation for measure 7 and 8.

Month	Baseline (kWh)	Measure 7 (kWh)	Measure 8 (kWh)
1	638,165	636,332	621,666
2	566,942	565,181	560,573
3	669,196	667,888	654,747
4	653,532	652,223	643,184
5	739,281	737,909	736,129
6	723,314	721,973	721,029
7	881,724	880,367	879,434
8	799,616	798,222	796,018
9	752,193	750,841	740,467
10	715,404	714,045	709,559
11	668,029	666,762	653,786
12	628,319	626,076	625,892
Total	8,435,715	8,417,819	8,342,484
Energy saving (%)	--	0.2	1.11
ROI (year)	--	37.3	0

CONCLUSION

In this study, the building energy software EnergyPlus was used to calculate annual energy consumptions of the hospital building under various energy saving measures. The feasibilities of energy saving measures have been evaluated in terms of annual savings and pay-back period of the investment. The pay-back periods of the investment of energy saving measures such as replacement of a high-performance chiller of 250 RT, increasing outlet chilled water temperature of chillers by 1°C, decreasing inlet cooling water temperature of chillers by 1°C, and increasing room temperature of regular spaces by 1°C are less than 5 years. Those analyses could be beneficial to establish a feasible retrofit plan for energy performance of this hospital building.

ACKNOWLEDGEMENTS

The authors would like to thank the Bureau of Energy of the Ministry of Economic Affairs of Taiwan for sponsoring this research work.

REFERENCES

Ascione, F., Bianco, N., De Masi, R.F., de Rossi, F., De Stasio, C. and Vanoli, G.P. 2016. Energy audit of health care facilities: dynamic simulation of energy performances and energy-oriented refurbishment of system and equipment for microclimatic control. American Journal of Engineering and Applied Sciences, in press.

ASHRAE, 2011. Procedures for commercial building energy audit.

Bonnema, E., Studer, D., Parker, A., Pless, S. and Torcellini, P. 2010. Large hospital 50% energy savings: technical support document. National Renewable Energy Laboratory, U.S. Department of Energy.

Bureau of Energy of Taiwan. 2015. Energy audit annual report for non-productive industries.

Chen, P.H. and Kan, M.S. 2014. Integrating energy simulation in energy saving analysis of Taiwan's green hospital buildings. In 2014 Proceedings of the 31st International Symposium on Automation and Robotics in Construction and Mining, Sydney, Australia. 561-567.

Rahman, M.M., Rasul, M.G. and Khan, M.M.K. 2010. Energy conservation measures in an institutional building in sub-tropical climate in Australia. Applied Energy 87: 2994-3004.

Saipi, N., Schuss, M., Pont, U. and Mahdavi, A. 2014. Comparison of simulated and actual energy use of a hospital building in Austria. Advanced Materials Research 899: 11-15.

Shchuchenko, V., Lie, B. and Harsem, T.T. 2013. Influence of building form of hospital on its energy performance. European Scientific Journal 3: 286-295.

Taipower company. 2011. Report of electricity development of Taipower company in Taiwan.

U.S. Department of Energy. 2014. EnergyPlus engineering reference, version 8.1.

AUTHOR INDEX

American Society of Heating, Refrigerating and Air-Conditioning Engineers (ASHRAE)
1791 Tullie Circle
N.E. Atlanta, GA 30329
USA

ISBN 978-1-7138-0598-4

9 781713 805984

CURRAN ASSOCIATES INC.
proceedings.com